AELIAN

ON THE CHARACTERISTICS
OF ANIMALS

II

BOOKS VI—XI

AELIAN

ON THE CHARACTERISTICS
OF ANIMALS

WITH AN ENGLISH TRANSLATION BY
A. F. SCHOLFIELD
FELLOW OF KING'S COLLEGE, CAMBRIDGE

IN THREE VOLUMES
II

BOOKS VI—XI

CAMBRIDGE, MASSACHUSETTS
HARVARD UNIVERSITY PRESS
LONDON
WILLIAM HEINEMANN LTD
MCMLIX

Printed in Great Britain

CONTENTS

v

ERRATA (Vol. I)

Page xiv, line 7 from bottom : *for* boredom, *read* boredom. (full
        stop for comma)
 ,, xxvi, line 6, and elsewhere : *for* viz. *read* viz
 ,,   19, margin : *for* Parro *read* Parrot
 ,, 109, note *a* : *for* χρινοῦν *read* χρυσοῦν
 ,, 165, margin : *for* conjuga *read* conjugal
 ,, 185, line 1 : *for* un *read* un-
 ,, 191, margin of ch. 28 : *for* ' Perseus *read* ' Perseus '
 ,, 257, note 5 : *for* 1875 *read* 1876
 ,, 335, last line : *for* knowledge. *read* knowledge, (comma for
        full stop)

AELIAN

ON THE CHARACTERISTICS
OF ANIMALS

SUMMARY

3

SUMMARY

4

SUMMARY

5

SUMMARY

6

SUMMARY

7

BOOK VI

ΑΙΛΙΑΝΟΥ

ΠΕΡΙ ΖΩΩΝ ΙΔΙΟΤΗΤΟΣ

ς

1. Δέονται μὲν ἄνθρωποι λόγου τοῦ προτρέψον-
τος καὶ ἀναπείσοντος ἀγαθοὺς εἶναι καὶ τὴν μὲν
δειλίαν διώξοντος, τὸ δὲ εὐθαρσὲς παρασκευά-
σοντος, ἀθληταὶ μὲν ἐς τὰ στάδια, στρατιῶται δὲ
ἐς τὰ ὅπλα· τὰ δὲ ζῷα οὐ δεῖται τῆς ἔξωθεν
ἐπιρρώσεως, ἑαυτοῖς δὲ παροξύνει τὴν ἀλκήν, καὶ
ἑαυτὰ ἀνίστησι καὶ ἐγείρει. ὁ γοῦν σῦς μέλλων
ἐς μάχην ἰέναι πρὸς ταῖς λείαις πέτραις τοὺς
ὀδόντας ὑποθήγει. τοῦτό τοι καὶ Ὅμηρος τῷ
ζῴῳ μαρτυρῶν δῆλός ἐστι. καὶ μέντοι καὶ ὁ
λέων τῇ ἀλκαίᾳ ἑαυτὸν ἐπεγείρει μαστίζων, καὶ
βλακεύειν καὶ ἐλινύειν οὐκ ἐπιτρέπει.[1] καὶ τοῦτο
δὲ ὁ ποιητὴς εἰδὼς ᾄδει περὶ τοῦ λέοντος. οἱ δὲ
ἐλέφαντες τῇ προβοσκίδι ἑαυτοὺς παίουσιν ἐς τὸν
ἀγῶνα ἐξάπτοντες, ὅταν τούτου ᾖ καιρός, καὶ οὐ
δέονται τοῦ προσάσοντος καὶ ἐροῦντος οὐχ ἕδρας
ἔργον οὐδ' ἀμβολᾶς, οὐδὲ μὴν τὰ Τυρταίου μέτρα
ἀναμένουσι. ταῦρος δὲ ἡγεμὼν τῆς ἀγέλης ὅταν
ἡττηθῇ ἡγεμόνος ἄλλου, ἑαυτὸν ἀποκρίνει ἐς
χῶρον ἕτερον, καὶ ἑαυτῷ γίνεται γυμναστής, καὶ
ἀθλεῖ πᾶσαν ἄθλησιν κονιόμενος καὶ τοῖς δένδροις
τὰ κέρατα προσανατρίβων[2] καὶ τῇ τε ἄλλη

---

[1] ἐπιτρέπων.

10

AELIAN
ON THE CHARACTERISTICS OF ANIMALS
BOOK VI

1. Men have need of the spoken word to stimulate and persuade them to be good, to banish cowardice, to gather courage: athletes, with a view to running; soldiers, with a view to fighting. Animals however Animal courage need no extraneous encouragement but stimulate their prowess for themselves and rouse and incite themselves. For instance, the Boar when intending to do battle, whets his tusks on smooth rocks. Homer, you know, gives clear evidence touching the animal [*Il*. 13. 471]. Further, the Lion rouses himself by lashing himself with his tail and allows no idling and no repose. And this the poet knew when he spoke of the Lion [*Il*. 20. 170]. And Elephants inflame themselves for the fight, whenever the occasion arises, by beating themselves with their trunk: they need no one to sing to them and say ' This is no time for sitting still or for delaying ' [Bacc. *fr.* 11 Jebb], still less do they wait for the poems of Tyrtaeus. And when a Bull that is the leader of a herd The Bull in defeat is defeated by another leader, he departs to some other place and becomes his own trainer and practises every method of fighting, scattering the dust over himself and rubbing his horns against tree-

---

² *Reiske*: προσανατρίβων καὶ θηλειῶν βοῶν ἀπεχόμενος.

ἑαυτὸν ἐς τὴν ἄσκησιν τῆς ἀλκῆς ῥυθμίζων καὶ
οὖν καὶ ἀφροδίτης ἀπεχόμενος καὶ σωφρονῶν ὡς
Ἴκκος ὁ Ταραντῖνος, ὅνπερ οὖν ὑμνεῖ Πλάτων ὁ
Ἀρίστωνος[1] παρὰ τὸν τῆς ἀθλήσεως χρόνον
πάντα[2] συνουσίας ἀμαθῆ καὶ ἄπειρον διαμεῖναι
ἁπάσης. καὶ Ἴκκῳ μὲν ὄντι ἀνθρώπῳ καὶ
Ὀλυμπίων ἐρῶντι καὶ Πυθίων καὶ κλέους αἰσθα-
νομένῳ καὶ δόξης γλιχομένῳ οὐδὲν μέγα ἦν
κεκολασμένως καθεύδειν καὶ σωφρόνως· τὰ γὰρ
ἆθλά οἱ κλεινὰ καὶ ἦν καὶ ἐδόκει, κότινος Ὀλυμπι-
κὸς καὶ Ἰσθμικὴ πίτυς καὶ δάφνη Πυθική, καὶ
ζῶντα μὲν περιβλέπεσθαι, ἀποθανόντα[3] δὲ εὐφη-
μεῖσθαι. καὶ μέντοι καὶ τὸν κιθαρῳδὸν Ἀμοιβέα
ἀκούω γῆμαι μὲν ὡραιοτάτην γυναῖκα, ἀποσχέ-
σθαι δὲ αὐτῆς παρὰ πάντα τὸν χρόνον, παρ' ὃν
ἀγωνιούμενος ἐς τὰ θέατρα ᾔει. Διογένης δὲ ὁ
τῆς τραγῳδίας ὑποκριτὴς τὴν ἀκόλαστον κοίτην
ἀπείπατο παντελῶς πᾶσαν. Κλειτόμαχος δὲ ὁ
παγκρατιαστὴς καὶ κύνας εἴ ποτε εἶδε μιγνυμέ-
νους, ἀπεστρέφετο, καὶ ἐν συμποσίῳ δὲ εἰ λόγον
ἀκόλαστον ἤκουσε καὶ ἀφροδίσιον, ἐξαναστὰς
ἀπηλλάττετο. ἀνθρώπους δὲ ὄντας ποιεῖν ταῦτα
ἢ ὑπὲρ τοῦ κερδᾶναι χρήματα ἢ ὑπὲρ τοῦ φήμης
καὶ κλέους μεταλαχεῖν, οὐ πάνυ τι[4] θαυμαστόν·
ταύρῳ δὲ νικῶντι ταῦρον ἀντίπαλον ποῖα μὲν
κηρύγματα, ὦ παῖ Ἀρίστωνος, ἀποκηρύττουσι,[5]
ποῖα δὲ ἆθλα ἀποκρίνουσιν;

2. Τὰ ἄλογα καὶ τῶν συνήθων σφίσι γενομένων
ἀπέχεσθαι φιλεῖ καὶ φείδεσθαι πολλάκις. ἀκούω
γοῦν τὸν λόγον ἐκεῖνον. πάρδαλιν ἐκ νηπίου

---

[1] Π. ὁ τοῦ Ἀ. Jac, H.   [2] Ges : πάντη.

trunks and fitting himself in other ways to display his strength, and particularly abstaining from sexual acts and living continently like Iccus of Tarentum, whom Plato the son of Ariston celebrates [*Legg*. 8. 839 E] as refraining from all sexual commerce during the entire period of the Games. Now to Iccus, who was a man and who loved the Olympic and Pythian games and who understood what glory was and who longed for fame, it was no great matter to restrain himself and to spend the nights continently. For to him the prizes meant glory—the wild olive of Olympia, the Isthmian pine, and the Pythian laurel, admiration in his lifetime, and after death an honoured name. Again, the harper Amoebeus, I am told, married a woman of surpassing beauty but had no intercourse with her when he was going to the theatre in order to compete there. And Diogenes the actor in tragedies eschewed absolutely all licentious unions. And Clitomachus the pancratiast, if ever he saw dogs coupling, would turn away; and if at a wine party he heard some licentious and bawdy story, would get up and leave. There is nothing surprising that being men they should behave so, either in order to make money or to achieve renown and fame. But, O son of Ariston, when a bull overcomes his adversary, what proclamation announces his victory, and what prizes do men award him?

The continence of athletes

2. Brute beasts are in the habit of not molesting their companions and of frequently sparing them. For instance, I have heard the following story. A hunter had a Leopard which he had tamed from its

A tame Leopard

---

³ *Schn* : καὶ ἀποθανόντα. ⁴ οὐ πάντῃ.
⁵ *Reiske* : ὑπο-.

AELIAN

θηρατὴς ἀνὴρ ἡμερώσας εἶχεν, οἷα δήπου φίλην
ἢ ἐρωμένην ἀγαπῶν καὶ περιέπων ἰσχυρῶς.
οὐκοῦν ἔριφον αὐτῇ φέρων ζῶντα ἐδίδου, τροφὴν
ἐν ταὐτῷ καὶ ἡδονήν τινα ἐπινοήσας τῷ θηρίῳ ἐν
τῷ διασπᾶν τὸν ἔριφον, ἀλλὰ μὴ δοκεῖν ἐσθίειν
κενέβρειόν τε καὶ θνησείδιον. καὶ δὴ κομισθέντος
⟨τοῦ⟩[1] ἐρίφου ἡ δὲ ἐγκρατῶς ἔσχε, δεομένη
ἀπόσιτος εἶναι διὰ πλησμονήν. ἔδρασε δὲ καὶ τῇ
δευτέρᾳ τοῦτο· ἐδεῖτο γὰρ ἔτι ὡς φαρμάκου τοῦ
λιμοῦ. τῆς δὲ τρίτης ἐπιστάσης ἡμέρας ἐπείνη
μὲν καὶ συνήθως ἐπεδείκνυτο τοῦτο τῷ φθέγματι,
οὐ μὴν τοῦ ἐρίφου γενομένου δύο ἡμερῶν ἑαυτῇ
φίλου ἔτι προσήψατο, ἀλλὰ ἐκεῖνον μὲν εἴασεν,
ἄλλον δὲ ἔλαβεν. ἄνθρωποι δὲ καὶ ἀδελφοὺς
προύδοσαν καὶ τοὺς γειναμένους καὶ φίλους
ἀρχαίους, καὶ πολλοὶ πολλάκις.

3. Ἡ ἄρκτος ὅτι τίκτει σάρκα ἄσημον εἶτα τῇ
γλώττῃ διαρθροῖ αὐτὴν καὶ οἱονεὶ διαπλάττει,
ἄνω που λέλεκται. ὃ δὲ οὐκ εἶπον ἤδη, τοῦτο
εἰρήσεται νῦν, καὶ μάλα ἐν καλῷ. χειμῶνος μὲν
ἀποτίκτει, καὶ φωλεύει τεκοῦσα, καὶ ὑφορωμένη
τοὺς κρυμοὺς τὴν ἐπιδημίαν τοῦ ἦρος προσμένει,
οὐδ' ἂν πρὶν ἢ πληρωθῆναι τρεῖς μῆνας ἐξαγάγοι
ποτὲ τὰ βρέφη. ὅταν δὲ αἴσθηται ἑαυτῆς πεπλησ-
μένης, ὑφορωμένη τοῦτο ὡς νόσον, ζητεῖ φωλεόν.
ἔνθεν τοι καὶ[2] κέκληται τῇ ἄρκτῳ φωλεία τὸ
πάθος. εἶτα ἐσέρχεται οὐ βαδίζουσα, ἀλλὰ ὑπτία,
ἀφανίζουσα τοῖς θηραταῖς τὰ ἴχνη· ἑαυτὴν γὰρ
ἐπισύρει κατὰ τὰ νῶτα. καὶ παρεσελθοῦσα ἡσυ-

1 ⟨τοῦ⟩ add. H.   2 ἐντεῦθέν τοι.

earliest days and which he loved and tended assiduously as though it were his friend or darling. Now he brought a kid and gave it to the Leopard alive, thinking to provide it at once with food and with the pleasure of tearing the kid to pieces, and supposing that it would refuse to eat dead meat. In fact when the kid was brought the Leopard controlled itself: being full-fed it needed to abstain from food. And it did the same on the second day, for it still needed the medicine of starvation. But when the third day came it began to grow hungry and, as usual, showed that it was by the sound of its voice; for all that, it still would not touch the kid which had been its friend for two days, but left it alone, though it accepted another one.

Men however have betrayed even their brothers and their parents and old friends; there have been many and frequent cases.

3. I have described in some earlier passage [a] how The Bear the Bear produces some shapeless flesh and then licks it into shape and, so to say, moulds it. But what I have not already mentioned I will mention now, and this is a suitable occasion. It gives birth in the winter time, and having done so, hibernates; and as it dreads the frosts it awaits the coming of spring, and would never bring its cubs out until three full months have passed. But when it perceives that it is pregnant it dreads this as though it were some sickness, and seeks for a lair. (Hence the Bear's hibernation is called its ' lair period.') Then it enters, not on its feet but lying down, thus effacing its tracks for those who hunt it, for it drags itself along on its back. And

[a] See 2. 19.

χάζει, καὶ τρόπον τινὰ τὴν ἕξιν ῥινᾷ, καὶ δρᾷ
τετταράκοντα ἡμερῶν αὐτό. καὶ λέγει μὲν Ἀριστοτέλης ὅτι ἄρα δὶς ἑπτὰ ἡμερῶν ἀκίνητος μένει
καὶ ἀτρεμεῖ, τῶν δὲ ἄλλων στρέφεται μόνον.
ἄσιτος δὲ ἄρα διαμένει τῶν τετταράκοντα πασῶν
καὶ ἄτροφος, ἀπόχρη δὲ αὐτῇ τὴν δεξιὰν περι-
λιχμᾶσθαι. ἐκ δὲ τῆς συντήξεως τῆς ἄγαν
συνέπτυκται τὸ ἔντερον αὐτῇ καὶ συνῆλθεν. ὅπερ
εἰδυῖα, ὅταν προέλθῃ, τοῦ καλουμένου ἄρου τοῦ
ἀγρίου ἐσθίει· τὸ δὲ ἄρα φυσῶδες ὂν διίστησιν
αὐτῇ τὸ ἔντερον, καὶ εὐρύνει αὐτό, καὶ ἀποφαίνει
τροφῆς δεκτικόν. ὅταν δὲ αὖ πάλιν ᾖ πεπλη-
ρωμένη, μυρμήκων ἐσθίει, καὶ κενοῦται ῥᾷστα.
κενώσεις μὲν δὴ φυσικαὶ τῶν ἄρκτων καὶ πληρώ-
σεις ἐς δέον [1] εἴρηταί μοι μήτε ἰατρῶν μήτε
συγκραμάτων,[2] ὦ ἄνθρωποι, δεόμεναι.

4. Οἱ δράκοντες ὅταν ὀπώρας μέλλωσι γεύεσθαι,
τῆς πικρίδος καλουμένης ῥοφοῦσι τὸν ὀπόν·
ὀνίνησι δὲ ἄρα αὐτοὺς αὕτη πρὸς τὸ μὴ φύσῃς
τινὸς ὑποπίμπλασθαι. μέλλοντες δέ τινα ἐλλοχᾶν
ἢ ἄνθρωπον ἢ θῆρα, τὰς θανατηφόρους ῥίζας
ἐσθίουσι καὶ τὰς πόας μέντοι τὰς τοιαύτας. οὐκ
ἦν δὲ ἄρα οὐδὲ Ὅμηρος αὐτῶν τῆς τροφῆς
ἀμαθής. λέγει γοῦν ὅπως ἄνδρα [3] μένει περὶ τὸν
φωλεὸν εἰλούμενος, προεμπλησθεὶς σιτίων πολλῶν
φαρμακωδῶν καὶ κακῶν.

5. Οἱ ἔλαφοι τὰ κέρατα ἀποβαλόντες ἐσδύονται[4]
παρελθόντες ἐς τὰς λόχμας, τοὺς ἐπιόντας σφίσι

---

[1] εἰς δέον ἐς τοσοῦτον.   [2] Weigel : συγγραμμάτων.

having entered, it rests, and in some way reduces its figure; and this it does for forty days. Aristotle however says [*HA* 600 b 2] that the Bear remains motionless and does not stir for fourteen days, and for the remainder she just turns. So she passes the entire forty days without food or nourishment: it is enough for her to lick her right paw. And owing to excessive colliquescence her intestines become wrinkled up and compressed. Knowing this, as soon as she emerges she eats some of the plant called 'wild arum';[a] and as this induces flatulence, it opens up her gut, widens it, and renders it capable of admitting food. And when she has filled herself out once more, she eats some ants and obtains an easy evacuation. I have now sufficiently described how Bears empty and fill their bodies by natural means without any need, my fellow men, of doctors or of concoctions.

4. When Snakes intend to eat fruit they swallow the juice of the herb called *picris*.[b] It helps to prevent them from being filled with wind. And when they intend to lie in wait for a human being or an animal, they eat poisonous roots and herbs too of the same description. So it seems that Homer too was aware of what they ate. For instance, he tells [*Il.* 23. 93] how a Snake waits for a man, lying coiled up near its lurking-place, after it has taken its fill of much poisonous, deadly provender. *The Snake, its diet of poison*

5. When Deer have cast their antlers they go and hide in coverts and so protect themselves against at- *The Stag and its antlers*

---

[a] Cuckoo-pint.   [b] See 1.35 n.

---

[3] *Valck*: ἀνά.   [4] *Cobet*: ἐσδύν- MSS H.

φυλαττόμενοι, καὶ εἰκότως· ἔρημοι γὰρ τῶν
ἀμυντηρίων ὄντες ἀφῃρῆσθαι καὶ τὴν ἀλκὴν
πεπιστεύκασιν ἐν τῷ τέως. λέγονται δὲ καὶ
φυλάττεσθαι μή ποτε ἄρα νεαροῖς οὖσιν αὐτῶν τοῖς
στελέχεσιν [1] εἶτα προσπίπτουσα ἡ ἀκτὶς πρὶν ἢ
παγῆναι καὶ τοὺς καλουμένους χόνδρους λαβεῖν ἡ
δὲ τὴν σάρκα ὑποσήψῃ.

6. Οἱ παριόντες ἐς πόλεμον ἵπποι ὑπόπτους [2]
ἔχουσι καὶ τάφρων πηδήσεις καὶ ἅλλεσθαι βόθρον
καὶ διαβῆναι σταυροὺς καὶ σκόλοπας καὶ τὰ
τοιαῦτα. πάρεστι δὲ καὶ Ὁμήρου λέγοντος ἀκούειν ὑπὲρ τῶν τοιούτων

ὣς Ἕκτωρ ἀν᾿ ὅμιλον ἰὼν ἐλλίσσεθ᾿ [3] ἑταίρους,
τάφρον ἐποτρύνων διαβήμεναι. οὐδέ οἱ ἵπποι
τόλμων ὠκύποδες, μάλα δὲ χρεμέτιζον ἐπ᾿ ἄκρῳ
χείλει ἐφεσταότες· ἀπὸ γὰρ δειδίσσετο τάφρος
εὐρεῖ᾿, οὔτ᾿ ἄρ᾿ ὑπερθορέειν σχεδὸν οὔτε περῆσαι
ῥῃδίη.

7. Ἐν τῇ Αἰγύπτῳ περὶ τὴν λίμνην τὴν καλουμένην Μοίριδος,[4] ὅπου Κροκοδίλων πόλις, κορώνης
τάφος δείκνυται, καὶ τὴν αἰτίαν ἐκείνην Αἰγύπτιοί
φασι. τῷ βασιλεῖ τῷ τῶν Αἰγυπτίων (Μάρης δὲ
οὗτος ἐκαλεῖτο) ἦν κορώνης θρέμμα πάνυ ἥμερον,
καὶ τῶν ἐπιστολῶν ἃς ἐβούλετό οἱ κομισθῆναί ποι
θᾶττον ἐκόμιζεν αὕτη, καὶ ἦν ἀγγέλων ὠκίστη,
καὶ ἀκούσασα ᾔδει ἔνθα ἰθῦναι χρὴ τὸ πτερόν, καὶ
τίνα χρὴ παραδραμεῖν χῶρον, καὶ ὅπου ἤκουσαν

[1] Reiske : ἕλκεσιν.   [2] Gill : ὑπόπτως.
[3] εἰλίσσεθ᾿.   [4] μύριδος.

tackers; and rightly so, for as they are without means of self-defence they are convinced that they have for the time being lost their strength. It is said also that, while the stumps are still fresh and before they have hardened and the young horns, called *chondroi*, have begun to form, they take care that the sun's rays shall not fall upon them and cause the flesh to putrefy.

6. When Horses march to battle they become sus- The Horse picious at having to jump trenches, at having to leap in battle over pits and to pass through stakes and palisades and the like. And one finds Homer saying about such matters [*Il.* 12. 49]

'Thus Hector passing through the throng im- plored his comrades, urging them to cross the trench. But even his swift horses dared not, but neighed loudly as they stood upon the sheer brink, for the yawning trench dismayed them, not easy to leap from close up, nor to cross.'

7. In Egypt near the lake Moeris as it is called, The Crow of where is Crocodilopolis, the tomb of a Crow is King Mares pointed out. The Egyptians give the following reason. The King of Egypt (Mares [a] was his name) possessed a remarkable Crow which was quite tame. Any despatches that he wished to have delivered any- where this Crow would speedily carry; and it was the swiftest of messengers : having heard its destination, it knew where it must direct its flight to, which spot it must pass, and where it must pause on arrival. In

[a] Mares (or Marres) is the Greek form of 'Moeris', the nickname given to King Amenemhet III; see Hdt. 2. 101 with How–Wells's note.

AELIAN

ἀναπαύσασθαι. ἀνθ' ὧν ἀποθανοῦσαν ὁ Μάρης
ἐτίμησεν αὐτὴν καὶ στήλῃ καὶ τάφῳ.

8. Ἴδιον δὲ καὶ ὄνομα τῆς κατὰ τροφὴν κομιδῆς
ἕκαστα τῶν ζῴων κέκτηται. πωλοδαμνικὴ γοῦν
κληθείη τις ἂν καὶ σκυλακοτροφικὴ καὶ ἐλεφαντο-
κομία καὶ λεοντοτροφία καὶ ὀρνιθοτροφία καὶ τὰ
τοιαῦτα.

9. Σοφὰ [1] δὲ ἄρκτου ἦν ἄρα ἐκεῖνα. ἐὰν διώκηται μετὰ τῶν αὐτῆς σκυλακίων, προωθεῖ αὐτὰ ἐς
ὅσον δύναται· ὅταν δὲ συνίδῃ ὅτι ἀπεῖπε, τὸ μὲν
κατὰ τοῦ νώτου φέρει, τὸ δὲ κατὰ τοῦ στόματος,
καὶ δένδρου λαβομένη ἀναπηδᾷ· καὶ τὸ μὲν
ἔχεται τοῦ νώτου τοῖς ὄνυξι, τὸ δὲ ἐν τοῖς ὀδοῦσι
φέρεται ἀναθεούσης αὐτῆς. ταύρῳ δὲ λιμώτ-
τουσα ὅταν ἐντύχῃ, κατὰ μὲν τὸ καρτερὸν καὶ ἐξ
εὐθείας οὐ μάχεται, προσπαλαίει δέ, καὶ τοῦ
τένοντος λαβομένη κλίνει, καὶ ἅμμα σφίγγει.[2] ὁ
δὲ πιέζεται καὶ μέμυκε, καὶ τελευτῶν ἀπεῖπε καὶ
κεῖται, καὶ ἐκείνη ἐμπίπλαται.

10. Μαθεῖν δὲ ἀγαθὰ ζῷα καὶ ταύτῃ κατέγνω-
μεν. ἐπὶ τῶν Πτολεμαίων οἱ Αἰγύπτιοι τοὺς
κυνοκεφάλους καὶ γράμματα ἐδίδασκον καὶ ὀρχεῖ-
σθαι καὶ αὐλεῖν καὶ ψαλτικήν. καὶ μισθὸν κυνοκέφαλος ἐπράττετο ὑπὲρ τούτων, καὶ τὸ διδόμενον
ἐς φασκώλιον ἐμβαλὼν ἐξηρτημένον ἔφερεν, ὡς οἱ
τῶν ἀγειρόντων δεινοί. ὅτι δὲ Συβαρῖται καὶ
ὄρχησιν ἵππους ἐπαίδευσαν, πάλαι κεκήρυκται.

[1] σοφία.   [2] ἅμα ἐσθίει.

20

reward for these services Mares honoured it when dead with a monument and a tomb.

8. Every animal has a special word to denote the care spent on its upbringing. For example, one might speak of the 'breaking in' of horses, the 'rearing' of hounds, the 'grooming' of elephants, the 'rearing' of lions, the 'rearing' of birds, and so forth.

The care of animals

9. Now here the Bear shows its clever tricks. If it is pursued together with its cubs it pushes them along in front as far as it is able. But when it realises that they are exhausted, it carries one on its back and another in its mouth, then laying hold of a tree, climbs up. And one cub clings to its back with its claws, while the other is carried in the teeth of the Bear as it mounts. If when famished it comes across a bull, it does not engage in a straightforward battle of strength, but wrestles with it and seizing its neck brings it down and tightens its clench. And while the bull is being crushed it bellows, until at last it gives up and lies dead; and the Bear takes its fill.

The Bear and its cubs

10 (i). Here is further evidence to show that animals are apt at learning. Under the Ptolemies the Egyptians taught baboons their letters, how to dance, how to play the flute and the harp. And a baboon would demand money for these accomplishments, and would put what was given him into a bag which he carried attached to his person, just like professional beggars. It has long been noised abroad that the people of Sybaris have even taught horses how to

Docility of certain animals

21

ἐλεφάντων δὲ τὸ εὐπειθὲς ἐς τὰ μαθήματα καὶ τὸ
ῥᾴδιον ἀνωτέρω εἶπον. κύνες δὲ ἄρα καὶ τὰ οἴκοι
ὑπηρετεῖν τοῖς ἐκπαιδεύσασιν αὐτοὺς ἱκανοί, καὶ
ἀπόχρη πένητι δοῦλον κύνα ἔχειν. ἦσαν δὲ ἄρα
καὶ τῶν τοιούτων ἄδουλοι, ὥσπερ οὖν Ἀράβων
μὲν οἱ Τρωγλοδύται, Λιβύων δὲ οἱ Νομάδες, καὶ
τῶν Αἰθιόπων ὅσον [1] λιμνόβιόν ἐστι, πέρα τῆς ἐκ
τῶν ἰχθύων τροφῆς μεμαθηκὸς σιτεῖσθαι οὐδὲ ἕν.

Μέμνηται δὲ ὧν πάσχει τὰ ζῷα, καὶ δεῖταί γε
τέχνης τῆς ἐς τὴν μνήμην οὐ Σιμωνίδου, οὐχ
Ἱππίου, οὐ Θεοδέκτου, οὐκ ἄλλου τινὸς τῶν ἐς
τόδε τὸ ἐπάγγελμα καὶ τήνδε τὴν σοφίαν κεκηρυγ-
μένων. ἔνθα γοῦν ἀφῃρέθη ἡ βοῦς τὸν μόσχον,
ἐνταῦθα ἐλθοῦσα ὠδύρατο μυκηθμῷ συντρόφῳ τὸ
πάθος. καὶ βόες μέντοι ὑπὸ ζυγὸν ἰέναι μέλλοντες
οἱ μὲν μειδιῶσιν, οἱ δὲ ἐπὶ πόδα ἀναχωροῦσιν.
ἵππος δὲ ὅταν ἀκούσῃ ψαλίων κρότον καὶ χαλινοῦ
κτύπον, καὶ προστερνίδιά τε καὶ προμετωπίδια
θεάσηται, φριμάττεται ἐνταῦθα, καὶ τὰς ὁπλὰς
σκιρτῶν ἐπικροτεῖ καὶ ἐνθουσιᾷ, ἥ τε τῶν ἱπποβο-
σκῶν βοὴ ἐγείρει αὐτόν, καὶ τὰ ὦτα ὤρθωσεν
αὐτὸς καὶ τοὺς μυκτῆρας διέστησε μνήμῃ δρόμου
καὶ συνηθείας ἴυγγι ἀμάχῳ.

11. Τίκτει δὲ ἔλαφος παρὰ τὰς ὁδούς, καὶ
ἔοικέ γε σοφίᾳ τοῦτο δρᾶν· δέδοικε γὰρ τὰ θηρία
καὶ τὰς ἐξ αὐτῶν ἐπιβουλάς, τοὺς δὲ ἀνθρώπους
θαρρεῖ. καὶ ἐκείνων μὲν πεπίστευκεν ἀσθενεστέρα
οὖσα, τούτους δὲ ἀποδρᾶναι δύνασθαι οὐκ ἀμφιβάλ-
λει. καταπιανθεῖσα δὲ οὐκ ἂν ἔτι τέκοι παρὰ τὰς

----

[1] ὅσον τό.

dance.[a] Of the ease with which elephants can be
induced to learn I have spoken above.[b] Now dogs
are capable of managing household affairs for those
who have trained them, and for a poor man it is
enough to have a dog as slave. There are after all
people who are without slaves even of this kind,
among the Arabs for instance the Troglodytes, among
the Libyans the Nomads, and among the Ethiopians
all the lake-dwellers, people who have never learnt
to eat anything other than fish.

(ii). Animals retain the memory of their experi- Memory in
ences and have no need of those mnemonic systems animals
devised by Simonides, by Hippias, and by Theodectes,
or by any other of those who have been extolled for
their profession and their skill in this matter. For
instance, a cow goes to the spot where her calf was
taken from her and mourns for it, lowing as is her
wont. Some oxen too when about to be yoked express their pleasure, others draw back. And a horse
on hearing the clash of curb-chain and the clang of
bit, and seeing chest-plates and frontlets, begins to
snort and makes his hoofs ring as he prances, and is
in an ecstasy. And the shouting of the stablemen
stimulates him and he pricks up his ears and dilates
his nostrils as he remembers his galloping and yearns
irresistibly for his wonted exercise.

11. The Deer produces its young by the roadside The Deer
and appears to do so from a wise precaution, because and its
it dreads wild beasts and their designs, but has no young
fear of human beings: it knows full well that it is
weaker than the former, but has no doubt that it can
escape from the latter. But when it has grown fat it

<hr>

[a] See 16. 23.   [b] See 2. 11.

AELIAN

ὁδούς· οἶδε γὰρ ὅτι δραμεῖν ἐστι νωθεστέρα.
τίκτει οὖν ἐν τοῖς ἄγκεσι καὶ ἐν τοῖς δρυμοῖς καὶ
ἐν τοῖς αὐλῶσι.

12. Ἡ χερσαία χελώνη διατραγοῦσα ὀριγάνου
παρ᾽ οὐδὲν ποιεῖται τὸν ἔχιν. ἐὰν δὲ ἀπορήσῃ
αὐτοῦ,[1] πηγάνου ἐμφαγοῦσα ὥπλισται πρὸς τὸν
ἐχθρόν. ἐὰν δὲ ἑκατέρου ἀτυχήσῃ, ἀνῄρηται.

13. Ὁ ἔλαφος, ὡς ἀκούω, τὰ παρόντα ἀγαπᾷ,
καὶ οὐκ ἐρᾷ πλειόνων, ἀλλὰ σωφρονεῖ περὶ τὴν
γαστέρα τῶν ἀνθρώπων μᾶλλον. περὶ γοῦν τὸν
Ἑλλήσποντόν ἐστι λόφος, καὶ νέμονται κατὰ τοῦδε
ἔλαφοι, καὶ τῶν ὤτων αὐτοῖς τὸ ἕτερον διέσχισται,
περαιτέρω δὲ οὐ χωροῦσι τοῦ λόφου, οὐδὲ νομῆς
ἐρῶσι ξένης, οὐδὲ λειμῶνας ποθοῦσιν ἑτέρους
πόας χρεία περιττοτέρας· ἀπόχρη δὲ ἄρα τὰ
παρόντα αὐτοῖς δι᾽ ἔτους ὅλου. τί πρὸς ταῦτα, ὦ
ἄνθρωποι, ὑμεῖς, οὓς οὐκ ἂν ἐμπλήσειέ ποτε ἕως
θανάτου

οὐδ᾽ ὅσα λάινος οὐδὸς ἀφήτορος ἐντὸς ἐέργει;

14. Ἡ ὕαινα, ὡς Ἀριστοτέλης λέγει, ἐν τῇ
ἀριστερᾷ χειρὶ ἔχει δύναμιν ὑπνοποιόν, καὶ
ἐνεργάζεται κάρον μόνον προσθιγοῦσα. πάρεισι
γοῦν ἐς τὰ αὔλια πολλάκις, καὶ ὅταν ἐντύχῃ τινὶ
καθεύδοντι, προσελθοῦσα ἡσυχῇ[2] τὴν ὑπνοποιὸν
ὡς ἂν εἴποις χεῖρα προσέθηκε τῇ ῥινί, ὁ δὲ ἄγχε-
ταί[3] τε καὶ πιέζεται.[4] καὶ ἐκείνη μὲν ὑπορύττει

---

[1] τοῦ. [2] Jac: ἡσυχάζει καί MSS, del. H.
[3] ἕλκεται. [4] πιέζεται καὶ ἀναισθήτῳ μᾶλλον ἔοικε.

24

would no longer give birth by the roadside, for it knows that it is too sluggish to run, and so it brings forth its young in glens, in thickets, and in ravines.

12. The Land Tortoise after eating some marjoram treats a viper with contempt. But if it lacks marjoram it arms itself against its enemy by consuming some rue. If however it fails to find either, it is killed. *Tortoise and Viper*

13. The Deer (so I am told) is content with what is before it and has no further wants, but is more frugal than man in its appetite. For instance, in the neighbourhood of the Hellespont there is a hill pastured by Deer, which have one of their ears cleft, and they do not stray beyond this hill, do not want strange food, desire no other meadows from any need of a larger amount of grass; so what is at hand is enough for them the whole year round. What have you, O men, to say to this, you whom *The Deer, its frugality*

'not even all the wealth contained within the Archer's [a] threshold of stone ' [Hom. *Il.* 9. 404]

would satisfy until the day of death?

14. The Hyena, according to Aristotle,[b] has in its left paw the power of sending to sleep and can with a mere touch induce torpor. For instance, it often visits stables, and when it finds any creature asleep it creeps softly up and puts what you might call its sleep-inducing paw upon the creature's nose, and it is suffocated and overpowered. Meantime the Hyena scoops out the earth beneath the head to such *The Hyena, its narcotic powers*

[a] Apollo.
[b] Not in any extant work; *fr.* 321 (Rose, p. 347).

τὴν γῆν τὴν ὑπὸ τῇ κεφαλῇ ἐς τοσοῦτον, ἐς
ὅσον ἀνέκλασεν ἐς τὸν βόθρον καὶ τὴν φάρυγγα
ὑπτίαν ἀπέφηνε καὶ γυμνήν· ἐνταῦθα δὲ ἡ ὕαινα
ἐνέφυ καὶ ἀπέπνιξε καὶ ἐς τὸν φωλεὸν ἀπάγει.
τοῖς [1] κυσὶ δὲ ἐπιτίθεται ἡ αὐτὴ τὸν τρόπον
ἐκεῖνον. ὅταν ᾖ πλήρης ὁ τῆς σελήνης κύκλος,
κατόπιν λαμβάνει τὴν αὐγήν, καὶ τὴν αὑτῆς
σκιὰν ἐπιβάλλει τοῖς κυσί, καὶ παραχρῆμα αὐτοὺς
κατεσίγασε, καὶ καταγοητεύσασα ὡς αἱ φαρμακί-
δες εἶτα ἀπάγει σιωπῶντας, καὶ κέχρηται ὅ τι καὶ
βούλεται τὸ ἐντεῦθεν αὐτοῖς.

15. Ἔρωτα δελφῖνος ἐν Ἰασῷ ἐς μειράκιον
καλὸν πάλαι ᾀδόμενον ἄμοιρον μνήμης τῆς ἐξ
ἐμοῦ ἀπολιπεῖν οὔ μοι δοκεῖ, καὶ διὰ ταῦτα
εἰρήσεται.[2] τὸ γυμνάσιον τὸ τῶν Ἰασέων ἐπίκει-
ται τῇ θαλάττῃ, καὶ οἵ γε ἔφηβοι μετὰ τοὺς
δρόμους καὶ τὰς κονίστρας κατιόντες ἐνταῦθα
ἀπολοῦνται [3] κατά τι ἔθος ἀρχαῖον. διανηχομένων
οὖν αὐτῶν ἑνὸς τοῦ τὴν ὥραν ἐκπρεπεστάτου [4]
ἐρᾷ δελφὶς ἔρωτα δριμύτατον. καὶ τὰ μὲν πρῶτα
πλησίον γενόμενος ἐφόβησέ τε καὶ ἐξέπληξεν
αὐτόν, εἶτα μέντοι τῇ συνηθείᾳ φιλίαν τινὰ καὶ
εὔνοιαν ἐς ἑαυτὸν ἐκ [5] τοῦ παιδὸς ἰσχυρὰν ἐπηγά-
γετο. ἀθύρειν γοῦν μετ' ἀλλήλων ὑπήρξαντο, καὶ
πῇ μὲν ἡμιλλάσθην παρανηχομένω τε καὶ ἐρίζοντε,
πῇ δὲ ὁ παῖς ἀναβαίνων ὡς πῶλον ἱππότης,
ὑπονηχομένου τοῦ ἐραστοῦ γαῦρος ἐφέρετο. καὶ
ἦν τοῖς Ἰασεῦσι καὶ τοῖς ξένοις τὸ πραττόμενον
ἀξιόζηλον. προῄει μὲν γὰρ τὰ παιδικὰ ὁ δελφὶς

[1] καὶ τοῖς.   [2] εἰρήσεται ὁ ἔρως.
[3] ἀπολούονται.   [4] ἐκπρεποῦς.

a depth as makes the head bend back into the hole,
leaving the throat uppermost and exposed. Thereupon it fastens on to the animal, throttles it, and
carries it off to its lair. And it attacks dogs in the
following manner. When the moon's disc is full, the
Hyena gets the rays behind it and casts its own
shadow upon the dogs and at once reduces them to
silence, and having bewitched them, as sorceresses do,
it then carries them off tongue-tied and thereafter
puts them to such use as it pleases.

15. The story of a Dolphin's love for a beautiful
boy at Iassus [a] has long been celebrated, and I am
determined not to leave it unrecorded; it shall
accordingly be told.

The gymnasium at Iassus is situated close to the
sea, and after their running and their wrestling the
youths in accordance with an ancient custom go
down there and wash themselves. Now while they
were swimming about, a Dolphin fell passionately in
love with a boy of remarkable beauty. At first when
it approached, it frightened the boy and completely
scared him; later on however, through constant
meeting, it even led the boy to conceive a warm
friendship and kindly feelings towards it. For instance, they began to sport with one another; and
sometimes they would compete, swimming side by
side in rivalry, sometimes the boy would mount, like
a rider on a horse, and be carried proudly along on
the back of his lover. And to the people of Iassus
and to strangers the event seemed marvellous. For

Dolphin and boy at Iassus

---

[a] Town on SW coast of Caria.

⁵ καὶ ἐκ.

φέρων ἐπὶ πλεῖστον τῆς θαλάττης καὶ ἐς ὅσον τῷ
παιδὶ εἶχεν ὀχουμένῳ καλῶς·[1] εἶτα ὑπέστρεφεν
καὶ ἦγε τοῦ αἰγιαλοῦ πλησίον, καὶ ἀλλήλων
διαλυόμενοι ὁ μὲν ἐς τὸ πέλαγος, ὁ δὲ ἐς τὰ
οἰκεῖα ἐπανῄεσαν. ἀπήντα δὲ ὁ δελφὶς ἐς τὸν
καιρὸν τῆς τῶν γυμνασίων ἀφέσεως, ὅ τε παῖς
ἥδετο τῇ προσδοκίᾳ τῇ τοῦ φίλου καὶ τῇ σὺν
αὑτῷ παιδιᾷ, καὶ πρὸς τῷ κάλλει τῷ φυσικῷ
περίβλεπτος ἦν, οἷα δήπου μὴ μόνον τοῖς ἀνθρώποις, ἀλλὰ καὶ τοῖς ἀλόγοις δοκῶν ὡραιότατος.
οὐ μέντοι μετὰ μακρὸν καὶ οὗτος ὁ ἀντέρως[2]
ἡττήθη τοῦ φθόνου. ἔτυχε γοῦν ὁ παῖς πλείω
γυμνασάμενος, καὶ καμὼν ἑαυτὸν τῷ ὀχοῦντι κατὰ
τὴν γαστέρα ἐπιβάλλει, καί πως ἔτυχεν ἡ τοῦ
ζῴου ἄκανθα ἡ κατὰ τοῦ νώτου ὀρθὴ οὖσα, καὶ
τῷ ὡραίῳ τὸν ὀμφαλὸν κεντεῖ. εἶτά τινες φλέβες
ὑπορρήγνυνται, καὶ αἵματος ἔπειτα ῥοὴ πολλή,
καὶ ὁ παῖς ἐνταῦθα ἀποθνῄσκει. ὅπερ οὖν ὁ
δελφὶς συναισθόμενος ἐκ τοῦ βάρους (ἐπέκειτο
γὰρ οὐ συνήθως κοῦφος, ἅτε μὴ τῷ πνεύματι
ἑαυτὸν ἐλαφρίζων) καὶ θεασάμενος πορφυροῦν ἐκ
τοῦ αἵματος τὸ πέλαγος, τὸ πραχθὲν συνῆκεν καὶ
ἐπιβιῶναι τοῖς παιδικοῖς οὐκ ἐτόλμησε. πολλῇ
τοίνυν τῇ ῥύμῃ[3] χρησάμενος, ὥσπερ οὖν ῥόθιον[4]
σκάφος, εἶτα ἑαυτὸν ἐς τοὺς αἰγιαλοὺς ἑκὼν[5]
ἐξέβρασε, καὶ τὸν νεκρὸν συνεξήνεγκε, καὶ ἔκειντο
ἄμφω ὁ μὲν τεθνεώς, ὁ δὲ ψυχορραγῶν. Λάϊος

---

[1] Schn : καλῶς εἰς τοσοῦτον.   [2] ἀντερῶν.
[3] Jac : ῥώμῃ.   [4] Ῥοδίων.   [5] ἕλκων.

the Dolphin would go a long way out to sea with its
darling on its back and as far as it pleased its rider;
then it would turn and bring him close to the beach,
and they would part company and return, the Dolphin to the open sea, the boy to his home. And the
Dolphin used to appear at the hour when the gymnasium was dismissed, and the boy was delighted to
find his friend expecting him and to play together.
And besides his natural beauty, this too made him
the admired of all, namely that not only men but
even dumb animals thought him a boy of surpassing
loveliness.

In a little while however even this mutual affection
was destroyed by Envy.[a] Thus, it happened that
the boy exercised himself too vigorously, and in an
exhausted state threw himself belly downwards on to
his mount, and as the spike on the Dolphin's dorsal
fin chanced to be erect it pierced the beautiful boy's
navel. Whereupon certain veins were severed;
there followed a gush of blood; and presently the
boy died. The Dolphin perceiving this from the
weight—for the boy lay heavier than usual, as he
could not lighten himself by breathing—and seeing
the surface of the water crimson with blood, realised
what had happened and could not bear to survive its
darling. And so with all the gathered force of a ship
dashing through the waves it made its way to the
beach and deliberately cast itself upon the shore,
bringing the dead body with it. And there they both
lay, the boy already dead, the Dolphin breathing its
last. (But Laïus,[b] my good Euripides, did not act

[a] I.e. divine envy; cp. Soph. Ph. 776.
[b] Laïus, King of Thebes, loved Chrysippus, the son of
Pelops. See Nauck TGF p. 632.

AELIAN

δὲ ἐπὶ Χρυσίππῳ, ὦ καλὲ Εὐριπίδη, τοῦτο οὐκ
ἔδρασε, καίτοι τοῦ τῶν ἀρρένων ἔρωτος, ὡς λέγεις
αὐτὸς καὶ ἡ φήμη διδάσκει, Ἑλλήνων πρώτιστος
ἄρξας. ἀμειβόμενοι δὲ [1] Ἰασεῖς τὴν φιλίαν ἐκείνων τὴν ἰσχυράν, ἀπέφηναν τάφον κοινὸν ὡραίου
μειρακίου καὶ δελφῖνος ἐρωτικοῦ, καὶ στήλην
ἐπέστησαν. καλὸς παῖς ἱππεύων ἐπὶ δελφῖνος ἦν.
καὶ νόμισμα δὲ ἀργύρου καὶ χαλκοῦ εἰργάσαντο,
καὶ ἐνέθλασαν σημεῖον τὸ ἀμφοῖν πάθος, καὶ
μνήμῃ παρέδοσαν ἔργον ⟨τοῦ⟩ [2] τοσούτου θεοῦ
τιμῶντες οἱ ἐκεῖθι. [3] πυνθάνομαι δὲ καὶ ἐν τῇ
Ἀλεξάνδρου πόλει κατὰ τὸν Πτολεμαῖον τὸν
δεύτερον ἐρασθῆναι δελφῖνα ἔρωτα παραπλήσιον
καὶ ἐν Δικαιαρχίᾳ τῆς Ἰταλίας. ἅπερ οὖν εἰ
Ἡρόδοτος ἔγνω, οὐκ ἂν ἐμοὶ δοκεῖν ἐθαύμασε τῶν
ἐπ᾽ Ἀρίονι [4] τῷ Μηθυμναίῳ ἧττον αὐτά.

16. Λιμοῦ μέλλοντος ἐπιδημεῖν αἰσθητικῶς ἔχουσι κύνες καὶ βόες καὶ ὖς καὶ αἶγες καὶ ὄφεις καὶ
ζῷα ἄλλα, καὶ λοιμοῦ δὲ ἀφιξομένου συνίησι
πρώτιστα καὶ σεισμοῦ. προγινώσκει δὲ καὶ ὑγίειαν [5] ἀέρων καὶ εὐφορίαν καρπῶν. καὶ λόγου
μὲν οὐ μετείληχε τοῦ καὶ σώζειν καὶ ἀποκτείνειν
δυναμένου, τῶν γε μὴν προειρημένων οὐ διαμαρτάνει.

17. Ἐν τῇ τῶν καλουμένων Ἰουδαίων γῇ ἢ
Ἰδουμαίων ᾖδον οἱ ἐπιχώριοι καθ᾽ Ἡρώδην τὸν
βασιλέα ἐρασθῆναι μείρακος ὡρικῆς δράκοντα

---

[1] δὲ καί.　　　　　　　　[2] ⟨τοῦ⟩ add. H.
[3] κεῖθι.　　　　　　　　　[4] Ἀρίωνι.
[5] Schn : ὑγείαν.

30

so in the case of Chrysippus, although, as you yourself
and the common report tell me, he was the first among
the Greeks to inaugurate the love of boys.) And
the people of Iassus to requite the ardent friendship
between the pair built one common tomb for the
beautiful youth and the amorous Dolphin, with a
monument at the head. It was a handsome boy
riding upon a Dolphin. And the inhabitants struck
coins of silver and of bronze and stamped them with a
device showing the fate of the pair, and they com-
memorated them by way of homage to the operation
of the god[a] who was so powerful.

And I learn that at Alexandria also, in the reign of
Ptolemy II,[b] a Dolphin was similarly enamoured; at
Puteoli also, in Italy. So, had these facts been
known to Herodotus, I think they would have sur-
prised him no less than what happened to Arion of
Methymna.[c]

16. Dogs, oxen, swine, goats, snakes, and other Prophetic
animals have a presentiment of an impending famine; powers of
they are the first too to know when a pestilence or an animals
earthquake is approaching. They can foretell fair
weather and the fertility of the crops. Though devoid of reason, which can be a man's salvation or his
destruction, they are not mistaken at any rate in the
matters mentioned above.

17. In the country of those known as Judaeans or Serpent in
Edomites the natives of the time of Herod the King love with
used to tell of a Serpent of enormous size being girl

[a] The God of Love.
[b] Ptolemy II, Philadelphus, 308–246 B.C.
[c] See Hdt. 1. 23–4.

μεγέθει μέγιστον, ὅσπερ οὖν ἐπιφοιτῶν εἶτα μέντοι
τῇ προειρημένῃ συνεκάθευδε σφόδρα ἐρωτικῶς.
οὐκοῦν ἡ μεῖραξ τὸν ἐραστὴν οὐκ ἐθάρρει, καίτοι
προσέρποντα ὡς ἐνῆν πρᾴτατά τε καὶ ἡμερώτατα.
ὑπεξῆλθεν οὖν, καὶ διέτριψε μῆνα, οἷα δήπου
λήθην τοῦ δράκοντος ἕξοντος κατὰ τὴν τῆς
ἐρωμένης ἀποδημίαν. τῷ δὲ ἄρα ἡ ἐρημία ἐπέ-
τεινε τὸ πάθος, καὶ ἐφοίτα μὲν ὁσημέραι καὶ
νύκτωρ· οὐ μὴν ἐντυγχάνων ᾗ ἠβούλετο, ὡς
ἐραστὴς ἀτυχῶν ἐν τῷ πόθῳ καὶ ἐκεῖνος ἤλγει.
ἐπεὶ δὲ ἡ ἄνθρωπος ὑπέστρεψεν αὖθις, ὁ δὲ
ἀφικνεῖται, καὶ περιβαλὼν τῷ λοιπῷ σώματι, τῇ
οὐρᾷ τὰς κνήμας τῆς ἐρωμένης πεφεισμένως
ἔπαιεν, ὑπεροφθείς τε καὶ μηνίων δῆθεν. οὔκουν[1]
ὁ καὶ τοῦ Διὸς ἄρχων αὐτοῦ καὶ τῶν θεῶν τῶν
ἄλλων οὐδὲ τῶν ἀλόγων ὑπερορᾷ, ἀλλ' ὅπως ἔχει
πρὸς αὐτὰ καὶ διὰ τούτων καὶ δι' ἄλλων ἀποδεί-
κνυται.

18. Οἱ ὄφεις ἑαυτοῖς συνεγνωκότες τὸν στό-
μαχον λεπτὸν καὶ μακρὸν ἔχουσιν, ὅμως ὄντες
ἀδηφάγοι καὶ παμβορώτατοι, ὡς Ἀριστοτέλης
λέγει, ἀνίστανται ὀρθοὶ καὶ ἐπ' ἄκρας τῆς οὐρᾶς[2]
ἑστᾶσι, καὶ ἡ τροφὴ κατολισθάνει αὐτοῖς, καὶ ἐς
τὸν ὄγκον τοῦ σώματος ἀποχωρεῖ· ἄποδες δὲ
ὄντες εἶτα ἕρπουσιν ὤκιστα. ἤδη δὲ καὶ ἀκοντίων
δίκην ἑαυτόν τις μεθίησι καὶ ἐπιφέρεται, καὶ τό
γε ὄνομα ἐξ οὗ δρᾷ ἔχει· κέκληται γὰρ ἀκοντίας.

enamoured of a lovely girl: he used to visit her and later even slept with her like an ardent lover. Now the girl was terrified of her lover, although he slid up to her as softly and gently as he could. So she escaped from him and remained away for a month, supposing that the Serpent in consequence of his darling's absence would forget her. But loneliness augmented his misery, and every day and night he used to haunt the place. Since however he did not find the object of his desire, he too felt all the pains of a disappointed lover. But when the girl came back once more, he arrived and, encircling her with the rest of his body, with his tail gently lashed her legs, presumably in anger at finding himself despised. So he [a] that is above even Zeus himself and the other gods does not overlook even brute beasts, but by these and by other acts manifests his relations towards them.

18. Snakes, conscious that they have a narrow, The Snake, elongated gullet, despite the fact that they are its voracity greedy and exceedingly voracious, as Aristotle says and speed [HA 594 a 18], rise upright and stand upon the tip of their tail, so that food slides down into them and passes into the bulk of their body. And having no feet they crawl at a great speed. Indeed one snake launches itself and flies with the speed of a javelin; and its name is derived from its action, for it is called *Acontias* (the Javelin-snake).

[a] The God of Love.

---

¹ οὐκοῦν.　　　² *Schn*: τὰς οὐράς.

AELIAN

19. Τῶν ἐν ᾠδαῖς τε καὶ μούσαις ὀρνίθων
οὐδεὶς διαλέληθεν, ἀλλ' ἴσμεν χελιδόνας καὶ
κοσσύφους καὶ τὸ ⟨τῶν⟩[1] τεττίγων φῦλον, καὶ
κίτταν λάλον καὶ βομβοῦσαν ἀκρίδα καὶ πάρνοπα
ὑποκρίζοντα καὶ μὴ σιωπῶσαν τρωξαλλίδα, ἀλκυό-
νας τε ἐπὶ τούτοις καὶ ψιττακούς· τῶν δὲ ἐνύδρων
ὀλολυγὼν οὐ σιωπᾷ. φθέγγεται δὲ αὐτῶν τὰ μὲν
γοερὰ καὶ θηλύφωνα, τὰ δὲ ὄρθια καὶ διάτορα·
καὶ τὰ μὲν ἀπὸ τῶν κλάδων ἐπὶ τοὺς κλάδους
μεταθέοντα ᾄδει, ὥσπερ οὖν οἴκους ἐξ οἴκων
ἀμείβοντα,[2] τὰ δὲ ἐν τοῖς λειμῶσι κατάδει,
οἱονεὶ πανηγυρίζοντα, καὶ βίον ὡς ἂν εἴποις
ἀνθηρὸν καὶ ἁβρὸν διαιτώμενα τὴν ἦρος ἐπιδημίαν μελῳδίαις ἔγωγ' ἂν φαίην εὐφημεῖ. κύκνων
δὲ πέρι καὶ ὅτου θεῶν θεράποντές εἰσιν ἀνωτέρω
εἶπον. ἡ κίττα δ' οὖν καὶ τῶν ἄλλων φωνημάτων
μιμηλότατόν ἐστι, τοῦ δὲ ἀνθρωπικοῦ πλέον.
ἰδιάζει δὲ ταῖς μιμήσεσι τῶν τοιούτων ὅ τε ἄνθος
καλούμενος[3] καὶ ἡ σάλπιγξ καὶ ἡ ἴυγξ καὶ ὁ κόραξ.
καὶ ὁ μὲν ἄνθος ὑποκρίνεται χρεμέτισμα ἵππου,
τὴν σάλπιγγα δὲ ἡ ὁμώνυμος, καὶ τὸν πλάγιον ἡ
ἴυγξ αὐλόν· βούλεται δὲ τῶν ὄμβρων μιμεῖσθαι
τὰς σταγόνας ὁ κόραξ.

---

[1] ⟨τῶν⟩ add. H.
[2] οἴκους . . . ἀμείβοντα] ἐπὶ τοὺς οἴκους ἐκ τῶν προτέρων
οἴκων ἀμείβοντες διὰ τὴν τρυφὴν καὶ τὴν τοῦ βίου θρύψιν.

34

19. Not one of the birds that sing and make melody has escaped observation, but we know that swallows, blackbirds, and the tribe of cicadas sing, that the jay is talkative, that the cricket buzzes,[a] the locust makes a light strumming, the grasshopper is not silent, and moreover that halcyons and parrots are vocal, while among aquatic creatures the croak of the male frog is not silent. And of these some utter a plaintive feminine note, others a note shrill and piercing; and some sing as they hurry from branch to branch, as though they were changing house, while others carol in the meadows as though they were holding festival, and while leading an existence that is, as it were, all flowers and delicacy, hail (so I would say) with their music the coming of spring. Touching swans and the god whose ministers they are I have spoken above.[b] Now the jay can imitate all other sounds but especially the human voice. And the buff-backed heron, as it is called, and the *salpinx* (trumpet) [c] and the wryneck and the raven are peculiarly fitted to imitate the following sounds. The buff-backed heron represents the neighing of a horse; the salpinx, the instrument whose name it bears; and the wryneck, the cross-flute; while the raven tries to imitate the sound of raindrops.

The song of Birds

ability to imitate other sounds

[a] Ἀκρίς elsewhere in Ael. is a *locust*; it can hardly bear this meaning here. I have ventured to render it ' cricket,' signifying the ' field-cricket,' *Acheta* or *Gryllus campestris*.
[b] See 2. 32; 5. 34.
[c] Thompson does not cite this passage in his *Glossary*, s.v. σάλπιγξ, which cannot here = ὀρχίλος, a wren. Gossen (§ 192) suggests the Roller, *Coracias garrulus*.

---

[3] ὁ καλ-.

20. Σκορπίων μὲν ὁ ἄρρην ἐστὶ χαλεπώτατος,
ὁ δὲ θῆλυς δοκεῖ πραότερος. ἀκούω δὲ αὐτῶν
γένη ἔνδεκα [1]· λευκὸν εἶναι καὶ αὖ πάλιν πυρρόν
τινα, ⟨καὶ⟩[2] καπνώδη [3] ἄλλον ⟨καὶ⟩ μέλανα ἐπὶ
τούτοις· πέπυσμαι δὲ καὶ χλωρὸν καὶ γαστρώδη
τινά καὶ καρκινώδη [4] ἄλλον· τόν γε μὴν χαλεπώ-
τατον φλογώδη ᾄδουσι.[5] παρείληφα δὲ ἄρα φήμῃ
καὶ πτερωτοὺς καὶ δικέντρους τινάς· καί που
ἑπτὰ ἔχων σφονδύλους ὤφθη τις. σκορπίος δὲ
οὐκ ᾠὰ ἀλλὰ ζῷα ἀποτίκτει. χρὴ δὲ εἰδέναι ὅτι
καί φασί τινες οὐκ ἐκ τῆς πρὸς ἀλλήλους ὁμιλίας
γίνεσθαι τὴν ἐπιγονὴν τοῖς ζῴοις τοῖσδε, ἀλλ᾽ . . .[6]
ἐς τὰ καύματα ἄγαν τίκτειν σκορπίους. ἐγχρίσας
δὲ ἕκαστος αὐτῶν τὸ κέντρον ὁποῖα ἐργάζεται καὶ
ἀναιρεῖ τίνα τρόπον ἀλλαχόθεν εἴσεσθε.

21. Ἐν Ἰνδοῖς, ὡς ἀκούω, ἐλέφας καὶ δράκων
ἐστὶν ἔχθιστα. οὐκοῦν οἱ μὲν ἐλέφαντες ἀποσπῶν-
τες τῶν δένδρων τοὺς κλάδους, ἐκείνους νέμονται.
ὅπερ οὖν εἰδότες οἱ δράκοντες ἐπ᾽ αὐτὰ μὲν
ἀνέρπουσιν, τὸ δὲ ἥμισυ σφῶν αὐτῶν τὸ οὐραῖον
τῶν δένδρων περιβάλλουσι τῇ κόμῃ, τὸ δὲ ἐς τὴν
κεφαλὴν προϊὸν ἥμισυ μεθῆκαν καλωδίου δίκην
ἀπηρτημένον. καὶ ὁ μὲν προσῆλθεν ἀποδρέψασθαι

---

[1] γ. ἐννέα MSS, del. H (1864), but (1858) γ. ἔνδεκα which
Wellmann also reads.
[2] ⟨καὶ⟩ . . . ⟨καὶ⟩ add. H.
[3] καπνοειδῆ.
[4] καρκινοειδῆ . . . φλογοειδῆ.
[5] καλοῦσι.
[6] Lacuna.

---

[a] Steier (art. Spinnentiere, RE 3 A 1801) identifies four of
them thus : λευκός, the young of most scorpions; πυρρός,

20. The male Scorpion is exceedingly ferocious, The
but the female seems to be of a milder temper. And Scorpion: various
I have heard that there are eleven kinds:[a] one is kinds
white, while another is red, another smoke-colour,
there is also a black kind; I have learnt also that
there is one kind that is green, another pot-bellied,
and another that resembles a crab. But it is commonly said that the fiercest is the fiery-coloured one.
I have also learned by report that there are Scorpions
with wings and others with a double sting, and somewhere one has been seen with seven vertebrae. The
Scorpion is not oviparous but viviparous. And it
should be known that some say that the offspring of
these creatures are not produced by mating but . . .
heat causes Scorpions to be exceedingly prolific.
And how they all inflict their sting, and the effect this
produces, and how they kill, you will learn from
another source.

21. In India, I am told, the Elephant and the Elephant
Python (?) are the bitterest enemies. Now Elephants and Python
draw down the branches of trees and feed upon them.
And the Pythons, knowing this, crawl up the trees
and envelop the lower half of their bodies in the
foliage, but the upper portion extending to the head
they allow to hang loose like a rope. And the
Elephant approaches to pluck the twigs, whereat the
Python springs at its eyes and gouges them out.

Buthus occitanus ; μέλας, *Androctonus afer* (cp. 15. 26 ; 17. 40);
χλωρός, if equivalent to the μελίχλωρος of Nic. *Th.* 797, may be
Androctonus (Buthus) australis. The πτερωτός is perhaps the
harmless insect *Panorpa communis*. There are no scorpions
'with two stings' or 'with seven vertebrae.' Καρκινοειδής is
perhaps the Crab-spider, *Thomisius onustus* ; see J. H. Fabre,
Life of the Spider, 181. See also Gossen §§ 42–4.

τῶν ἀκρεμόνων ὁ ἐλέφας, ὁ δὲ δράκων ἐμπηδᾷ
τοῖς ὀφθαλμοῖς καὶ ἐξορύττει, εἶτα τῷ τραχήλῳ
περιερπύσας, †τείνων† ¹ τῷ οὐραίῳ μέρει καὶ
σφίγγων θατέρῳ ἀπάγχει τὸ θηρίον ἀήθει βρόχῳ
καὶ καινῷ.

22. Ἔχθιστα δὲ τῷ μὲν λέοντι πῦρ καὶ ἀλεκ-
τρυών, ὕαινα δὲ τῇ παρδάλει, σκορπίῳ ² δὲ
ἀσκαλαβώτης·³ νάρκη γοῦν τὸν σκορπίον καταλαμβάνει προσαχθέντος οἱ τοῦ ζῴου τοῦ προειρημένου. δράκοντα δὲ ἐλέφας ὀρρωδεῖ· ὑποζύγιον
δὲ πᾶν τὴν μυγαλῆν οὐ θαρρεῖ, ἀστακὸς δὲ
πολύποδα. καὶ μέντοι ⟨καὶ⟩⁴ προωθούμενοι ἐκ
τῶν τεγῶν οἱ κύνες, οὐκ ἂν αὐτοὺς ῥίψειας· τοῦ
γάρ τοι κινδύνου δεδοίκασι τὸ μέγεθος.

23. Οἷα δὲ ἄρα σοφίσματα καὶ τοῖς σκορπίοις
ἡ φύσις ἔοικε δοῦναι καὶ τοῖσδε ἴδια. οἱ Λίβυες
τὸ πλῆθος αὐτῶν ὑφορώμενοι καὶ τὰ ⁵ τεχνάσματα
μηχανὰς αὐτοῖς μυρίας ⁶ ἀντεπινοοῦσι ⁷ κοῖλα
ὑποδήματα φοροῦντες καὶ ὑψηλοὶ καὶ μετέωροι
καθεύδοντες καὶ τῶν τοίχων ἀναστέλλοντες τὰς
κειρίας ⁸ καὶ τῶν κλινῶν τοὺς πόδας ἐς ὑδρίας
ὕδατος πεπληρωμένας ἐντιθέντες, καὶ οἴονται τὸ
λοιπὸν ἐν ἀδείᾳ τε καὶ εἰρήνῃ καθεύδειν πολλῇ.
οἱ δὲ ὁποῖα παλαμῶνται. σκορπίος εἰ λάβοιτο
ὁπόθεν ἑαυτὸν ἐξαρτήσει κατὰ τοῦ ὀρόφου,⁹
ἔχεται τούτου ταῖς χηλαῖς καὶ μάλα ἐγκρατῶς

¹ τείνων corrupt.
² σκορπίος MSS, σκορπίοις, Reiske.
³ Reiske : ἀσκαλαβώτῃ.
⁴ ⟨καὶ⟩ add. H.   ⁵ Jac : πάντα τά.

Next the snake winds round the Elephant's neck,
and [as it clings to the tree?][a] with the lower part
of its body, it tightens its hold with the upper part
and strangles the Elephant with an unusual and
singular noose.

22. To the lion fire and a cock are utterly hateful; Enmities
to the leopard a hyena, to the scorpion a gecko. and fears of
Thus, if the aforesaid creature is brought near to a animals
scorpion, the latter is seized with numbness. And
the elephant shrinks from the python; and every
beast of burden dreads the shrew-mouse; the lobster, the octopus. Furthermore if you were to try
to push dogs off the roof, you would not succeed in
throwing them down: they are afraid of the great
danger involved.

23. What ingenuity, peculiar to their kind, Nature The
seems to have imparted to Scorpions! The people Scorpion in
of Libya dreading their numbers and their machina- Libya
tions, devise endless schemes to counter them: they
wear high boots; they sleep in beds raised high
above the ground, setting their bed-cords away
from the walls; they place the feet of their beds in
vessels full of water, and imagine that they wil
thereafter sleep without fear and in peace. But what
tricks do the Scorpions devise! If a Scorpion can
find some spot in the roof to which he can hang,
he clings to it firmly with his claws and lets down

[a] The text is corrupt and the translation is conjectural.

---

[6] *Reiske* : μηχαναῖς αὐτοὺς μυρίαις.
[7] ἀντεπινοοῦσι φυλαττόμενοι.
[8] *Jac* : χεῖρας.   [9] τὸν ὄροφον.

AELIAN

καὶ καθῆκε τὸ κέντρον. οὐκοῦν ὁ δεύτερος
κάτεισιν[1] ἐκ τῆς στέγης, καὶ διὰ τοῦ πρώτου
καθέρπει,[2] καὶ τοῦ κέντρου τοῦ ἐκείνου ἔχεται καὶ
αὐτὸς ταῖς χηλαῖς, καὶ τό γε ἑαυτοῦ[3] μετέωρον
εἴασε κέντρον· καὶ ὁ τρίτος ἐκεῖθεν ἔχεται, καὶ ὁ
τέταρτος ἐκ τοῦ τρίτου, καὶ ὁ πέμπτος κατὰ
στοῖχον, καὶ οἱ ἐπ᾽ ἐκείνοις διὰ τῶν πρώτων
καθέρποντες. εἶτα ὁ τελευταῖος ἔπαισε τὸν καθεύδοντα καὶ διὰ τῶν ἀνωτέρω ἀνέρπει,[4] καὶ ὁ μετ᾽
ἐκεῖνον καὶ ὁ κάτωθεν τρίτος καὶ οἱ λοιποί,
ἔστε[5] οἱ πάντες ἀλλήλων ἀπελύθησαν οἷα δήπου
λύσαντες ἄλυσιν.

24. Δολερὸν χρῆμα ἡ ἀλώπηξ. ἐπιβουλεύει
γοῦν τοῖς χερσαίοις ἐχίνοις τὸν τρόπον τοῦτον.
ὀρθοὺς αὐτοὺς καταγωνίσασθαι ἀδύνατός ἐστι. τὸ
δὲ αἴτιον, αἱ ἄκανθαι ἀνείργουσιν αὐτήν. ἡ δὲ
ἡσύχως καὶ πεφεισμένως ⟨ἔχουσα⟩[6] τοῦ ἑαυτῆς
στόματος ἀνατρέπει αὐτοὺς καὶ κλίνει ὑπτίους,
ἀνασχίσασά τε ἐσθίει ῥᾳδίως τοὺς τέως φοβερούς.
τὰς δὲ ὠτίδας ἐν τῷ Πόντῳ θηρεύουσιν οὕτως.
ἀποστραφεῖσαι αὐταὶ καὶ ἐς γῆν κύψασαι τὴν
κέρκον ἀνατείνουσιν ὥσπερ οὖν τράχηλον ὄρνιθος·
αἱ δὲ ἀπατηθεῖσαι προσίασιν[7] ὡς πρὸς ὄρνιν
ὁμόφυλον, εἶτα πλησίον γενόμεναι τῆς ἀλώπεκος
ἁλίσκονται ῥᾷστα ἐπιστραφείσης[8] καὶ ἐπιθεμένης
κατὰ τὸ καρτερόν. τὰ σμικρὰ δὲ ἰχθύδια θηρῶσι
πάνυ σοφῶς. παρὰ τὴν ὄχθην τὴν τοῦ ποταμοῦ
ἔρχονται καὶ τὴν οὐρὰν καθιᾶσιν ἐς τὸ ὕδωρ· τὰ

[1] κάτεισι μέν.　　[2] ἔρπει.
[3] αὐτοῦ.　　[4] ἔρπει.
[5] ἔστ᾽ ἄν.　　[6] ⟨ἔχουσα⟩ add. H.

40

his sting. Then a second descends from the roof, crawls down over the first, and with his claws holds fast to his sting and lets his own dangle in the air. Then a third holds on to that, and a fourth on to the third, and a fifth in a line, while those that follow crawl down over the preceding ones. Then the last Scorpion strikes the sleeper; crawls up again over the one above; after him the next; then the third from the bottom; then the rest, until the entire lot are disconnected, just as if they had undone a chain.

24. The Fox is a crafty creature. For instance, it plots against Hedgehogs in the following way. It cannot overcome them by a direct attack, the reason being that their prickles prevent it; and so, gingerly and taking great care of its mouth, it turns them over and lays them on their back and after ripping them open, easily devours those whom till then it dreaded. *The Fox and Hedgehogs*

And this is the way that Foxes hunt Bustards in Pontus. They reverse themselves and put their head down upon the ground and stick their tail up, like a bird's neck. And the bustards are taken in and approach, supposing it to be some bird of their own kind; then when they come close up, they are easily caught by the Fox, which turns upon them and attacks them violently. *and Bustards*

Their manner of catching very small fishes is extremely dexterous. They move along the bank of a stream and trail their tails in the water. And the *and small Fish*

---

7 *Schn*: προΐασιν.
8 ἐπιστραφείσης τῆς ἀλώπεκος.

AELIAN

δὲ προσνέοντα ἐνίσχεταί τε καὶ ἐμπαλάσσεται [1]
τῷ δάσει τῷ τῶν τριχῶν. αἱ δὲ αἰσθόμεναι τοῦ
μὲν ὕδατος ἀναχωροῦσιν, ἐλθοῦσαι δὲ ἐς τὰ ξηρὰ
χωρία διασείουσι τὰς οὐράς, καὶ ἐκπίπτει τὰ
ἰχθύδια, καὶ ἐκεῖναι δεῖπνον ἁβρότατον ἔχουσιν.
οἱ δὲ Θρᾷκες τῆς τῶν ποταμῶν πήξεως τῆς οὐ
σφαλερᾶς ποιοῦνται γνώμονα τήνδε τὴν θήρα.
καὶ ἐὰν διαδράμῃ τὸν κρύσταλλον μὴ ἐνδιδόντα
μηδὲ εἴκοντα τοῖς ἐκείνης βήμασι, θαρροῦσι καὶ
ἕπονται. πεῖραν δὲ αὐτὴ ποιεῖται τοῦ μὴ σφαλε-
ροῦ πόρου τὸν τρόπον τοῦτον. παραβάλλει τὸ
οὖς τῷ [2] κρυστάλλῳ· κἂν μὲν αἴσθηται μὴ
ὑπηχοῦν κάτωθεν τὸ ῥεῦμα μηδὲ ὑποψοφοῦν
ἡσυχῆ ἐς βάθος, ἡ δὲ ὡς ἑστῶτι τῷ κρυστάλλῳ
θαρρεῖ, διαθεῖ τε ἀτρέπτως [3]· εἰ δὲ μή, οὐκ ἂν
ἐπιβαίη.

25. Οἱ ποιηταὶ μὲν τὴν παῖδα τὴν τοῦ Ἴφιδος
σέβουσι, καὶ τά γε θέατρα ὑπ' αὐτῶν ἐμπέπλησται
ὑμνούντων τὴν ἡρωίνην ἐκείνην, ἐπεὶ τὰς ἄλλας
ὑπερεπήδησε τῇ σωφροσύνῃ, τὸν ἑαυτῆς γαμέτην
προτιμήσασα τοῦ βίου· τὰ δὲ ζῷα ὑπερβολὴν
φιλοστοργίας οὐ παραλέλοιπεν. ὁ γοῦν Ἠριγόνης
κύων ἐπαπέθανε τῇ δεσποίνῃ, καὶ ὁ Σιλανίωνος
καὶ ἐκεῖνος τῷ [4] δεσπότῃ, καὶ οὔτε πρὸς βίαν οὔτε
σὺν κολακείᾳ ἀπέστη τοῦ τάφου. Δαρείῳ δὲ τῷ
τελευταίῳ βασιλεῖ τῶν Περσῶν ἐν τῇ πρὸς

---

[1] ἐμπλάσσεται.   [2] *Schn* : ἐπὶ τῷ.
[3] πρώτῃ.   [4] ἐπὶ τῷ.

---

[a] Evadne, see above 1. 15.

fish swim up and are immeshed and entangled in the
thick hairs. When the Foxes notice this, they withdraw from the water and go to dry ground where they
shake their tails thoroughly: the little fishes tumble
out, and the Foxes make a delicious meal.

The people of Thrace use this animal as an indica- Fox tests
tor of whether a frozen river is safe to cross. And if strength of
the Fox runs across without the ice bending or giving
way beneath its tread, they make bold to follow. The
Fox tests the safety of the transit in the following
manner: it puts its ear down to the ice, and if it hears
no sound of the flow beneath and no murmur in the
depths, it has no fear, the ice being solid, and it
races over without hesitation. Otherwise it would
not set foot upon it.

25. Poets pay homage to the daughter of Iphis,[a]
and the theatres are packed when they celebrate this
famous heroine, since she excelled all other women
in her chaste resolve, reckoning her husband more
precious than her own life.

But animals have not been wanting in inordinate Devotion of
affection. For instance, the hound of Erigone [b] died Dogs to
upon the body of its mistress: also the hound of
Silanio [c] upon the body of its master, and neither
force nor blandishment could move it from the grave.
And when Darius, the last King of Persia,[d] was struck

[b] Daughter of Icarius, hanged herself on finding her father
slain.

[c] Tzetzes, repeating the story (*Chil.* 4. 200), adds that he
was a Roman general. More than that I have been unable
to discover.

[d] Darius III, *c.* 380-330 B.C., defeated at Issus and Gau-
gamela by Alexander and finally murdered by his own
followers.

AELIAN

Ἀλέξανδρον μάχῃ βληθέντι ὑπὸ Βήσσου καὶ κειμένῳ, πάντων τὸν νεκρὸν ἀπολιπόντων, ὁ κύων ὁ ὑπ᾽ αὐτῷ[1] τραφεὶς μόνος παρέμεινε πιστός, τὸν οὐκέτι τροφέα μὴ προδοὺς ὡς ἔτι ζῶντα. τοιοῦτόν τι ὑπὲρ τῶν τοῦ Κύρου φίλων τοῦ νεωτέρου Ξενοφῶν ὁ Γρύλλου[2] νεανιευόμενος δῆλός ἐστι λέγων, τοὺς ὁμοτραπέζους μόνους οἳ[3] συμπαραμεῖναι καὶ συναπολέσθαι, καὶ τὸν εὐνοῦχον, ὃς ἦν οἱ σκηπτοῦχος τὴν τιμήν, ὄνομα δὲ Ἀρταπάτης, ἑαυτὸν ἐπαποκτεῖναι τῷ νεκρῷ, ἀτιμάσαντα[4] τὸν βίον ἐσόμενόν οἱ ἔρημον Κύρου. καὶ Λυσιμάχῳ δὲ τῷ βασιλεῖ κύων κοινοῦ τοῦ τέλους ἑκὼν μετέλαβε σωθῆναι δυνάμενος.

26. Ἡ πιθήκη ὑπό τινων ὀρειβάτης κέκληται, ὑπ᾽ ἄλλων γε μήν, ὡς ἀκούω, ὑλοδρόμος,[5] καὶ ἐν τοῖς δένδροις γίνεται καὶ ἔχει τρίχας· κέκληται δὲ ὑπ᾽ ἐνίων καὶ ψύλλα. ἐντέτμηται μὲν οὖν τὴν γαστέρα ἡσυχῇ, ὡς εἰπεῖν ὅτι λίνῳ διακέκοπται. δάκνει δὲ σφαλερώτατα δήγματα, καὶ παρέπεται τρόμος τῷ δηχθέντι, καὶ περὶ τὴν καρδίαν ἄλγημα ἰσχυρὸν ἐπιγίνεται, καὶ τὰ οὖρα ἐμφράττεται, καὶ ὁ ἕτερος πόρος γίνεται καὶ ἐκεῖνος ἄπορος. ἔοικε δὲ τοῖς προειρημένοις ἀντίπαλος ὁ καρκίνος ὁ ποτάμιος εἶναι βρωθείς.

---

[1] αὐτοῦ.
[2] ὁ Γ. A, de Stefani : Ξ. ὁ τοῦ Γ. most MSS, H.
[3] οἱ οἴκτῳ MSS, οἱ ὀκτὼ Gron, comp. Xen. An. 1.8.27.
[4] μετ᾽ ἐκεῖνον ἀτιμάσαντα.
[5] ὑπόδρομος.

44

by Bessus in the battle against Alexander and lay dead, all forsook the corpse, only the dog which had been reared under his care remained faithfully at his side, unwilling to abandon, as though he was still alive, the man who could no longer tend him. Xenophon the son of Gryllus is clearly using the high-flown language of youth when he relates [*An.* 1. 8. 27] a similar tale of the friends of Cyrus the Younger,[a] how his table-companions alone stood fast at his side and were slain along with him, while the eunuch who held the office of sceptre-bearer and was called Artapates, slew himself upon the corpse, not caring to live henceforward without Cyrus. And the hound of King Lysimachus[b] of its own free will shared his death although its life might have been saved.

26. The Monkey-spider has by some been called 'the mountain-ranger,' but by others (I am told) 'the wood-runner.'[c] It is born on trees and is hairy. It has also by some been called 'the flea.' Its belly has a slight incision, so that one might say it had been cut in two by a thread. It inflicts the most dangerous bites, and they are attended by a trembling on the part of the victim; there ensues a sharp pain in the region of the heart; the urine is stopped; and the other passage also becomes blocked. It seems that the remedy for these afflictions is to eat a river-crab.

The Monkey-spider

---

[a] Cyrus, see 1. 59 note *c*.

[b] General of Alexander the Great, became King of Thrace, defeated in battle by Seleucus, 281 B.C.

[c] May be identical with the wolf-spider of Arist. *HA* 622 b 30, or more probably the malmignatte.

27. Αἰλούρων ὁ μὲν ἄρρην ἐστὶ λαγνίστατος, ὁ
δὲ θῆλυς φιλότεκνος, φεύγει δὲ τὴν πρὸς τὸν
ἄρρενα ὁμιλίαν· ἀφίησι γὰρ τὸν θορὸν θερμότατόν
τε καὶ προσεοικότα πυρί, καὶ κάει τῆς θηλείας τὸ
ἄρθρον. εἰδὼς οὖν ὁ ἄρρην τοῦτο τὰ κοινὰ
βρέφη διαχρῆται, ἡ δὲ παίδων ἑτέρων πόθῳ
ἑαυτὴν παρέχει συνελθεῖν γλιχομένῳ. φασὶ δὲ
τοὺς αἰλούρους πάντα ὅσα δυσώδη ἐστὶ μισεῖν τε
καὶ βδελύττεσθαι. ταύτῃ τοι καὶ τὸ σφέτερον
περίττευμα ἀφιέναι πρότερον βόθρον ὀρύξαντας,
ἵνα ἀφανίσωσιν αὐτὸ τῆς γῆς ἐπιβαλόντες.

28. Ἀκολαστότατον τῶν ἰχθύων τὸν πολύποδα
εἶναί φασι καὶ ἐς τοσοῦτον λαγνεύειν, ἐς ὅσον
αὐτῷ ἡ [1] πᾶσα τοῦ σώματος ῥώμη ἐκρυεῖσα
ἀσθενῆ ἀπέφηνε καὶ νήξασθαι μὲν ἥκιστον, τροφὴν
δὲ μαστεῦσαι ἀδύνατον, καὶ διὰ ταῦτα ἄλλους
τρέφειν. τὰ γοῦν σμικρὰ τῶν ἰχθυδίων καὶ τὰς
καλουμένας καρκινάδας καὶ τοὺς καρκίνους ἐπιφοι-
τᾶν τε ἅμα καὶ κατεσθίειν αὐτόν. λέγουσι δὲ
ταύτην εἶναι τὴν αἰτίαν τοῦ μὴ δύνασθαι πολύποδα
ἐνιαυτοῦ βιῶναι χρόνον μακρότερον. καὶ θῆλυς
δὲ ἄρα πολύπους ἀναλίσκεται ῥᾳδίως τίκτων
πολλάκις.

29. Φύλαρχος μέμνηται παῖδα ἰσχυρῶς φιλόρ-
νιθα ἀετοῦ νεοττὸν λαβεῖν δῶρον καὶ τρέφειν
τροφῇ ποικίλῃ καὶ κηδεμονίαν αὐτῷ προσφέρειν
πᾶσαν· οὐ γάρ τί που ὡς ἄθυρμα ἐς παιδιὰν
ἔτρεφε τὸν ὄρνιν, ἀλλὰ ἐρωμένου δίκην ἢ ἀδελφοῦ
νεωτέρου, οὕτως ἄρα ὁ παῖς τοῦ ἀετοῦ προμηθῶς
εἶχεν. προϊὼν [2] δὲ ⟨ὁ⟩ [3] χρόνος ἐς φιλίαν αὐτοὺς
46

27. The Tom-cat is extremely lustful, but the The Cat
Female cat is devoted to her kittens and tries to avoid
sexual intercourse with the male, because the semen
which he ejaculates is exceedingly hot and like fire,
and burns the female organ. Now the Tom-cat
knowing this, makes away with their kittens, and the
Female in her yearning for other offspring yields to
his lust. They say that Cats hate and abhor all foul-
smelling objects, and that is why they dig a hole
before they discharge their excrement, so that they
may get it out of sight by throwing earth upon it.

28. They say that the Octopus is the most in- The Octopus
continent of fish and copulates until all the strength
of its body is drained away, leaving it weak, incapable
of swimming, and unable to seek for food; in consequence of which it provides food for others, thus:
small fishes, and what are known as ' hermit-crabs,'
and crabs come about it and devour it. And they
say that this is the reason why the Octopus cannot
live for more than a year. And as to the female, it
is soon exhausted by giving birth so frequently.

29. Phylarchus records how a youth who was Eagle and
deeply devoted to birds was presented with an boy
eaglet, and how he fed it on a variety of foods and
tended it with all possible care. He reared the bird
not as a plaything to sport with, but as a favourite or
as a younger brother, so full of thought was the
youth for the Eagle. As time passed it lit the flame

---

[1] καὶ ἥ. [2] καὶ προϊών.
[3] ⟨ὁ⟩ add. H.

AELIAN

ἐξῆψεν ἀλλήλων ἰσχυράν. συνηνέχθη δὲ καμεῖν
τὸ σῶμα τὸν παῖδα. καὶ τὸν ἀετὸν παραμένειν
καὶ νοσηλεύειν τὸν τροφέα, καὶ καθεύδοντος μὲν
ἡσυχάζειν, ἐγρηγορότος δὲ παρεστάναι, ἀσιτοῦντος
δὲ τροφὴν μὴ προσίεσθαι. ἐπεὶ δὲ καὶ τὸν βίον
ὁ παῖς κατέστρεψεν, ἠκολούθησε καὶ ὁ ἀετὸς μέχρι
τοῦ μνήματος· καομένου δὲ ἑαυτὸν ἐς τὴν πυρὰν
ἐνέβαλεν.

30. Ὁ ἰχθὺς ὁ ὄνος τὰ μὲν ἄλλα ὅσα ἐντὸς
προσπέφυκεν οὐ πάνυ τι [1] τῶν ἑτέρων διεστῶτα
κέκτηται, μονότροπος δέ ἐστι καὶ σὺν ἄλλοις
βιοῦν οὐκ ἀνέχεται. ἔχει δὲ ἄρα ἰχθύων μόνος
οὗτος ἐν τῇ γαστρὶ τὴν καρδίαν καὶ ἐν τῷ ἐγκεφάλῳ
λίθους, οἵπερ οὖν ἐοίκασι μύλαις τὸ σχῆμα.
Σειρίου δὲ ἐπιτολῇ φωλεύει μόνος, τῶν ἄλλων ἐν
ταῖς κρυμωδεστάταις φωλεύειν εἰθισμένων.

31. Οἱ θηρῶντες τοὺς παγούρους [2] ἐπ᾽ αὐτοὺς
ἐμηχανήσαντο τὴν μουσικὴν δέλεαρ. φωτιγγίῳ
γοῦν (ὄνομα δὲ ὀργάνου τοῦτο) αἱροῦσιν αὐτούς.
οἱ μὲν γὰρ ἐν τοῖς φωλεοῖς δεδύκασιν, οἱ δὲ
ὑπάρχονται τοῦ μέλους. καὶ ἀκούσαντες οἱ πάγου-
ροι πείθονται ὡς ὑπό τινος ἴυγγος προελθεῖν τῆς
θαλάμης, εἶτα ὑπὸ τῆς ἡδονῆς ἑλκόμενοι προΐασι
καὶ ἔξω τῆς θαλάττης, οἱ δὲ [3] αὐλοῦντες ἐπὶ πόδα
ἀναχωροῦσι. καὶ ἐκεῖνοι ἕπονται καὶ ἐπὶ τῆς γῆς
ἁλίσκονται.

32. Οἱ δὲ τῇ Μαρείᾳ λίμνῃ προσοικοῦντες τὰς
θρίσσας θηρῶσι τὰς ἐκεῖθι ᾠδῆς μέλει τορωτάτῳ [4]

[1] οὐ πάντη.

48

of a strong mutual friendship. It happened that the youth fell sick, and the Eagle stayed at his side and nursed its keeper: while he slept, the bird remained quiet; when he woke, it was there; if he took no food, it refused to eat. And when the youth at last died, the Eagle also followed him to the tomb, and as the body burned it threw itself on to the pyre.

30. The Hake though not differing widely from other fish in its inward parts, is nevertheless solitary in its habits and cannot endure to live with other fish. It is the only fish that has its heart in its belly and stones in its brain resembling millstones. At the rising of the Dog-star [a] it alone lurks in its den, while other fish are in the habit of doing so in the very frostiest seasons. *The Hake*

31. Those who hunt Crabs have hit upon the device of luring them with music. At any rate they catch them by means of a flageolet (this is the name of an instrument). Now the Crabs have gone down into their hiding-places, and the men begin to play. And at the sound, as though by a spell, the Crabs are induced to quit their den, and then captivated with delight even emerge from the sea. But the flute-players withdraw backwards and the Crabs follow and when on the dry land are caught. *The Crab and music*

32. Those who live by the lake of Marea [b] catch the Sprats there by singing with the utmost shrillness, *The Sprat and music*

[a] About mid-July.
[b] Near the westernmost mouth of the Nile.

---

[2] παγούρους τὸν τρόπον τοῦτον αἱροῦσιν.
[3] τε.   [4] *Jac*: γοερωτάτῳ.

AELIAN

καὶ κρότῳ ὀστράκων ὁμορροθοῦντι πρὸς τὸ μέλος·
αἱ δὲ ὥσπερ ὀρχούμεναι ὑπὸ τῷ μέλει πηδῶσι,
καὶ ἐμπίπτουσι τοῖς θηράτροις, ἅπερ οὖν αὐτοῖς
περιπέπταται,[1] καὶ λαμβάνουσιν οἱ Αἰγύπτιοι
θήραν εὔοψον σὺν χορείᾳ τε καὶ παιδιᾷ.

33. Αἰγυπτίους ἐγὼ πυνθάνομαι μαγείᾳ τινὶ
ἐπιχωρίῳ τοὺς ὄρνιθας ἐκ τοῦ οὐρανοῦ καταφέ-
ρειν· τῶν δὲ φωλεῶν τοὺς ὄφεις ἐπαοιδαῖς τισι
καταγοητεύσαντες εἶτα μέντοι προάγουσι ῥᾷστα.

34. Ὁ κάστωρ ἀμφίβιόν ἐστι ζῷον, καὶ μεθ᾽
ἡμέραν μὲν ἐν τοῖς ποταμοῖς καταδὺς διαιτᾶται,
νύκτωρ δὲ ἐπὶ τῆς γῆς ἀλᾶται, οἷς ἂν περιτύχῃ
τούτοις τρεφόμενος. οὐκοῦν ἐπίσταται τὴν αἰτίαν
δι᾽ ἣν ἐπ᾽ αὐτὸν οἱ θηραταὶ σὺν προθυμίᾳ τε καὶ
ὁρμῇ τῇ πάσῃ χωροῦσι, καὶ ἐπικύψας καὶ δακὼν
ἀπέκοψε τοὺς ἑαυτοῦ ὄρχεις, καὶ προσέρριψεν
αὐτοῖς, ὡς ἀνὴρ φρόνιμος λῃσταῖς μὲν περιπεσών,
καταθεὶς δὲ ὅσα ἐπήγετο ὑπὲρ τῆς ἑαυτοῦ σωτηρίας, λύτρα δήπου ταῦτα ἀλλαττόμενος. ἐὰν δὲ
ᾖ πρότερον ἐκτεμὼν[2] καὶ σωθεὶς εἶτα πάλιν
διώκηται, ὁ δὲ ἀναστήσας ἑαυτὸν καὶ ἐπιδείξας
ὅτι τῆς αὐτῶν σπουδῆς οὐκ ἔχει τὴν ὑπόθεσιν, τοῦ
περαιτέρω καμάτου παρέλυσε τοὺς θηρατάς· ἧττον
γάρ τοι[3] τῶν κρεῶν ἐκείνοις φροντίς ἐστι. πολλάκις δὲ καὶ ἔνορχοι ὄντες, ὡς ὅτι πορρωτάτω
ἀποσπάσαντες[4] τῷ δρόμῳ, εἶτα ὑποστείλαντες τὸ
σπουδαζόμενον μέρος, πάνυ σοφῶς καὶ πανούργως
ἐξηπάτησαν, ὡς οὐκ ἔχοντες ἃ κρύψαντες εἶχον.[5]

---

[1] προσπέπταται.

accompanying their song with the clash of castanets. And the fishes, like women dancing, leap to the tune and fall into the nets spread for their capture. And through their dancing and frolics the Egyptians obtain an abundant catch.

33. I am informed that the Egyptians bring birds Egyptian down from the sky by some magic peculiar to them. magic And they have certain spells to bewitch snakes and draw them without any difficulty from their lurking-places.

34. The Beaver is an amphibious creature : by day The Beaver it lives hidden in rivers, but at night it roams the land, feeding itself with anything that it can find. Now it understands the reason why hunters come after it with such eagerness and impetuosity, and it puts down its head and with its teeth cuts off its testicles and throws them in their path, as a prudent man who, falling into the hands of robbers, sacrifices all that he is carrying, to save his life, and forfeits his possessions by way of ransom. If however it has already saved its life by self-castration and is again pursued, then it stands up and reveals that it offers no ground for their eager pursuit, and releases the hunters from all further exertions, for they esteem its flesh less. Often however Beavers with testicles intact, after escaping as far away as possible, have drawn in the coveted part, and with great skill and ingenuity tricked their pursuers, pretending that they no longer possessed what they were keeping in concealment.

---

² ⟨ἑαυτὸν⟩ ἐκτεμών Reiske.   ³ ἔτι.
⁴ ἑαυτοὺς ἀποσπάσαντες.   ⁵ ἔσχον.

35. Βούπρηστις ζῷόν ἐστιν, ὅπερ οὖν ἐὰν βοῦς καταπίῃ, πίμπραται καὶ ῥηγνύμενος ἀπόλλυται οὐ μετὰ μακρόν.

36. Αἱ κάμπαι ἐπινέμονται τὰ λάχανα, τάχα δὲ καὶ διαφθείρουσιν αὐτά. ἀπόλλυνται δὲ αὗται, γυνὴ τὴν ἐπιμήνιον κάθαρσιν καθαιρομένη εἰ διέλθοι μέση τῶν λαχάνων.

37. Εἶεν δ' ἂν βουσὶν ἔχθιστα οἶστρος καὶ μύωψ. καὶ ὁ μὲν οἶστρος κατὰ τὰς μυίας τὰς μεγίστας ἐστί, καὶ ἔχει στερεὸν καὶ μέγα κέντρον, καὶ ἦχόν τινα βομβώδη ἀφίησι καὶ τραχύν· ὁ δὲ μύωψ τῇ κυνομυίᾳ προσείκασται, βομβεῖ δὲ τοῦ οἴστρου μᾶλλον, ἔλαττον δὲ ἔχει τὸ κέντρον.

38. Τῶν ὑπ' ἀσπίδος δηχθέντων οὐ μνημονεύεται οὐδεὶς ἐξάντης τοῦ κακοῦ γεγονέναι. ἔνθεν τοι καὶ τοὺς βασιλεῖς ἀκούω τῶν Αἰγυπτίων ἐπὶ τῶν διαδημάτων φορεῖν πεποικιλμένας ἀσπίδας, τῆς ἀρχῆς αἰνιττομένους τὸ ἀνίκητον δὴ ἐκ τῆς τοῦ ζῴου μορφῆς τοῦ προειρημένου. γίνονται δὲ καὶ πενταπήχεις ἀσπίδες. καὶ μέλαιναι μὲν αἱ πλεῖσται ἢ τεφραῖαι, ἴδοις δ' ἂν καὶ πυρρὰν ἀσπίδα. οἱ δηχθέντες δὲ ὑπ' ἀσπίδος οὐ περαιτέρω βιοῦσι τετάρτης ὥρας, πνιγμὸς δὲ αὐτοὺς καὶ σπασμὸς διώκει καὶ λυγμός, ὥς φασιν. ἀκούω δὲ τὸν ἰχνεύμονα τῆς ἀσπίδος τὰ ᾠὰ ἀφανίζειν, οἱονεὶ τοῖς ἑαυτοῦ παισὶν ὑπεξαίροντα [1] τοὺς μέλλοντας ἀντιπάλους. Λίβυσσαν δὲ ἄρα ἀσπίδα καὶ ἀποφαίνειν τυφλοὺς τῷ φυσήματι λέγει τις λόγος.

52

35. The *Buprestis* (cow-inflater) is a creature which, if swallowed by a cow, causes it to swell and presently to burst and die.

The Buprestis

36. Caterpillars feed upon vegetables and in a short while destroy them. But they in turn are destroyed if a woman with her monthly courses upon her walks through the vegetables.

The Caterpillar

37. The worst enemies of cattle are the Gadfly and the Horsefly. The Gadfly is the size of the very largest flies, and its sting is powerful and long, and it makes a harsh buzzing sound. But the Horsefly is like the dogfly: its buzz is louder than the Gadfly, but its sting is smaller.[a]

The Gadfly

The Horsefly

38. Among all those who have been bitten by an Asp there is no record of a single man having escaped disaster. That is why (I am told) the Kings of Egypt wear asps embroidered upon their crowns, hinting through the figure of the aforesaid creature at the invincibility of their rule. There are Asps as much as five cubits long; the majority are black or of an ashy hue; and one may even see a red one. Those who have been bitten by an Asp do not live for more than four hours and are assailed by choking and convulsions and retching, so they say. But I am told that the Ichneumon destroys the eggs of the Asp with intent to do away with the future enemies of its own young. And there is a story that the Libyan Asp even blinds men with its breath.

The Asp

its bite fatal

The Ichneumon

---

[a] Comp. 4. 51.

---

[1] ὑπεξαιροῦντα Cobet.

39. Εἶτα οὐ χρὴ θαυμάσαι τὴν φύσιν τῇ τε
ἄλλῃ καὶ ἐνταῦθα μέντοι; τῶν ἀρρένων . . .¹ οἱ
πατέρες τοὺς πλείστους νεβροὺς ἀναιροῦσιν, ἵνα
μὴ πληθύωνται εἶτα μέντοι καὶ τὰς μητέρας
ἀναβαίνωσι· μίασμα γὰρ καὶ ἐν τοῖς ἀλόγοις καὶ
ἐναγὲς ἔργον δοκεῖ τοῦτο δήπου. Κύρῳ δὲ καὶ
Παρυσάτιδι, ὦ Πέρσαι, καὶ καλὰ ταῦτα καὶ
ἔνδικα ἐδόκει· καὶ ἐφίλει Κῦρος τὴν μητέρα
κακῶς, καὶ ἐφιλεῖτο ὑπὸ τῆς μητρὸς φιλίαν
ὁμοίαν. καὶ ταῦτα μέντοι †σωμένης†,² οἱ δὲ
ἄνθρωποι πάντων μὲν ἐπιθυμοῦντες, μηδενὸς δὲ
φειδόμενοι.

40. Νῆσος ἐν τῷ Πόντῳ Ἡρακλεῖ ἐπώνυμος
ἐκτετίμηται. οὐκοῦν ὅσον μυῶν ἐστιν ἐνταῦθα
σέβει τὸν θεόν, καὶ πᾶν ὅσον ἀνεῖται αὐτῷ, τοῦτο
πιστεύει τῷ θεῷ κεχαρισμένον ἀνεῖσθαι ³ καὶ οὐκ
ἂν προσάψαιτο αὐτοῦ. οὐκοῦν καὶ ἄμπελος τῷ
θεῷ κομᾷ, καὶ τετίμηται ὡς ἀνάθημα αὐτῷ μόνῳ,
καὶ φυλάττουσιν οἱ θεραπευτῆρες τοῦ δαίμονος ἐς
τὰς θυσίας τοὺς βότρυς. ὅταν οὖν ἐς ἀκμὴν αἱ
ῥᾶγες ἔρχωνται, οἱ δὲ ἀπολείπουσι τὴν νῆσον οἱ
μύες, ἵνα μὴ μείναντες ἄκοντες γοῦν προσάψωνται
ὧν θιγεῖν οὐκ ἄμεινον· εἶτα τῆς ὥρας διαδραμού-
σης οἱ δὲ ἐς ἤθη τὰ οἰκεῖα ὑποστρέφουσι. καὶ
μυῶν μὲν Ποντικῶν ἀγαθὰ ταῦτα· Ἵππων δὲ καὶ

¹ *Lacuna* : ⟨τῶν ἀγρίων ὄνων⟩ τῶν ἀρρένων *Jac, comp.* Opp.
Cyn. 3. 201–6, Plin. *HN* 8.108.
² κεκολασμένως or πεφεισμένως *conj. H.*
³ ἀφεῖσθαι.

54

39. Now does not Nature claim our admiration for this reason especially, besides others? Of the males . . . the sires destroy most of the male fawns to prevent their multiplying and then mounting their dams. Even among brute beasts, I fancy, such an act is regarded as bringing defilement and a curse. But Cyrus and Parysatis, you men of Persia, thought it a fine and legitimate action. And Cyrus conceived a vile passion for his mother, a passion which his mother reciprocated. ⟨While animals are moderate in their desires?⟩ [a] men desire everything and stop at nothing.

Animals abhor incest

40. There is an island [b] in the Black Sea named after Heracles which has been highly honoured. Now all the Mice there pay reverence to the god, and every offering that is made to him they believe to have been made to gratify him and would not touch it. And so the vine grows luxuriantly in his honour and is reverenced as an offering to him alone, while the ministers of the god preserve the clusters for their sacrifices. Accordingly when the grapes reach maturity the Mice quit the island so that they may not, by remaining, even involuntarily touch what is better not touched. Later when the season has run its course they return to their own haunts. This is a merit in the Pontic Mice. But Hippon,[c] Diagoras,

Heracles revered by Mice

---

[a] I have given what may have been the sense of the passage.

[b] Unidentified.

[c] Hippon of Samos, Pythagorean philosopher, 5th cent. B.C., satirised by Cratinus as an atheist.—Diagoras of Melos, called 'the atheist,' incurred the enmity of the Athenians by his attacks on their religion and withdrew from Athens, 411 B.C. —Herostratus of Ephesus burnt the temple of Artemis, 356 B.C.

AELIAN

Διαγόρας καὶ Ἡρόστρατος καὶ ὁ λοιπὸς τῶν
θεοῖς ἐχθρῶν κατάλογος πῶς ἂν ἐφείσαντο τῶν
βοτρύων ἢ ἀναθημάτων ἄλλων οἱ καὶ τὰ τῶν
θεῶν ὀνόματα καὶ ἔργα ἀμωσγέπως συλᾶν προῃρη-
μένοι;

41. Ἔθος τοῦτο Αἰγύπτιον. ὅταν κατὰ τὴν
Αἴγυπτον ὕσῃ (ῥανίσι δὲ ὕει λεπταῖς), μύες
παραχρῆμα τίκτονται. οὐκοῦν κατὰ τὰς ἀρούρας
πλανώμενοι οὗτοι λυμαίνονται τοῖς ληίοις ὑποτέμ-
νοντες τοὺς στάχυς καὶ ὑποκείροντες, ἤδη μέντοι
καὶ τοὺς σωροὺς τῶν δραγμάτων κεραΐζοντες
λυποῦσι τοὺς Αἰγυπτίους. καὶ διὰ ταῦτα πάγαις
τε αὐτοὺς ἐλλοχῶσι [1] καὶ θριγκοῖς ἀναστέλλουσι
καὶ τάφροις ἀνείργουσι καὶ κάουσιν [2] ἐν ταύταις
πῦρ. οἱ τοίνυν μύες οὐδὲ [3] τὴν ἀρχὴν πρὸς
⟨τὰς⟩ [4] πάγας προσφοιτῶσιν, ἐῶσι δὲ αὐτὰς
ἑστάναι ἄλλως· τοῖς δὲ θριγκοῖς καίτοι [5] λε-
λειωμένοις ὑπὸ τῆς χρίσεως ἐπαναβαίνουσιν,
ἁλτικώτατοί τε [6] ὄντες εἶτα μέντοι ὑπερπηδῶσι
τὰς τάφρους. ὅταν οὖν ἀπαγορεύσαντες οἱ Αἰγύπτιοι τὰς μηχανὰς καὶ τὰς ἐπιβουλὰς ὡς ἀχρήστους
ἀπολίπωσι καὶ ἐκτραπέντες ἐπί τε δέησιν ἔλθωσι
καὶ ἱκετείας τὰς πρὸς τοὺς θεούς, ἐνταῦθα δήπου
τὴν ἐκ τῶν θεῶν μῆνιν ὀρρωδοῦσιν οἱ μύες, καὶ
ἔς τι ὄρος ἀναχωροῦσι τάξιν πλαισίου φυλάττοντες.
οἱ μὲν οὖν νεώτατοι πρῶτοι, οὐραγοῦσι δὲ οἱ
μέγιστοι, καὶ τοὺς ὑπολειπομένους ἐπιστραφέντες
εἶτα μέντοι ἕπεσθαί σφισιν ἐκβιάζονται. ἐὰν δὲ
οἱ νεώτατοι κάμνοντες ὑποστῶσι, καὶ τὸ ἑπόμενον

[1] πάγαις . . . ἐλλοχῶσι] πάγας τε αὐτοῖς ἐλλοχῶντες ἱστᾶσι.

56

and Herostratus, and all the rest in the tale of
heaven's enemies, how would they have kept their
hands off the grapes or other offerings—men who
preferred by one means or another to rob the gods
of their names and functions.

41. This is what commonly happens in Egypt.
When it rains in Egypt (the raindrops are minute)
Mice are produced forthwith. Now they roam the
ploughlands and damage the standing crops by cutting away and nibbling the ears of corn from below,
and actually ravage the stacked sheaves and cause
the Egyptians much trouble. On that account the
people try to trap them, to exclude them by building
walls, to keep them off by digging trenches in which
they light fires. Now the Mice go nowhere near the
traps but allow them to remain useless. And although the walls have been rendered smooth with a
wash of mortar, they climb up them and then, being
exceedingly nimble, jump over the trenches. And
so the Egyptians abandon their traps and schemes as
ineffectual and turn from them to prayers and suppli-
cations to the gods. Whereupon the Mice, I fancy,
are in dread of the wrath of heaven and retreat in the
formation of a hollow square to some mountain.
Now the youngest go in front and the oldest bring
up the rear, and if any are left behind, the latter turn
and force them to follow. If however the youngest
ones halt from exhaustion, the entire lot behind them

The Mouse in Egypt

---

² ἀναστέλλουσι . . . ἀνείργουσι . . . κάουσιν] ἀναστέλλοντες
. . . ἀνείργοντες . . . καίοντες.
³ *Jac*: οὔτε. ⁴ ⟨τάς⟩ *add. H.*
⁵ *Reiske*: καὶ τοῖς. ⁶ *Reiske*: γε.

AELIAN

ἵσταται πᾶν, ὡς ἐν δυνάμει στρατιωτικῇ πέφυκε
γίνεσθαι. ὅταν δὲ ὑπάρξηται ἡ ἐκ τῶν πρώτων
κίνησις, ἐνταῦθα καὶ οἱ λοιποὶ ἕπονται. λέγουσι
δὲ καὶ οἱ τὸν Πόντον οἰκοῦντες τὰ αὐτὰ καὶ
ἐκεῖθι τοὺς μῦς δρᾶν. πεπίστευται δὲ καὶ ἐξ
οἰκίας ἁπάσης, ἥτις μέλλει πεσεῖσθαι, ἢ ποδῶν
ἔχουσι μετοικίζεσθαι πάντας. ἰδιότης δὲ ἄρα
μυῶν καὶ ἐκείνη. ἐπειδὰν ἀκούσωσι γαλῆς τρι-
ζούσης [1] ἢ συρίττοντος ἔχεως, ἐκ τῆς μυωπίας τῆς
μιᾶς τὰ ἑαυτῶν βρέφη ἄλλο ἄλλῃ μετοικίζουσιν.

42. Λόγον δὲ Ἰταλὸν τῇ Συβαριτῶν πόλει
συνακμάσαντος ἔργου μνημονεύοντα καὶ φοιτή-
σαντα ἐς ἐμὲ εἰπεῖν οὐ χεῖρόν ἐστι. τὴν ἡλικίαν
ἀντίπαις, αἰπόλος τὸ ἐπιτήδευμα, ὄνομα Κρᾶθις,
ἐς ὁρμὴν ἀφροδίσιον ἐμπεσὼν τῇ τῶν αἰγῶν ἰδεῖν
ὡραιοτάτῃ μίγνυται, καὶ τῇ ὁμιλίᾳ ἥσθη, καὶ εἴ
ποτε ἐδεῖτο ἀφροδίτης ὡς αὐτὴν ἐφοίτα, καὶ εἶχεν
ἐρωμένην αὐτήν· καὶ μέντοι καὶ οἷα λαμβάνειν
ἠδύνατο δῶρα, τοιαῦτα ὁ ἐραστὴς αἰπόλος τῇ
ἐρωμένῃ τῇ προειρημένῃ προσέφερε, καὶ κυτίσου [2]
ποτὲ τοὺς [3] ὡραιοτάτους ἀκρεμόνας καὶ μίλακος
πολλάκις καὶ σχίνου τραγεῖν παρέσχε, τὸ στόμα
ἀποφαίνων αὐτῷ, εἰ δεηθείη φιλῆσαι, εὐῶδες
αὐτῆς. ἀλλὰ καὶ στιβάδα ἐγκαθεύδειν ὡς νύμφῃ
παρεσκεύασεν ἁβροτάτην τε καὶ μαλθακήν. οὔκουν ἀμελῶς ταῦτα ἐθεάσατο ὁ τῆς ἀγέλης
ἡγεμὼν τράγος, ἀλλὰ αὐτὸν ἔπεισι ζηλοτυπία.
καὶ κατέκρυπτε μὲν τέως τὸν θυμόν, καθήμενον
δὲ αὐτὸν ποτε ἐλλοχᾷ καὶ καθεύδοντα· ἦν δὲ ἄρα
ἐμβαλὼν τὸ πρόσωπον ἐς τὸν κόλπον. ὡς οὖν

_____
[1] Jac : τρυζούσης.

58

halt also, as is customary for an armed force. And
when the front rank begins to move, then the remainder follow. And the inhabitants of Pontus say
that the Mice there do the same. And it is believed
that whenever a house is threatening to fall, all the
Mice will change house as fast as their legs can carry
them. Now here is another peculiar trait of Mice:
whenever they hear the squeak of a marten or the
hiss of a viper they transfer their young from one
hole to a number of different holes.

42. An Italian story, which records an event that The story
occurred when affairs were at their prime in the city ^{of Crathis}
of Sybaris, has reached me and is worth relating.

A mere boy, a goatherd by occupation, whose name
was Crathis, under an erotic impulse lay with the
prettiest of his goats, and took pleasure in the union,
and whenever he wanted sexual pleasure he would
go to her; and he kept her as his darling. Moreover
the amorous goatherd would bring to his loved one
aforesaid such gifts as he could procure, offering her
sometimes the loveliest twigs of tree-medick, and
often bindweed and mastic to eat, so making her
mouth fragrant for him if he should want to kiss her.
And he even prepared for her, as for a bride, a leafy
bed ever so luxurious and soft to sleep in. But the
he-goat, the leader of the flock, did not observe these
proceedings with indifference, but was filled with
jealousy. For a time however he dissembled his
anger and watched for the boy to be seated and
asleep; and there he was, his face dropped forward

---

² κισσοῦ or κυτίσας.
³ ποτὲ τούς] ποτὲ καὶ σχίνου τούς.

AELIAN

εἶχε δυνάμεως τὴν κεφαλὴν προσήραξε,[1] καὶ
διέθρυψέν οἱ τὸ βρέγμα. διαρρεῖ τοίνυν ἐς τοὺς
ἐπιχωρίους τὰ πραχθέντα, καὶ τῷ μὲν οὐκ ἀφανῆ
τάφον ἀνέστησαν, ἐξ αὐτοῦ δὲ τὸν ποταμὸν
Κρᾶθιν ὠνόμασαν. γίνεται δὲ ἐκ τῆς ὁμιλίας τῆς
πρὸς τὴν αἶγα παιδίον, καὶ ἦν αἲξ τὰ σκέλη, τὸ
πρόσωπον ἄνθρωπος. τοῦτον καὶ ἐκθεωθῆναι λόγος
ἔχει, καὶ θεὸν ὑλαῖόν τε καὶ ναπαῖον νομισθῆναι
τὸν αὐτόν. μετειληχέναι δὴ καὶ ζηλοτυπίας τὰ
ζῷα ὁ τράγος διδάσκει.

43. Σύριγγας μὲν Αἰγυπτίας ᾄδουσιν ⟨μὲν⟩[2] οἱ
συγγραφεῖς, ᾄδουσι δὲ καὶ λαβυρίνθους τινὰς
Κρητικοὺς ἐκεῖνοί τε αὐτοὶ καὶ τὸ τῶν ποιητῶν
φῦλον· μυρμήκων δὲ ἐν γεωρυχίᾳ ποικίλας[3]
ἀτραποὺς καὶ ἑλιγμοὺς καὶ περιόδους οὔπω ἴσασι.
σοφίᾳ δὲ ἄρα τὴν ὑπόγειον οἰκονομίαν τήνδε
ἀπεργάζονται[4] σκολιωτάτην, τοῖς ἐπιβουλεύουσί
σφισι τῶν θηρίων τὴν πάροδον δύσπορον ἢ καὶ
παντελῶς ἄπορον ἀποφαίνοντες. τὴν δὲ γῆν ἣν
ἐξορύττουσιν, ἀλλὰ[5] ταύτην ὑπὲρ τοῦ στομίου
περιβαλόντες οἱονεὶ τείχη τινὰ καὶ προβλήματα
ἐργάζονται, ὡς μὴ τὸ ὕδωρ τὸ ἐξ οὐρανοῦ κα-
ταρρέον[6] εἶτα ῥᾳδίως[7] αὐτοὺς[8] ἐπικλύσαν ἢ
ἀπολέσῃ πάντας ἢ τούς γε πλείστους. αἱμασιὰς
δέ τινας μέσας διειργούσας ἀπ᾽ ἀλλήλων τοὺς
χηραμοὺς διατειχίζουσι καὶ μάλα ἐντέχνως, εἶεν
δ᾽ ἂν οἱ χῶροι τρεῖς, ὡς ἐν οἰκίᾳ σοβαρᾷ. καὶ τὸν
μὲν ἀποφαίνουσιν ἀνδρῶνα[9] εἶναι, ἐν ᾧ διαιτῶνται
οἱ ἄρρενες καὶ ὅσον σὺν αὐτοῖς θῆλυ· τὸν δὲ

---

[1] τῇ κεφαλῇ προσέρρηξε.　　[2] ⟨μὲν⟩ add. H.
[3] ποικίλας τε.　　[4] Reiske : ὑπεργάζονται.

on his chest. So with all the force at his command
the he-goat dashed his head against him and smashed
the fore-part of his skull.

The event reached the ears of the inhabitants, and
it was no mean tomb that they erected for the boy;
and they called their river ' the Crathis ' after him.
From his union with the she-goat a baby was born
with the legs of a goat and the face of a man. The
story goes that he was deified and was worshipped
as a god of the woods and vales. From the goat we
learn that animals have indeed their share of jealousy.

43. Historians celebrate the underground passages Ants and
of the Egyptians; they also with the company of their nests
poets celebrate certain labyrinths in Crete. They
have yet to learn of the elaborate tracks with their
mazy windings dug by Ants in the earth. Now in
their wisdom these make their underground dwelling
so very tortuous as to render access difficult or totally
impossible for such creatures as have designs upon
them. And the soil which they excavate they put
around the mouth, forming as it were walls and
barriers, so that the rain which descends from the sky
may not easily flood them and destroy all or at any
rate most of them. And with consummate skill they
build partitioning walls, as you might say, to separate
their cells from one another, and, as in some fine
house, there will be three divisions: the first they
design for the ' men's quarters,' in which the males
live and any females that are with them; the second,

---

5 ἀλλὰ καί.
7 Reiske : ῥ. εἶτα.
9 ὅσον ἀνδρῶνα.

6 Cobet : καταθέον MSS, H.
8 Ges : αὐτῶν.

AELIAN

ἕτερον, ἔνθα ἀποτίκτουσι κύουσαι μύρμηκες,
οἱονεὶ γυναικῶνα· τρίτον δ' ἕτερον θησαυρόν τε
καὶ σιρὸν ἀποκρίνουσι τοῖς ἠθροισμένοις σπέρμασι.
καὶ οὔτε Ἰσχόμαχος ἐνταῦθα οὔτε Σωκράτης
ὑπὲρ τῆς ἀξιοζήλου οἰκονομίας σπουδάζοντες
διδάσκουσιν αὐτά. οἱ δὲ μύρμηκες προϊόντες
ὥσπερ οὖν ἐπισιτίσασθαι, τοῖς μεγίστοις ἕπονται·
οἱ δὲ ἄγουσι στρατηγῶν δίκην. καὶ ἐς τὰ λήια
ἐλθόντες οἱ μὲν ἔτι νεαροὶ τὴν καλάμην ὑφεστᾶσιν,
οἱ δὲ ἡγεμόνες ἀνέρπουσι, καὶ τοὺς καλουμένους
οὐραχοὺς τῶν καρπίμων διατραγόντες τῷ δήμῳ
τῷ κάτω ῥίπτουσι· οἱ δὲ περιελθόντες τοὺς μὲν
ἀθέρας ἀποκόπτουσιν, ἐκλέπουσι δὲ τὰς τὸν πυρὸν
στεγούσας τε καὶ περιαμπεχούσας θυλακίδας. καὶ
μήτε ἀλοητοῦ δεόμενοι μήτε ἀνδρῶν λικμῆσαι
δυναμένων μήτε μὴν ἐπειγομένων πνευμάτων
ἀποκρῖναι καὶ διαστῆσαι τάς τε ἄχνας καὶ τὸν
καρπόν, ἀνθρώπων ἀρούντων τε καὶ σπειρόντων
τροφὰς ἔχουσι μύρμηκες. σοφὸν δὲ καὶ ἐκεῖνο
προσακήκοα, ὅτι ἄρα τοὺς τεθνεῶτας μύρμηκας
οἱ προσήκοντες ἐν ταῖς τῶν πυρῶν κηδεύουσι
θυλακίσιν, ὡς πατέρας ἢ πᾶν τὸ φίλιον ἐν ταῖς
σοροῖς οἱ ἄνθρωποι.

44. Ἵππος εἰ τυγχάνοι κηδεμονίας, ἀμείβεται
τὸν εὐεργέτην εὐνοίᾳ τε καὶ φιλίᾳ. καὶ ὁποῖος
μὲν ἦν ὁ Βουκεφάλας ἐς Ἀλέξανδρον διαρρεῖ
πανταχόσε ὁ λόγος, καὶ οὔ μοι λέγειν αὐτὸν
ἥδιόν ἐστι. καὶ τὸν Ἀντιόχου δὲ ἵππον τὸν
τιμωρήσαντα τῷ δεσπότῃ καὶ ἀποκτείναντα τὸν
Γαλάτην ὥσπερ οὖν ἀπέσφαξε τὸν Ἀντίοχον ἐν τῇ

62

in which the pregnant ants bring forth their young—
the 'women's quarters,' as it might be; and the
third they set apart as a treasury and a pit for the
seeds they have collected. And no Ischomachus,[a]
no Socrates, with their interest in the management
of a household on admirable lines, is there to teach
them these things. When Ants go abroad to collect
food, they follow the biggest ones, and these lead the
way, like generals. And as soon as they reach the
crops the young ones stand at the foot of the stalks
while the leaders crawl up and having eaten through
what are called the 'rhacillae' of the fruitful ears,
throw the ears down to the crowd below. And these
go about and cut off the chaff and peel off the capsules
that protect and envelop the wheat. They need no
threshing, no men who can winnow, nor even 'rushing winds' [Hom. *Il.* 5. 501] to separate and sunder
the chaff and the grain, yet the Ants possess the food
of men who plough and sow.

I have also heard the following example of their
cleverness: their relations bury dead ants in the
capsules of wheat, just as men bury their parents or
all whom they love in coffins.

44. If a Horse receives careful attention, he repays
his benefactor by being good-natured and friendly.
How Bucephalus bore himself to Alexander is a story
that is current everywhere and would give me no
pleasure to repeat. I shall also pass over the horse
of Antiochus [b] which avenged his master by killing

The Horse's
devotion to
its master

---

[a] Ischomachus in Xenophon's *Oeconomicus* (chs. 7–end)
propounds a system of domestic economy that wins the
approval of Socrates.

[b] Antiochus Soter, founder of the Seleucid dynasty, reigned
280–261 B.C.; fell in battle against the Gauls.

AELIAN

μάχῃ (ὄνομα δὲ τῷ Γαλάτῃ Κεντοαράτης ἦν) ἐῶ
καὶ τοῦτον. Σωκλῆς δὲ ἄρα (οὐ γάρ τί που
πολλοὶ τόνδε μοι δοκοῦσιν ἐγνωκέναι) Ἀθηναῖος
μὲν ἦν, καλὸς δὲ καὶ ἐδόκει καὶ ἐπεφύκει. οὗτος
οὖν ἐπρίατο ἵππον ὡραῖον μὲν καὶ αὐτόν, ἐρωτικὸν δὲ ἰσχυρῶς καὶ οἷον σοφώτερον ἢ κατὰ τοὺς
ἄλλους [1] ἵππους. οὐκοῦν ἐρᾷ τοῦ δεσπότου δριμύ-
τατα, καὶ προσιόντος ἐφριμάττετο καὶ ἐπικροτοῦν-
τος ἐφρυάττετο, καὶ ἀναβαίνοντος ἑαυτὸν παρεῖχεν
εὐπειθῆ, καὶ παρεστῶτος κατὰ πρόσωπον ὁ δὲ
ὑγρὸν ἑώρα. καὶ ταῦτα μὲν ἐρωτικὰ ὄντα ἤδη
ὅμως τερπνὰ ἐδόκει· ἐπεὶ δὲ ἦν ὥς τι καὶ δρασείων ἐς τὸ μειράκιον προπετέστερος, καὶ διέρρει
λόγος ὑπὲρ ἀμφοῖν ἀτοπώτερος, ὁ Σωκλῆς οὐκ
ἐνεγκὼν τὸ ἀπόφημον, ὡς ἐραστὴν ἀκόλαστον
μισήσας ἀπημπόλησε τὸν ἵππον. ὁ δὲ οὐ φέρων
τὴν ἐρημίαν τὴν ἀπὸ τοῦ καλοῦ, ἑαυτὸν τοῦ ζῆν
ἀπήλλαξε λιμῷ βιαιοτάτῳ.

45. Νοοῦσι δὲ ἄρα ἀτταγᾶς μὲν ἀλεκτρυόνι
ἔχθιστα, ἀλεκτρυὼν δὲ αὖ πάλιν ἀτταγᾷ, καὶ
κορώνῃ κίρκος καὶ ἐκείνη ἐκείνῳ, καὶ πελαγίῳ
ἱέρακι ὁ κόραξ καὶ κόρακι ἐκεῖνος, τρυγόνι τε
κόραξ καὶ κίρκος, καὶ μέντοι καὶ ἡ τρυγὼν
ἑκατέρῳ. πέπυσμαι δὲ καὶ πελαργὸν νυκτερίδα
μισεῖν, καὶ ἐκείνην ἀντιμισεῖν ὡς πολέμιον, πελε-
κᾶνα δὲ μὴ νοεῖν φίλα ὄρτυγι. καὶ ἀμοιβὴν τοῦ
μίσους ἀκούω εἶναι.

46. Ἀποκτίννυσι δὲ ἀετὸν μὲν τὸ καλούμενον
σύμφυτον, τὴν δὲ ἶβιν ὑαίνης χολή, σκορόδου

[1] ἄλλους ὁρᾶν.

the Gaul (his name was Centoarates) who slew Antiochus on the battlefield. Socles then, about whom not many seem to know, was an Athenian who was esteemed, and indeed was, a comely boy. Now he bought a horse, handsome too like its master but of a violently amorous disposition and with a far sharper eye than other horses. Hence it conceived a passionate love for its master, and when he approached, it would snort; and if he patted it, it would neigh; when he mounted, it would be docile; when he stood before it, it would cast languishing glances at him. These actions already savoured of love, but were thought pleasing. When however the horse, becoming too reckless, seemed to be meditating an assault upon the boy, and tales about the pair of a too monstrous nature began to circulate, Socles would not tolerate the slander, and in his detestation of a licentious lover sold the horse. But the animal could not bear to be separated from the beautiful boy and ended its days by a rigorous starvation.

45. The Francolin entertains the bitterest hatred for the Cock, and the Cock on its side for the Francolin; likewise the Falcon for the Crow, and vice versa; and the Raven for the Sea-hawk, and the Sea-hawk for it; the Raven and the Falcon for the Turtle-dove, and the Turtle-dove for both. I have learnt also that the Stork abhors the Bat, and the Bat in return abhors it as an enemy; and the Pelican, I am told, is not friendly disposed to the Quail, and their hatred is mutual. *Birds and their enmities*

46. To the Eagle the herb called comfrey is fatal; to the Ibis the gall of the Hyena; to the Starling the *Substances fatal to Birds*

65

AELIAN

σπέρμα τὸν ψᾶρα, χαραδριὸν ἄσφαλτος, τὸν δὲ
ἔλανον¹ ὁ καλούμενος ποταμογείτων. ἔλανος δὲ
αἰθυίας χολὴν οὐχ ὑπομένει. κίρκος δὲ καὶ λάρος
καὶ τρυγὼν καὶ κόσσυφος καὶ τὸ γυπῶν ἔθνος
σίδην² κοπεῖσαν εἰ διατράγοιεν, ἀπολώλασι.
κέδρου τὸν καλαμοδύτην ἀπόλλυσι φύλλα, ἄνθος
δὲ ἄγνου τὸν μελαγκόρυφον, κόρακα δὲ εὐζώμου
σπέρμα. μύρῳ κάνθαρος ἀποθνήσκει, στέατι δορ-
κάδος ὁ ἔποψ. κορώνη δὲ λυκοβρώτου κρέως
λειψάνῳ περιτυχοῦσα ἀποθνήσκει. κορυδαλλὸς δὲ
νάπυος σπέρματι, γέρανος ἀμπέλου δάκρυον σπά-
σασα διεφθάρη.

47. Λαγὼ δὲ πέρι καὶ ἐνταῦθα ἔπεισιν εἰπεῖν
τοιαῦτά μοι. ἐς τὴν κοίτην τὴν συνήθη οὐ
πάρεισιν ὁ λαγὼς πρὶν ἢ ταράξαι τὰ ἴχνη, πῇ μὲν
ἐσιὼν πῇ δὲ ἐξιών, ἵνα ἀφανίσῃ τὴν ἐκ τῶν
θηρατῶν ἐς αὐτὸν ἐπιβουλήν, σοφίᾳ τινὶ φυσικῇ τὸ
θηρίον τοὺς ἀνθρώπους αἱμυλώτατα ἀπατῆσαν.

48. Μήτηρ δὲ ἄρα καὶ ἡ θήλεια ἵππος ἀγαθὴ
ἦν καὶ τοῦ πώλου τοῦ ἐξ αὐτῆς μεμνῆσθαι δεινή.
ὅπερ οὖν κατεγνωκὼς Δαρεῖος ὁ κάτω εἶτα μέντοι
ἐπήγετο ἐς τὰς μάχας ἐξ ὠδίνων ἵππους τὰ βρέφη
καταλιπούσας οἴκοι. τρέφονται δὲ καὶ ὀρφανοὶ
μητέρων οἱ πῶλοι γάλακτι ξένῳ, ὥσπερ οὖν καὶ
οἱ ἄνθρωποι. οὐκοῦν ὅτε ἡ τροπὴ τῆς μάχης τῆς
κατὰ τὸν Ἰσσὸν τὰ Περσῶν πιέζειν ὑπήρξατο, καὶ
ἐνικᾶτο Δαρεῖος, ἵππον ἀνέβη θῆλυν, φυγῆς δεόμενος καὶ σωτηρίας ὠκίστης. ἡ δὲ ἄρα τοῦ κατα-

¹ *Oud, Klein*: ἐχῖνον . . . ἐχῖνος MSS, 'corrupt' *H.*

66

seed of garlic; to the Stone-curlew bitumen; to the Kite pondweed, as it is called. And the Kite cannot endure the gall of the Shearwater. If a Falcon, or a Sea-mew, or a Turtle-dove, or a Blackbird, or the whole Vulture tribe eat a sliced pomegranate, they die. The leaves of the cedar are fatal to the Reed-warbler (?); the flower of the agnus-castus to the Marsh-tit; to the Raven the seed of the rocket. The Beetle is killed by perfume, and the Hoopoe by the fat of a gazelle. If a Crow comes upon the remains of flesh which a wolf has eaten, it is killed. A Lark is destroyed by mustard-seed, and a Crane if it drinks the gum from a vine.

47. It occurs to me at this point to speak of the The Hare Hare as follows. The Hare does not repair to its accustomed form until it has confused its tracks, here in entering, and there in leaving, in order to defeat the designs of huntsmen. It is by some kind of natural sagacity that it tricks men so very craftily.

48. It seems that the Mare is in fact a good mother The Mare and cherishes the memory of her foal. The younger and its love Darius had noted this; hence he would take into battle for its foal some mares that had lately foaled and had left their young at home. Foals that lose their dams are reared on the milk of a stranger, just as human beings are. Now when the changing fortune of the battle of Issus began to press the Persians, and Darius was being defeated, he mounted a Mare, being anxious to escape and to save himself with all possible speed.

---

[2] ῥοιᾶς σίδην.

AELIAN

λειφθέντος μνήμῃ, ὡς εἶχεν ἐπιθυμίας καὶ ποδῶν,
τὸν δεσπότην ὑμνεῖται τῆς ἀκμῆς τῶν ἐπικειμένων
κινδύνων ἐξαρπάσαι.

49. Ἡμίονος γέρων Ἀθήνησιν ὑπό γε τοῦ
δεσπότου τοῦ ἰδίου τῶν ἔργων ἀπολυθείς, ὡς
Ἀριστοτέλης λέγει, τοῦ μὲν φιλοπόνου καὶ
ἐθελουργοῦ καθ' ἡλικίαν ἑαυτὸν οὐκ ἀφῆκεν.
ἡνίκα γοῦν Ἀθηναῖοι κατεσκεύαζον τὸν Παρθε-
νῶνα, οὔτε ἐπισύρων οὔτε ἀχθοφορῶν ὅμως τοῖς
νέοις ὀρεῦσι προφορουμένοις τὴν ὁδὸν ἄκλητος καὶ
ἑκὼν οἱονεὶ παράσειρος ᾔει, δορυφορῶν ὡς ἂν
εἴποις καὶ παρορμῶν τὸ ἔργον τῇ βαδίσει τῇ
κοινῇ δίκην τεχνίτου παλαιοῦ τοῦ μὲν αὐτουργεῖν
ὑπὸ τοῦ γήρως ἀπολυθέντος, ἐμπειρίᾳ δὲ καὶ [1]
διδασκαλίᾳ ὑποθήγοντός τε ἅμα τοὺς νέους καὶ
ἐπαίροντος. ταῦτα οὖν μαθόντες ὁ δῆμος τῷ
κήρυκι ἀνειπεῖν προσέταξαν, εἴτε ἀφίκοιτο ἐς τὰ
ἄλφιτα, εἴτε ἐς τὰς κριθὰς παραβάλοι, μὴ ἀνείρ-
γειν, ἀλλ' ἐὰν σιτεῖσθαι ἐς κόρον, καὶ τὸν δῆμον
ἐκτίνειν τὸ ἀργύριον, τρόπον τινὰ ἀθλητῇ σιτή-
σεως ἐν Πρυτανείῳ [2] δοθείσης ἤδη γέροντι.

50. Κλεάνθην τὸν Ἄσσιον κατηνάγκασε καὶ
ἄκοντα εἶξαι καὶ ἀποστῆναι τοῖς ζῴοις τοῦ καὶ
ἐκεῖνα λογισμοῦ μὴ διαμαρτάνειν ἀντιλέγοντα
ἰσχυρῶς καὶ κατὰ κράτος ἱστορία τοιαύτη φασίν.
ἔτυχεν ὁ Κλεάνθης καθήμενος καὶ μέντοι καὶ

---

[1] καὶ παλαιᾷ διδασκαλίᾳ.

And the Mare, remembering the foal she had left
behind, is celebrated for having with the uttermost
eagerness and at full speed snatched her master away
from the critical moment of urgent danger.

49. At Athens an aged Mule was released from An aged
work by its master, so Aristotle tells us [*HA* 577 b 30], Mule
but declined to abandon its love of labour and its
willingness to work on the score of age.   Thus, at the
time when the Athenians were erecting the Par-
thenon, though it neither drew nor carried burdens,
yet it would unbidden and of its own free will walk
by the young mules as they went back and forth, like
a horse harnessed alongside a pair, acting as guard,
so to speak; and by treading a common path it en-
couraged their work, like some old craftsman whom
age has released from labour with his hands but
whose experience and knowledge are a stimulus and
incitement to the young.   Now when the people got
to hear of this they directed the herald to proclaim
that if it came in quest of barleymeal or approached
to get corn, it was not to be prevented but was to be
allowed to eat its fill, and that the populace would
defray the cost, as in the case of an athlete who in his
old age was given his meals in the Prytaneum.

50. The following story, they say, shows how Cleanthes
Cleanthes of Assos was forced against his will and and the
in spite of his vehement arguments to the contrary, to Ants
make a concession to animals and to allow that they
too are not destitute of reasoning power.   Cleanthes
happened to be seated and moreover was resting

---

² ἐκτίνειν ἐν Πρυτανείῳ MSS, *H*; ἐν Π. *transposed by Cobet.*

σχολὴν ἄγων μακροτέραν ἄλλως. οὐκοῦν μύρμηκες [1] περὶ τοῖς ποσὶν ἦσαν αὐτῷ πολλοί. ὁ δὲ
ἄρα ὁρᾷ ἐξ ἀτραποῦ τινος ἑτέρας νεκρὸν μύρμηκα
μύρμηκας ἄλλους κομίζοντας ἐς οἶκον ἑτέρων καὶ
ἑαυτοῖς οὐ συντρόφων, καὶ ἐπί γε τῷ χείλει τῆς
μυρμηκιᾶς ἑστῶτας αὐτῷ νεκρῷ, καὶ ἀνιόντας
κάτωθεν ἑτέρους καὶ συνιόντας [2] τοῖς ξένοις ὡς
ἐπί τινι εἶτα κατιόντας τοὺς αὐτούς, καὶ πλεονάκις
τοῦτο. καὶ τελευτῶντας σκώληκα οἱονεὶ λύτρα
κομίσαι, τοὺς δὲ ἐκεῖνον μὲν λαβεῖν, προέσθαι δὲ
ὅνπερ οὖν ἐπήγοντο νεκρόν. καὶ ἐκείνους ὑποδέξασθαι ἀσμένως ὡς υἱὸν κομιζομένους ἢ ἀδελφόν.
τί οὖν πρὸς ταῦτα Ἡσίοδος λέγει, λέγων ὅτι ἄρα
ὁ Ζεὺς τὰς φύσεις ἀπέκρινε, καὶ οὖν καὶ ‘ἰχθύσι
μὲν καὶ θηρσὶ καὶ οἰωνοῖς πετεηνοῖς’ ἔδωκεν
‘ἔσθειν ἀλλήλους, ἐπεὶ οὐ δίκη ἐστὶ μετ’ αὐτοῖς,
ἀνθρώποισι δ’ ἔδωκε δίκην’· ἀλλ’ οὐκ ἐρεῖ ταῦτα
ὁ Πρίαμος, εἴ γε καὶ ἐκεῖνος τὸν Ἕκτορα ἐλύ-
σατο πολλῶν κειμηλίων καὶ θαυμαστῶν παρὰ
τοῦ ἀνθρώπου ὁ ἄνθρωπος καὶ τοῦ ἥρωος καὶ
τοῦ Διὸς ἐκγόνου καὶ ἐκεῖνος ὢν ἔκγονος τοῦ
Διός.

51. Κατηγορεῖ τῆς διψάδος τὸ ἔργον αὐτὸ ἡμῖν
τὸ ὄνομα. καὶ ἔχεως μέν ἐστιν ὀλιγωτέρα τὸ
μέγεθος, ἀποκτεῖναι δὲ ὀξυτέρα· οἱ γάρ τοι τῷ
δήγματι προσπεσόντες ἐξάπτονταί τε ἐς δίψος
καὶ πιεῖν ἀναφλέγονται καὶ ἀμυστὶ σπῶσι καὶ
τάχιστα ῥήγνυνται. καὶ φησι μὲν Σώστρατος
λευκὴν εἶναι τὴν διψάδα, ἐπί γε μὴν τῆς οὐρᾶς

_____

[1] καὶ μύρμηκες.　　　　　[2] συνόντας.

quietly for some time. Now there were Ants about his feet in great numbers. So he observed how some were conveying a dead ant out of one track to a nest belonging to other ants not of their own kin. And they paused on the edge of the nest with the corpse while others came up from below and met the strangers seemingly with a view to some consultation; the same Ants then went down into the nest. And this happened several times until finally they brought up a worm, as it were a ransom. And the other party accepted it and surrendered the dead body which they had brought. And the Ants in the nest were glad to receive it, as though they were recovering a son or brother.

Now what answer can Hesiod make to this when he says [*OD* 277] that Zeus has made a distinction between various natures and has granted

' to fish on the one hand and to beasts and to winged fowl that they should devour one another, for among them there is no justice, but to mankind has he granted justice ' ?

But Priam will not admit this, since it was at the cost of many marvellous treasures that even he, a man and moreover a descendant of Zeus, redeemed Hector from the man who was also a hero and a descendant of Zeus.

51. The name of the *Dipsas* (thirst-provoker) declares to us what it does. It is smaller than the viper, but kills more swiftly, for persons who chance to be bitten burn with thirst and are on fire to drink and imbibe without stopping and in a little while burst. Sostratus declares that the Dipsas is white, though

The 'Dipsas' snake

ἔχειν γραμμὰς μελαίνας δύο. ἀκούω δὲ ὅτι καὶ
πρηστῆρας αὐτὰς καλοῦσί τινες, καύσωνας δὲ
ἄλλοι·[1] ὄχλον δὲ ἄρα ὀνομάτων ἐπαντλοῦσι τῷδε
τῷ θηρίῳ.[2] κέκληται δὲ καὶ μελάνουρος, ὥς φασι,
καὶ ἀμμοβάτης·[3] εἰ δὲ ἀκούσειας[4] κεντρίδα,
τὴν αὐτήν μοι λέγεσθαι νόει. δεῖ δὲ καὶ μῦθον
τῷδε τῷ ζῴῳ ἐπᾷσαί με ὅνπερ οὖν ἀκούσας
οἶδα,[5] ὡς ἂν μὴ δοκοίην ἀμαθῶς ἔχειν αὐτοῦ. τὸν
Προμηθέα κλέψαι τὸ πῦρ ἡ φήμη φησί, καὶ τὸν
Δία ἀγανακτῆσαι ὁ μῦθος λέγει καὶ τοῖς κατα-
μηνύσασι τὴν κλοπὴν δοῦναι φάρμακον γήρως
ἀμυντήριον. τοῦτο οὖν ἐπὶ ὄνῳ θεῖναι τοὺς
λαβόντας πέπυσμαι. καὶ τὸν μὲν προϊέναι τὸ
ἄχθος φέροντα, εἶναι δὲ ὥραν θέρειον, καὶ διψῶντα
τὸν ὄνον ἐπί τινα κρήνην κατὰ τὴν τοῦ ποτοῦ
χρείαν ἐλθεῖν. τὸν οὖν ὄφιν τὸν φυλάττοντα
ἀναστέλλειν αὐτὸν καὶ ἀπελαύνειν, καὶ ἐκεῖνον
στρεβλούμενον μισθὸν οἱ τῆς φιλοτησίας δοῦναι
ὅπερ οὖν ἔτυχε φέρων φάρμακον. οὐκοῦν ἀντίδοσις γίνεται, καὶ ὁ μὲν πίνει, ὁ δὲ τὸ γῆρας ἀποδύεται, προσεπιλαβὼν ὡς λόγος τὸ τοῦ ὄνου δίψος.
τί οὖν; ἐγὼ τοῦ μύθου ποιητής; ἀλλ' οὐκ ἂν
εἴποιμι, ἐπεὶ καὶ πρὸ ἐμοῦ Σοφοκλῆς ὁ τῆς
τραγῳδίας ποιητὴς καὶ Δεινόλοχος ὁ ἀνταγωνιστὴς
Ἐπιχάρμου καὶ Ἴβυκος ὁ Ῥηγῖνος καὶ Ἀριστίας
καὶ Ἀπολλοφάνης ποιηταὶ κωμῳδίας ᾄδουσιν
αὐτόν.

---

[1] ἄλλοι. γίνονται δὲ ἄρα ἐν Λιβύῃ τε καὶ Ἀραβίᾳ μᾶλλον.
[2] θηρίῳ καὶ ἄλλων.
[3] ἀμμοβάτης ὑπ' ἄλλων.
[4] ἀκούσαις καί.
[5] οἶδα, οὐ σιωπήσομαι τοῦτον.

it has two black stripes on its tail. And I have heard
that some people call these snakes *presteres* (inflaters);
others, *kausones* (burners). In fact they deluge this
creature with a host of names. It has also been called
melanurus (black-tail), so they say, and by others
ammobates (sand-crawler); and should you also hear
it also called *kentris* (stinger), you may take it from
me that the same snake is meant.

And it behoves me to repeat a story (which I know
from having heard it) regarding this creature, so that
I may not appear to be ignorant of it. It is said that
Prometheus stole fire, and the story goes that Zeus
was angered and bestowed upon those who laid information of the theft a drug to ward off old age. So
they took it, as I am informed, and placed it upon an
ass. The ass proceeded with the load on its back;
and it was summer time, and the ass came thirsting
to a spring in its need for a drink. Now the snake
which was guarding the spring tried to prevent it and
force it back, and the ass in torment gave it as the
price of the loving-cup the drug that it happened to
be carrying. And so there was an exchange of gifts:
the ass got his drink and the snake sloughed his old
age,[a] receiving in addition, so the story goes, the ass's
thirst.

What then? Did I invent the legend? I will
deny it, for before me it is celebrated by Sophocles,[b]
the tragic poet, and Dinolochus, the rival of Epi-
charmus, and Ibycus of Rhegium, and the comic poets
Aristias and Apollophanes.

Dipsas and Ass: an exchange of gifts

---

[a] Γῆρας is used in two senses: (i) old age, (ii) old
skin.
[b] Sophocles, in his Κωφοὶ Σάτυροι [*fr*. 362 P]. Of the following poets no fragment relating to this story survives.

52. Σοφὸν ἐλέφαντος ἔργον εἰ παραλίποιμι, φήσει μέ τις ἀγνοήσαντα οὐκ εἰπεῖν. ἔστι δὲ καὶ ἀκοῆς ἄξιον, καὶ διὰ ταῦτα ἀκούσωμεν αὐτοῦ. ὁ τῆς τούτου κομιδῆς ἐγχειρισθεὶς τὴν φροντίδα τῶν μὲν κριθῶν ὑφήρει, λίθους δὲ ὑποπάττων ἐκείνῳ μὲν ἄβρωτον τὸ πλεῖστον εἰργάζετο, ἀπέσωζε δὲ τὸν ὄγκον τοῦ μέτρου πρὸς τὸν ἐπισκοποῦντα ἀμφοῖν δεσπότην. καὶ τέως διελάνθανεν. οὐκοῦν ὁ ἐλέφας ἀθάρην ἰδὼν ἕψοντα τὸν ἐπίβουλόν οἱ, τῆς ἄμμου τῆς ἐν ποσὶ τῇ προβοσκίδι χύδην ἀναλαβὼν ἐνέβαλεν ἐς τὴν χύτραν καὶ ἠμύνατο ἀνθ' ὧν ἔπαθε δι' ὧν ἐποίησεν εὐμηχάνως.

53. Οἱ μὲν ἄλλοι κύνες καὶ ἑλεῖν καὶ ἀνιχνεῦσαι τὰ θηρία σοφοί, οἱ δὲ Αἰγύπτιοι φυγεῖν δεινότατοι. τὰ γοῦν ἐν τῷ Νείλῳ δεδιότας ἄγει μὲν αὐτοὺς τὸ δίψος πιεῖν, ἡσυχῇ δὲ καὶ ἐς κόρον πιεῖν τὸ δέος οὐ συγχωρεῖ. καὶ διὰ ταῦτα οὐ πίνουσιν ἐπικύψαντες, ὡς ἂν μή τι τῶν κάτωθεν ἀνερπύσαν εἶτα ἐξαρπάσῃ αὐτούς. οὐκοῦν τὴν μὲν ὄχθην παραθέουσι, λάπτουσι δὲ τῇ γλώττῃ, ἁρπάζοντες ὡς ἂν εἴποι τις ἢ καὶ νὴ Δία κλέπτοντες τὸ πῶμα.

54. Ἐχῖνος, οὐχ ὁ θαλάσσιος, ἀλλ' ὁ χερσαῖος, πολλὰ μὲν καὶ ἄλλα ὥς ἐστι πανοῦργος ἤδη μοι λέλεκται, ὃ δὲ οὐκ εἶπον αὐτοῦ δολερὸν ἔργον, τοῦτο εἰρήσεται τὰ νῦν. μέλλων ἁλίσκεσθαι ἑαυτὸν συνειλήσας ἄληπτον ἐργάζεται, εἶτα μέντοι καὶ πιέζει τὸ πνεῦμα καὶ ἀκίνητος ἀτρεμεῖ καὶ τὸν τεθνεῶτα ὑποκρίνεται.

---

[a] See 3. 10; 4. 17.

52. Were I to pass over a piece of cleverness on the part of an Elephant, someone will say that I failed through ignorance to record it. And it is really worth hearing, so let us hear it. The man who was entrusted with the care of its food was in the habit of purloining its corn, and by scattering stones underneath it he rendered most of the food uneatable, while preserving the bulk of the measure, so far as the master who supervised them both could see. And for a while he escaped detection. So the Elephant, observing the designing fellow as he was cooking some porridge, picked up with its trunk a mass of sand at its feet and flung it into the pot, thus adroitly avenging the treatment it had received at his hands. *An Elephant punishes dishonesty*

53. All other Dogs are clever at catching and tracking down wild animals; Egyptian Dogs however excel at running away. Thus, although they dread the creatures in the Nile, thirst compels them to drink, while their fear does not allow them to drink in peace as much as they want. For that reason they do not put their heads down and drink, for fear some creature from below may creep up and seize them; and so they run along the brink, lapping with their tongue and snatching or, one might say, positively stealing their drink. *The Dog in Egypt*

54. I have already *a* mentioned many other crafty tricks of the Land *Echinus* (hedgehog), not the Sea *Echinus* (sea-urchin), but one specimen of its guile which I failed to mention I will mention now. When it is likely to be caught it rolls itself up, which makes it impossible to handle; moreover it holds its breath and remains motionless and pretends to be dead. *The Hedgehog*

75

AELIAN

55. Αἱ λεπάδες, οὐκ ἂν αὐτὰς ἀποσπάσειας τῶν πετρῶν, οὐδὲ εἰ λάβοις δακτύλοις ⟨τοῖς⟩[1] τοῦ Μίλωνος, ὅσπερ οὖν ἐγκρατέστατα καὶ εὐλαβέστατα τὴν ῥοιὰν κατεῖχεν, ὡς μὴ αὐτὴν ἀφελέσθαι τῶν ἀντιπάλων τινὰ τῆς δεξιᾶς αὐτοῦ. ὅστις δὲ ἐπιχειρεῖ λεπάδα ἀποσπάσαι τῆς πέτρας, ᾗ προσέχεται, γελᾶται[2] μογῶν καὶ παρέχει[3] θυμηδίαν. ἀδυνατεῖ γοῦν ἐγκρατὴς οὗ σπεύδει γενέσθαι. ξυομένη δὲ σιδήρῳ σχίζεται ἀπὸ τῆς πέτρας ὀψέ.

56. Στρατεύονται δὲ ἄρα οἱ Λίβυες οὐ μόνον ἐπὶ τοὺς γείτονας, ἵνα αὐτῶν πλέον ἔχωσιν, ἀλλὰ καὶ ἐπὶ τοὺς ἐλέφαντας. καὶ ἴσασί γε ἐκεῖνοι τῆς ὁδοῦ τῆς ἐπ᾽ αὐτοὺς τὴν ὑπόθεσιν οὐδὲ ἕτερον εἶναι ἢ τοὺς ὀδόντας. οἱ τοίνυν πηρωθέντες τὸν ἕτερον ἐπὶ μετώπου ἑστᾶσι, τῶν λοιπῶν προβαλλομένων αὐτούς, ἵνα οἱ μὲν ὑποδέχωνται τὴν πρώτην ὁρμήν, οἱ δὲ ἀμύνωσιν ἀκεραίῳ τῇ τῶν ὀδόντων ῥώμῃ καὶ ἰσοπαλεῖ, ἴσως δὲ τῶν Λιβύων[4] καὶ καταφιλοσοφοῦντες καὶ ἐπιδεικνύντες αὐτοῖς ὅτι ἄρα οὐχ ὑπὲρ μεγάλου τοῦ ἄθλου κινδυνεύσοντες ἥκουσι. χρῶνται δὲ ἄρα τῷ μὲν τῶν ὀδόντων ὡς ὅπλῳ, καὶ τεθηγμένον αὐτὸν φυλάττουσι, τῷ δὲ ὡς σμινύῃ· καὶ γὰρ ἐν αὐτῷ ῥίζας ὀρύττουσι καὶ δένδρα ἐκμοχλεύσαντες ὑποκλίνουσιν.

57. Οὐ μόνον δὲ ἄρα ἦσαν ὑφαντικαὶ αἱ φάλαγγες καὶ εὔχειρες κατὰ τὴν Ἀθηνᾶν τὴν Ἐργάνην

---

[1] ⟨τοῖς⟩ add. H.   [2] Reiske : γελᾷ τε.
[3] Reiske : ἔχει.   [4] Ges : Ἰνδῶν.

55. You would not succeed in dislodging Limpets The Limpet
from the rocks, even were you to grasp them with the
fingers of a Milo [a] who clung with such strength and
tenacity to a pomegranate-tree that not one of his
opponents could wrench it from his right hand. But
anyone who undertakes to dislodge a Limpet from
the rock to which it is clinging is laughed at for his
pains and affords merriment to others. At all events
it is impossible for him to get what he wants. An
iron saw will at long last detach it from the rock.

56. It appears that the Libyans do not confine The
themselves to waging war upon their neighbours with Elephant
a view to gaining an advantage over them, but they and its
wage war upon Elephants also. And the latter are hunters
well aware that the purpose of their attack is nothing
else than to get their tusks. So those beasts that
have had one tusk mutilated stand in the front line,
the rest of the herd using them as a cover in order
that they may receive the first assault and that the
rest may help with the strength of their tusks undamaged and equal to the struggle. And perhaps
they are trying to convince the Libyans and to prove
to them that they are risking their lives for an in-
considerable reward. One of their tusks they use as
a weapon and keep sharpened; the other they use as
a mattock, for with it they dig up roots and lever up
and bend down trees.

57. It seems after all that Spiders are not only The Spider's
dexterous weavers after the manner of Athena the web

[a] Native of Crotona, 6th cent. B.C., proverbial for his
great strength, gained six Olympic and six Pythian victories
in wrestling.

τε καὶ Πηνῖτιν θεάν, πεφύκασι δὲ [1] καὶ γεωμετ-
ρίαν δειναί. τὸ γοῦν κέντρον φυλάττουσι καὶ τὸν
ἐξ αὐτοῦ κύκλον καὶ τὴν περιφέρειαν ἀκριβοῦσιν
ἰσχυρῶς, καὶ Εὐκλείδου δέονται οὐδὲ ἕν· κάθηνται
γὰρ ἐν τῷ κέντρῳ μέσῳ ἐλλοχῶσαι τὴν ἑαυτῶν
ἄγραν. εἰσὶ [2] δὲ ὡς [3] εἰπεῖν καὶ ὑφάντριαι γεν-
ναῖαι καὶ ἀκεστικὴν εὐπάλαμοι· καὶ ὅ τι ἂν
διαρρήξῃς ἐκείνων τῆς εὐπήνου τε καὶ εὐμίτου
σοφίας, αἱ δὲ ἀνακοῦνται, καὶ ἀπαθὲς καὶ ὁλόκλη-
ρον αὖθις ἀποδείκνυνται.

58. Ἄνευ δὲ λογιστικῆς οἱ φοίνικες συμβαλεῖν
ἐτῶν πεντακοσίων ἴσασιν ἀριθμόν, μαθηταὶ φύσεως
τῆς σοφωτάτης ὄντες, καὶ διὰ ταῦτά τοι μηδὲ [4]
δακτύλων δεδεημένοι ἢ ἄλλου τινὸς ἐς ἐπιστήμην
ἀριθμητικῆς. ὑπὲρ ὅτου δὲ ἴσασι τοῦτο καὶ
εἰδέναι ἀνάγκη αὐτούς, δημώδης ἐστὶν ὁ λόγος.
τὸν δὲ τῶν πεντακοσίων ἐτῶν χρόνον πληρούμενον
ἴσασιν Αἰγυπτίων ἤ τις ἢ οὐδείς, ὀλίγοι δὲ κομιδῇ
καὶ οὗτοι τῶν ἱερέων. οὗτοι [5] δ' οὖν [6] πρὸς
ἀλλήλους ὑπὲρ τούτων οὐ ῥᾳδίως συμβῆναι ἔχουσιν, ἀλλὰ οἱ μὲν ἐρεσχελοῦσι σφᾶς αὐτοὺς ἐρίζοντες
ὡς [7] οὐ νῦν ἀλλ' ἐς ὕστερον ὅδε ὁ θεῖος ὄρνις
ἀφίξεται ἢ ὡς ἐχρῆν ἥκειν· ὁ δὲ ἄλλως ἐκείνων
ἐριζόντων ἀποσημαίνεται δαιμονίως τὸν καιρὸν
καὶ πάρεστιν. οἱ δέ, θύειν ἀνάγκη αὐτοὺς καὶ

---

[1] πεφύκεσαν δὲ ἄρα.	[2] ἦσαν or ἔστι.
[3] ὡς ἰδόντι.	[4] μήτε.
[5] καὶ οὗτοι.	[6] Kaibel : γοῦν MSS, H.
[7] ἢ ὡς.	

Worker and goddess of the Loom, but that they are by nature clever at geometry.[a] Thus, they keep to the centre and fix with the utmost precision the circle with its boundary based upon it, and have no need of Euclid,[b] for they sit at the very middle and lie in wait for their prey. And they are, as you might say, most excellent weavers and adept at repairing their web. And any thread that you may chance to break of their skilled and delicate workmanship they repair and render sound and whole again.

58. The Phoenix knows how to reckon five hundred years without the aid of arithmetic, for it is a pupil of all-wise Nature, so that it has no need of fingers or anything else to aid it in the understanding of numbers. The purpose of this knowledge and the need for it are matters of common report. But hardly a soul among the Egyptians knows when the five-hundred-year period is completed; only a very few know, and they belong to the priestly order. But in fact the priests have difficulty in agreeing on these points, and banter one another and maintain that it is not now but at some date later than when it was due that the divine bird will arrive. Meantime while they are vainly squabbling, the bird miraculously guesses the period by signs and appears. And the priests are obliged to give way [c] and confess that they devote their time ' to putting the sun to

[a] Cp. Arist. *HA* 623 a 7 and D. W. Thompson (Eng. tr.) *ad loc.*

[b] Euclides of Alexandria, the famous geometer, *c.* 300 B.C.

[c] Lit. ' to offer sacrifice '; the word is used metaphorically of one who concedes a point, who admits that something is due to one in a stronger position than himself. See Headlam on Herodas 2. 71, Kaibel, *Hermes* 28 (1893) 53–4.

The Phoenix

AELIAN

ὁμολογεῖν ὅτι τὸν μὲν ἥλιον ἐν ταῖς λέσχαις κατα-
δύειν ἄγουσι σχολήν, οὐκ ἴσασι δὲ ὅσα ὄρνιθες.
ἐκεῖνα δέ, ὦ πρὸς τῶν θεῶν, οὐ σοφά, εἰδέναι ποῦ
μὲν Αἴγυπτός ἐστι, ποῦ δὲ καὶ Ἡλίου πόλις, ἔνθα
αὐτῷ πέπρωται ἥκειν, καὶ ὅπου ποτὲ τὸν πατέρα
καταθέσθαι χρὴ καὶ ἐν θήκαις τίσι; ταῦτα δὲ εἰ
μὴ δοκεῖ θαυμαστά, ἆρά γε τὰ ἀγοραῖα καὶ τὰ
ἐνόπλια καὶ τὰς ἄλλας τῶν ἀνθρώπων ἐς ἀλλήλους
τε καὶ κατ' ἀλλήλων ἐπιβουλὰς ἐροῦμεν σοφά;
ἐμοὶ μὲν οὐ δοκεῖ, ὦ Σίσυφον καὶ Κερκώπων καὶ
Τελχίνων ζηλωταὶ ἄνθρωποι. λέγω δὲ πρὸς τοὺς
ἀκριβοῦντας ταῦτα, πρός γε μὴν τοὺς ἀτελέστους
τοῖς προειρημένοις κακοῖς οὐ λέγω.

59. Τὸ δὲ ἐνθυμηματικὸν[1] καὶ διαλεκτικὸν καὶ
τὸ τοῦδε μᾶλλον ἢ τοῦδε αἱρετικὸν[2] εἰ καὶ τὰ ζῷα
οἶδεν, εἰκότως ἂν εἴποιμεν διδάσκαλον τῶν ὅλων
τὴν φύσιν ἄμαχον. ἐμοὶ γοῦν τις γευσάμενος
διαλεκτικῆς καὶ κυνηγεσίων ἀμωσγέπως ἐχόμενος
τοιαῦτα ἔλεγεν. ἦν θηρατικὴ κύων, ἦ δ' ὅς.
οὐκοῦν λαγὼ κατ' ἴχνια ᾔει. καὶ ὁ μὲν οὐχ
ἑωρᾶτό πω, μεταθέουσα δὲ ἡ κύων ἐντυγχάνει
που τάφρῳ, καὶ διαπορεῖ ἆρά γε ἐπὶ δεξιὰ ἄμεινον
ἢ ἐπὶ θάτερα διώκειν· ὡς δ' ἀποχρώντως ἐδόκει
σταθμήσασθαι, εἶτα εὐθύωρον ὑπερεπήδησεν. ὁ
φάσκων οὖν διαλεκτικός τε εἶναι καὶ θηρατικὸς

[1] ἐνθυμητικόν.   [2] Reiske: αἱρετόν.

80

rest with their talk ' [cp. Call. *ep.* 2 = *AP.* 7. 80];
but they do not know as much as birds. But, in
God's name, is it not wise to know where Egypt is
situated, where is Heliopolis whither the bird is
destined to come, and where it must bury its father
and in what kind of coffin ? [a] But if there is nothing
wonderful in all this, are we really to pronounce as
' wise ' affairs relating to the market, to armaments,
and men's other schemes for their mutual undoing ?
I think not, you men who rival Sisyphus [b] and the
Cercopes [c] and the Telchines.[d] I address myself to
those who perfect themselves in these matters, but
not to those who have not been initiated into the
aforesaid abominations.

59. If even animals know how to reason deduc-
tively, understand dialectic, and how to choose one
thing in preference to another, we shall be justified
in asserting that in all subjects Nature is an instruc-
tress without a rival. For example, this was told me
by one who had some experience in dialectic and
was to some degree a devotee of the chase. There
was a Hound, he said, trained to hunt ; and so it was
on the track of a hare. And the hare was not yet to
be seen, but the Hound pursuing came upon a ditch
and was puzzled as to whether it had better follow to
the left or to the right. And when it seemed to have
weighed the matter sufficiently, it leapt straight

The Dog, its reasoning power

[a] See Hdt. 2. 73.

[b] Sisyphus, mythical King of Corinth, became a byword for
deceitfulness and cruelty.

[c] Cercopes, mischievous dwarfs, who robbed Heracles;
changed by Zeus into monkeys.

[d] Telchines, under one aspect, were malignant demons with
the power of changing their shapes.

ταύτῃ πῃ συνάγειν τὴν ὑπὲρ τῶν λεχθέντων
ἐπειρᾶτο ἀπόδειξιν. ἐπιστᾶσα ἡ κύων ἐσκοπεῖτο
καὶ πρὸς ἑαυτὴν ἔλεγεν ʽἤτοι τῇδε ἢ τῇδε ἢ ἐκείνῃ
ὁ λαγὼς ἐτράπετο. οὔτε μὴν τῇδε οὔτε τῇδε·
ἐκείνῃ ἄρα.ʼ καὶ οὔ μοι ἐδόκει σοφίζεσθαι· τῶν
γὰρ ἰχνῶν μὴ ὁρωμένων ἐπὶ τάδε τῆς τάφρου,
κατελείπετο ὑπερπηδῆσαι τὸν λαγὼν αὐτήν. εἰκότως οὖν ἐπήδησε καὶ αὐτὴ κατʼ αὐτόν· ἰχνευ-
τικὴ γὰρ καὶ εὔρινος ἐκείνη [1] γε ἡ κύων ἦν.

60. Μασσαγέται μέν, ὡς Ἡρόδοτος λέγει, τὸν
φαρετρεῶνα πρό γε ἑαυτῶν κρεμάσαντες, εἶτα
μέντοι ὁμιλεῖ τῇ θηλείᾳ ὁ ἄρρην ἐμφανῶς, εἰ καὶ
ὁρῷεν αὐτοὺς οἱ πάντες, πεφροντικότες οὐδὲν
ἐκεῖνοί [2] γε. καμήλων δὲ ὁμιλία οὐκ ἄν ποτε
ἐμφανὴς γένοιτο, οὐδὲ ὁρώντων οἱονεὶ μαρτύρων·
ἀλλὰ εἴτε αἰδῶ φαμεν εἴτε φύσεως δῶρον ἀπόρ-
ρητον, ταῦτα Δημοκρίτῳ τε καὶ τοῖς ἄλλοις
καταλείπωμεν ἐλέγχειν τε καὶ τὰς αἰτίας οἴεσθαι [3]
λέγειν ἱκανοῖς ὑπὲρ τῶν ἀτεκμάρτων τε καὶ οὐ
συμβλητῶν. ἤδη δὲ καὶ ὁ νομεὺς ἀπαλλάττεταί
ποι, ὅταν αἴσθηται τῆς συμφοιτήσεως αὐτοῖς τῆς
πρὸς ἀλλήλους τὴν ὁρμήν, ὥσπερ οὖν ἀφιστάμενος
παριοῦσιν ἐς θάλαμον νύμφῃ τε καὶ νυμφίῳ.

61. Λυκοῦργος δὲ νομοθετεῖ νόμον φιλανθρω-
πότατον, ὡς ἐγῷμαι, θάκων τε καὶ ὁδῶν ἀφίστασθαι τοῖς πρεσβυτέροις τοὺς νεωτέρους αἰδοῖ

[1] Reiske : ἐκεῖνος ἢ ἐκείνη.   [2] ἐκείνων.
[3] οἴεσθαι del. H.

across. So the man who professed himself both dialectician and huntsman essayed to offer the proof of his statements in the following manner: The Hound paused and reflected and said to itself: ' The hare turned either in this direction or in that or went ahead. It turned neither in this direction nor in that; therefore it went ahead.' And in my opinion he was not being sophistical, for as no tracks were visible on the near side of the ditch, it remained that the hare must have jumped over the ditch. So the Hound was quite right also to jump over after it, for certainty that this particular Hound was good at tracking and keen-scented.

60. The Massagetae, according to Herodotus The Camel, [1. 216], hang up their quivers in front of themselves its modesty and then the man has commerce with the woman openly, even though all can see, though in fact they pay no attention.[a] Camels however would never couple in the open, nor if there were witnesses, so to say, looking on. But whether we are to call this modesty or a mysterious gift of Nature, let us leave it to Democritus and others to decide and suppose themselves competent to investigate and explain the causes of matters obscure and past conjecture. And even the herdsman at once takes himself off when he realises that the urge to couple is upon them, just as one withdraws when the bride and bridegroom are about to enter the marriage-chamber.

61. Lycurgus laid down a most humane law (as I The think), viz that younger men should give up their Elephant's respect for seats to, and leave the path for, their elders out of old age

[a] The statement is a travesty of Hdt. 1. 216.

χρόνου ἐς ὃν εὔχονται πάντες ἀφικέσθαι, ἐάνπερ
οὖν αὐτοῖς πεπρωμένον τοῦτο δήπου ᾖ. πῶς [1] δὲ
ὁ γενναῖος ὁ τοῦ Εὐνόμου δύναιτο ἂν τοῖς τῆς
φύσεως νόμοις ἁμιλλᾶσθαί τε καὶ ἀντικρίνεσθαι;
ἐπαΐουσι γοῦν τὸ τῶν ἐλεφάντων γένος, ὦ Λυκοῦρ-
γοί τε καὶ Σόλωνες καὶ Ζάλευκοι καὶ Χαρῶνδαι,
ὦνπερ οὖν ὑμεῖς νομοθετεῖτε οὐδὲ τὴν ἀρχήν, καὶ
ὅμως δρῶσι τοιαῦτα, καὶ τροφῆς ἀφίστανται τοῖς
πρεσβυτέροις οἱ νέοι, καὶ γήρᾳ παρειμένους
θεραπεύουσιν αὐτούς, καὶ κινδύνων ῥύονται, καὶ
ἐς ὀρύγματα ἐμπεσόντας οἶδε ἀνάγουσι, φρυγάνων
τινὰς ἀγκαλίδας καὶ φακέλους ἐμβαλόντες, οἷσπερ
οὖν ὡς ἀναβαθμοῖς χρώμενοι ἐκεῖνοι εἶτα ἀνίασι
γήρᾳ βαρεῖς ὄντες. ποῦ δαὶ [2] ἠλόησε πληγαῖς
πατέρα ἐλέφας; ποῦ δαὶ ἀπεκήρυξεν ὁ πατὴρ ὁ
ἐν τούτοις τὸν υἱόν; ὑμῖν δὲ ἴσως, ὦ ἄνθρωποι,
δοκῶ λέγειν μύθους τεχνίταις [3] (εἰ τἀληθῆ λέγειν
ἐθέλοιμεν) καὶ ποιηταῖς οὖσι τῶν μύθων τῶν
ἀπιστουμένων.

62. Φιλοδέσποτον μὲν [4] ὥς ἐστιν ὁ κύων,
τεκμηριοῖ καὶ τὰ ἤδη λεχθέντα, χρὴ δὲ ἄρα τάτ-
τειν ⟨σὺν⟩[5] αὐτοῖς καὶ ἐκεῖνό γε δήπου. Γέλων
ὁ Συρακόσιος καθεύδων βαθύτατα ἐδόκει διόβλητος
γεγονέναι. καὶ τὸ μὲν φάσμα ὄνειρος ἦν, ἐβόα
δὲ καίτοι καθεύδων [6] καὶ μάλα γε ὀξείᾳ καὶ
διατόρῳ τῇ φωνῇ. κύων οὖν ὑπ᾽ αὐτῷ τραφεὶς

---

[1] ποῦ.   [2] Reiske : δέ.
[3] Schn : τεχνίτας.   [4] μὲν οὖν.
[5] ⟨σὺν⟩ add. H.   [6] καθεύδων αὐτός.

respect for years which all pray they may attain, if
that chance to be their destiny. But how could the
noble son of Eunomus seek to rival and compete with
the laws of Nature? At any rate, you lawgivers, men
like Lycurgus,[a] Solon, Zaleucus, and Charondas, the
race of Elephants obeys laws which your legislation
does not even begin to touch. For all that, they
behave in the following manner: the young ones
give way to the elders in feeding; they wait upon
those that are weak with age; they guard them
from danger; when they fall into pits the young ones
drag them out by throwing in armfuls, so to say, and
bundles of dry sticks which the elders use as steps and
so climb out, though burdened with age. Where, I
should like to know, did an Elephant ever belabour its
sire with blows? Where, I ask, among Elephants did
a sire ever disinherit its son? But perhaps, my fellow
men, you who (if I am to speak the truth) fabricate and
invent incredible tales, think that I am telling tales.

62. What I have said above [b] proves that the Dog   Gelon and
certainly loves his master, and so I think I should   his dog
put the following story beside the rest. Gelon of
Syracuse [c] while fast asleep fancied that he had been
struck by Zeus.[d] But what he saw was only a dream;
yet, although asleep he cried aloud and at the top of
his voice. Whereupon a Dog which he kept, hear-

---

[a] Lycurgus, son of Eunomus and King of Sparta, perh.
9th cent. B.C., legislator *par excellence* of Sparta.—Zaleucus,
7th cent. B.C., drew up laws for Locri Epizephyrii.—Charondas
of Catana, perh. 6th cent. B.C., made laws for his city, for
Rhegium and other Chalcidian cities.

[b] See ch. 25.

[c] Tyrant of Syracuse, 485–78 B.C.

[d] *I.e.* by a thunderbolt. The story is repeated in *VH* 1. 13.

ἀκούσας φίλου καὶ συντρόφου φθέγματος, ὥς τι
τοῦ Γέλωνος ἐξ ἐπιβουλῆς παθεῖν κινδυνεύοντος,
ὡς εἶχεν ὁρμῆς ἀναθορὼν ἐπὶ τὴν στρωμνὴν καὶ
περιβὰς τὸν τροφέα, ὑλάκτει σφοδρότατα, οἷα δὴ
ἀμυνούμενος [1] τὸν ἐπιόντα. ἐξήγρετο τοίνυν ὁ
Γέλων καὶ ὑπὸ τοῦ δέους καὶ ὑπὸ τῆς ὑλακῆς
ἐκβαλὼν τὸν ὕπνον καίτοι βαθύτατον ὄντα.

63. Δράκων νήπιος νηπίῳ παιδί, τὸ γένος
Ἀρκάδι, κἀκεῖνος ἐπιχώριος γίνεται σύντροφος.
οὐκοῦν συνανιόντε [2] τὴν ἡλικίαν ὁ παῖς [3] ἦν
μειράκιον καὶ ὁ σύντροφος ὑπέρμεγας ἤδη ἦν.
καὶ ἀλλήλους μὲν ἐφίλουν, οἱ δὲ τῷ μειρακίῳ
προσήκοντες ὠρρώδουν τοῦ θηρίου τὸ μέγεθος·
τὸ γάρ τοι ζῷον τοῦτο, ὥκιστα μεγέθει μὲν
μέγιστον ἴδοις ἂν αὐτό, ὄψει δὲ φοβερώτατον.
καθεύδοντα οὖν σὺν τῷ παιδὶ ἐπί γε τῆς κλίνης
τῆς αὐτῆς ἀράμενοι ὡς ὅτι πορρωτάτω κομίζουσι,
καὶ ὁ μὲν ὑπανέστη ὁ παῖς, ὁ δὲ ἔμεινεν ὁ δράκων.
ὡς δὲ ὕλης ἐλάβετο καὶ τῶν ἐκεῖ φαρμάκων τῶν
συμφυῶν, διέτριβεν ἐνταῦθα ταῖς τῶν δρακόντων
τροφαῖς ἡδόμενος καὶ τὴν ἐρημίαν πρὸ τῶν
ἀστικῶν διατριβῶν καὶ τῶν ἐν τοῖς δωματίοις
προαιρούμενος ἐκεῖνος. διέρπων δὲ ὁ χρόνος τὸν
μὲν ἀπέφηνε νεανίαν, τὸν δὲ εἰργάσατο δράκοντα
ἤδη τέλειον. καί ποτε δι᾽ ἐρημίας ἰὼν ὁ Ἀρκὰς
ὁ τοῦ ζῴου τοῦ προειρημένου ἐραστὴς καὶ ἐρώμενος
λῃσταῖς περιτυγχάνει, καὶ παιόμενος ξίφει οἷα
εἰκὸς ἐβόα, τὰ μὲν ἀλγῶν, τὰ δὲ καὶ συμμάχους
παρακαλῶν. δράκων δὲ ἦν ἄρα ζῴων καὶ ἰδεῖν
ὀξυωπέστατος καὶ ὥκιστος ἀκοήν.[4] οὐκοῦν ἐκεῖ-

---

[1] ἀμυνόμενος.   [2] συνανιόντε τε or συνανιόντες.

ing the voice of its friend and comrade, as though
Gelon's life was in danger from a plot, leapt with all
its force on to the bed and stood over its master,
barking furiously, as though it would keep off the
assailant. So Gelon was roused and through fear
and the noise of barking threw off sleep though it was
of the deepest.

63. A young Snake was brought up along with a
child, an Arcadian born; the snake too was of the
country. So as the pair grew up the child became a
youth while his foster-brother had already become
enormous. And they were devoted to one another.
But the relatives of the youth were terrified at the
size of the monster. (You may see these creatures
attain in a very short time to an enormous size and
the most terrifying aspect.) And so while it was
asleep on the same bed with the boy, they picked it
up and took it as far away as possible. And the boy
rose up, but the Snake remained in that place. And
when it took to the forest and the drugs that grew
there, it lived there, enjoying the food of snakes and
preferring waste places to life in a city and confine-
ment in a room.

Time passed and turned one into a young man, the
other into a Snake now full-grown. And on one
occasion the Arcadian, the lover and the beloved of
the aforesaid creature, going through a lonely region,
fell in with brigands, and at a blow from a sword he
cried out, as was natural, both from pain and in order
to summon help. Now it seems that the Snake of all
creatures has the sharpest sight and the keenest

*Snake be-
friends boy*

---

[3] καὶ ὁ παῖς.   [4] ἀκοῇ.

AELIAN

νος, ἅτε αὐτῷ συντραφείς, τοῦ φθέγματος ἀκούει,
καὶ συρίσας ὀξύ, οἷον [1] ὠργισμένος, ἐξέπληξέ τε
ἐκείνους, τρόμος τε αὐτοὺς καταλαμβάνει, καὶ πᾶν
ὅσον ἦν κακοῦργον διασπείρονται ἄλλος ἄλλῃ, καὶ
μέντοι καὶ καταληφθέντας [2] τινὰς οἰκτίστῳ διαφθείρει θανάτῳ.[3] τοῦ γε μὴν παλαιοῦ τὰ τραύματα
καθήρας φίλου καὶ παρ᾽ [4] ὅσον ἔνθηρον ἦν τοῦ
τόπου [5] παραπέμψας, ᾤχετο ἀπιὼν ἔνθα αὐτὸν
ἐξέθεσαν, οὔτε μηνίσας ὑπὲρ τῆς ῥίψεως, οὔτε ὡς
οἱ κακοὶ τῶν ἀνθρώπων περιιδὼν ἐν κινδύνῳ ὄντα
τὸν τέως φίλτατον.

64. Ἡ ἀλώπηξ πονηρὸν ζῷόν ἐστιν, ἔνθεν τοι
καὶ κερδαλέην οἱ ποιηταὶ καλεῖν φιλοῦσιν αὐτήν·
πονηρὸν δὲ καὶ ὁ χερσαῖος ἐχῖνός ἐστι. καὶ ὁ μὲν
ἑαυτὸν συνειλήσας κεῖται, θεασάμενος ἤκουσαν
τὴν ἀλώπεκα, ἡ δὲ χανεῖν τε καὶ ἐνδακεῖν οὐ
δυναμένη, κᾆτα οὔρησεν αὐτοῦ ἐς τὸ στόμα· ὁ δὲ
ἀποπνίγεται, τοῦ [6] πνεύματος ἔνδον ἐκ τῆς συνει-
λήσεως κατεσχημένου καὶ ἐπιρρέοντός οἱ τοῦ
προειρημένου, καὶ μέντοι ⟨καὶ⟩[7] τὸν τρόπον τοῦτον κακὸν κακῇ περιελθοῦσα τὸν ἐχῖνον ἡ ἀλώπηξ
ᾕρηκεν αὐτόν. ἀνωτέρω δὲ θήρα λέλεκται ἄλλη.

65. Περὶ τὸ Κωνώπιον οὕτω καλούμενον (χῶρος
δὲ ἄρα τῆς Μαιώτιδός ἐστι) τοῖς ἀσπαλιευταῖς τε
καὶ θαλαττουργοῖς ἀνδράσιν οἱ λύκοι πιστῶς [8]
παραμένουσι, καὶ εἰ θεάσαιο, οὐκ ἂν εἴποις αὐτοὺς

---

[1] καὶ οἷον.　　　　[2] *Gill* : καταλειφθέντας.
[3] τῷ θανάτῳ.　　　[4] *Gow* : πᾶν MSS, 'corrupt' H.
[5] τοῦ τόπου] *Haupt*, τοῦτο MSS, 'corrupt' H.
[6] καὶ τοῦ.

ON ANIMALS, VI. 63–65

hearing. Accordingly this Snake, being the youth's foster-brother, heard his voice and hissing loudly as in anger, struck terror into the brigands, who were seized with trembling: the villains were all scattered in different directions, and what is more, some were overtaken by the Snake and perished miserably. But the Snake cleansed the wounds of its old friend, and after escorting him past that part of the region where wild beasts lurked, departed and went to the spot where the relations had exposed it: it showed no resentment at having been cast away, nor did it in the hour of danger, like base men, neglect one who had been its dearest friend.

64. The Fox is a rascally creature, hence poets are fond of calling it 'crafty.' The Hedgehog also is a rascal, for directly it sees the Fox approaching it rolls itself into a ball and lies still. And the Fox, unable to open his jaws and bite it, makes water into its mouth. And the Hedgehog is suffocated because its breathing is stopped through its being rolled up and because of the aforesaid stream. Moreover the Fox having thus tricked the Hedgehog, one scoundrel tricking another, catches it out.

I have earlier[a] described another method of capture.

65. In the neighbourhood of Conopeum as it is called (it is a district near the Maeotic lake[b]) Wolves are the faithful companions of the anglers and the fisherfolk, and were you to see them you would say

Fox and Hedgehog

Wolves and fishermen

---

[a] See ch. 24.　　　[b] Sea of Azov.

7 ⟨καί⟩ add. H.　　　8 δεινῶς.

89

κυνῶν οἰκουρούντων διαφέρειν. ἐὰν μὲν οὖν ἀπο-
λάχωσι τῆς ἄγρας τῆς θαλαττίου μοίρας οἶδε οἱ
λύκοι, εἰρηναῖα αὐτοῖς πρὸς τοὺς ἁλιέας καὶ
ἔνσπονδά ἐστιν· εἰ δὲ μή, διαξαίνουσιν αὐτῶν τὰ
δίκτυα καὶ ἀφανίζουσι, καὶ ἔδοσαν ὑπὲρ τῆς
σφετέρας ἀμοιρίας ζημίαν οἱ λύκοι αὐτοῖς.

ON ANIMALS, VI. 65

that they were no different from house-dogs. Now if these Wolves receive a share of the catch from the sea, there is a treaty of peace between them and the fishermen. Otherwise the Wolves rip up and destroy the nets, and for failing to give them a share inflict this damage upon the fishermen.

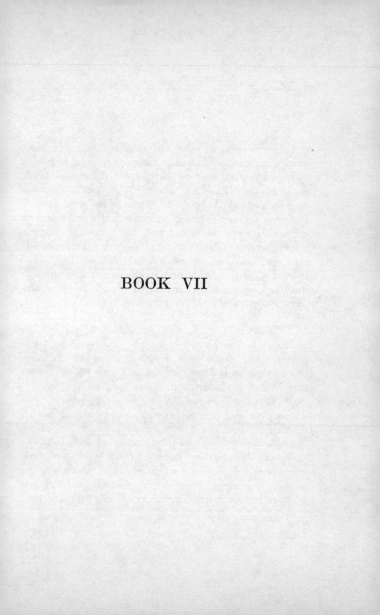

BOOK VII

Z

1. Πέπυσμαι δὲ ἄρα καὶ ἀριθμητικῆς τὰς βοῦς οὐκ ἀμοίρους εἶναι τὰς Σουσίδας. καὶ ὡς οὐκ ἔστιν ἄλλως κόμπος τὸ εἰρημένον, μάρτυς ὁ λόγος ὁ λέγων ἐν Σούσοις τῷ βασιλεῖ βοῦς ἐς τοὺς παραδείσους πολλὰς ἐς τὰ ἧττον ἐπίρρυτα ἀντλεῖν ἑκάστην κάδους ἑκατόν. οὐκοῦν ἢ τὸν ἐπινηθέντα αὐταῖς ἢ τὸν συντραφέντα ἐκ πολλοῦ μόχθον προθυμότατα ἐκτελοῦσι, καὶ οὐκ ἂν βλακεύουσάν τινα θεάσαιο· εἰ δὲ πέρα τῆς προειρημένης ἑκατοντάδος ἕνα γοῦν προσλιπαρήσειας κάδον ἀνιμήσασθαι, οὐ πείσεις οὐδὲ ἀναγκάσεις οὔτε παίων οὔτε κολακεύων. λέγει Κτησίας.

2. Ὑπὸ τοῖς ποσὶ τοῦ Ἄτλαντος (ὄρος δὲ ἄρα τοῦτο ὑμνεῖται καὶ ὑπὸ τῶν συγγραφέων καὶ μέντοι καὶ ὑπὸ τῶν ποιητῶν) νομαί τέ εἰσι θαυμασταὶ καὶ ὗλαι βαθύταται, καὶ τό γε δάσος αὐτῶν ἔοικεν ἄλσεσι πάνυ σκιεροῖς καὶ συνηρεφέσιν. ἐνταῦθα δήπου τοὺς ἤδη παλαιοὺς τῶν ἐλεφάντων φασὶν ἀφικνεῖσθαι, γήρᾳ βαρεῖς ὄντας· ἄγει δὲ αὐτοὺς ἄρα ἡ φύσις ὥσπερ οὖν ἐς ἀποικίαν, ἀναπαύσασα[1] ἤδη καὶ οἷον ὅρμον τινὰ καὶ λιμένα ποθητὸν ἀποφήνασα αὐτοῖς, ὅπου τοῦ βίου τοῦ σφετέρου τὸ λοιπὸν καταζήσουσι.[2] ἀνεῖται δὲ αὐτοῖς καὶ πηγὴ ποτίμου τε ὕδατος καὶ καθαροῦ μάλα ἄφθονος,[3] νομίζονταί τε ἱεροί, καὶ ἀφίενται

---

[1] ἀναπαύουσα.   [2] καταζῶσιν.   [3] ἀφθόνως.

BOOK VII

1. I have ascertained that the Cows in Susa are not unacquainted even with arithmetic. And that this is no idle boast the following story bears witness. In Susa the King has a large number of Cows of which each one draws one hundred buckets ⟨daily⟩ to water the drier places in his parks. Now they perform with the utmost zest the task which has either been heaped upon them or to which they have long been accustomed, and you would never see one of them idling. If however you were to urge them to draw so much as one bucket-ful in excess of the century, you will neither persuade nor compel them, whether by blows or by soft words, to do so. This is what Ctesias says.

2. At the foot of Atlas (this mountain is celebrated by historians and also by poets) there are marvellous pasture-lands and forests of the deepest, whose dense foliage is like that of groves all shady and over-arched. And that, you know, is where Elephants are said to resort in old age when heavy with years. And Nature leads them as it were to a colony, giving them rest at last and providing them with a desired anchorage and harbour, so to speak, where they can live out the rest of their life. And they have a spring of drinking-water pure and welling up abundantly; and they are regarded as sacred and are allowed to go unmolested; and they have an agree-

ἄσυλοι, καὶ παρά γε τῶν βαρβάρων τῶν τῇδε
εἰλήφασιν ἐς ἀθηρίαν σπονδάς, ᾄδονται[1] τε ὡς
ὑλαίοις τισὶ θεοῖς καὶ ναπαίοις τοῦ χώρου δεσπό-
ταις πάνυ μέλονται. διαρρεῖ δὲ ὑπὲρ αὐτῶν καὶ
ἐκεῖνος ὁ λόγος ὡς ἄρα τις τῶν βασιλέων τῶν
ἐπιχωρίων ἐπόθησε διὰ τὸ[2] κάλλος τῶν ὀδόντων[3]
καὶ τὸ μέγεθος ἀποκτεῖναί τινας αὐτῶν, ἵνα οἱ
γένηται κτῆμα ἐξαίρετον· εἶναι γὰρ διά τε
πολυετίαν καὶ πλῆθος χρόνου μέγιστον μέγιστα
ἐκείνων τῶν ζῴων ταῦτα τὰ ὅπλα. ὡς δὲ ἐσῆλθεν
ἤδε ἡ ἐπιθυμία αὐτόν, τριακοσίους λογάδας
ἐξέπεμψε κατακοντιοῦντας[4] τήνδε τὴν ἱερὰν
ἀγέλην. καὶ οἱ μὲν ᾗ ποδῶν εἶχον διανύσαντες
τὴν ὁδὸν ὡπλισμένοι καὶ δὴ τῷ χωρίῳ προσεπέλα-
ζον, λοιμὸς δὲ αὐτοὺς ἄφνω συλλαβὼν κατέστρωσε,
καὶ πλὴν ἑνὸς οἱ πάντες ἀπολώλασιν, ὅσπερ οὖν
ἐπανελθὼν τὸ πάθος διηγήσατο τῷ πέμψαντι καὶ
μάλα γε οἴκτιστον. οὕτω μὲν δὴ καὶ θεοφιλεῖς
ἐφωράθησαν ὄντες ἐλέφαντες.

3. Ζῷον ἔστι Παιονικόν, καὶ κέκληται μόνωψ,
καὶ ἔοικε ταύρῳ λασίῳ τὸ μέγεθος. οὗτος οὖν[5]
ὅταν διώκηται, ταραττόμενος ἀφίησι πυρῶδες καὶ
δριμὺ ἀποπάτημα, ὡς ἀκούω, ὅπερ οὖν εἰ προσπέσοι τῳ τῶν θηρατῶν ἀπέκτεινεν αὐτόν.

4. Ἴδιον δὲ ἦν ἄρα ταύρου καὶ τὸ εὐπειθές,
ἡμερωθέντος τε καὶ ἐς τὸ πρᾶον ἐκ τοῦ θηριώδους

---

[1] καὶ ᾄδονται.   [2] διά τε.
[3] ὀδόντων ἢ κεράτων.
[4] Reiske : κατακεντιοῦντας.

ment with the barbarians in those parts that they shall
not be hunted; and it is commonly said that they
are under the care of certain gods of the district who
are lords of wood and valley. And there is a story
current about them, as follows. A certain King of
that country was eager to kill some of them on
account of the splendour and size of their tusks, in
order to obtain a choice possession, for with the multitude of years and the lengthening of time these
weapons of these creatures become enormous.
So when this desire came upon him he despatched
three hundred picked men to shoot this sacred herd.
And all equipped they accomplished their journey
with the utmost speed, and were actually nearing
the spot when a pestilence suddenly seized them and
laid them low: all died save only one, and he returned and rendered to him who had sent them a
full account of the truly lamentable disaster. By
this means it was discovered that the Elephants
were beloved of the gods.

3. There is an animal in Paeonia *a* called *Monops*, The
and it is the size of a shaggy bull. Now when this 'Monops'
creature is pursued, in its agitation it voids a fiery
and acrid dung, so I am told; and should this happen
to fall on any of the hunters, it kills him.

4. It seems that a special characteristic of the Bull The Bull, its
is its docility, once it has been tamed and from being docility

---

a Paeonia, mountainous district N of Macedonia. The
animal was the Aurochs, now extinct.

---

⁵ οὖν ὃν μόνωπα καλοῦσιν οἱ Παίονες.

AELIAN

μεταβαλομένου.¹ μένουσι γοῦν καὶ ἐπὶ τῶν φερέ-
τρων ἀκίνητοι, εἴτε ὑπτίους αὐτοὺς ἐθέλοις
ἀτρεμεῖν εἴτε ἐπὶ στόμα, ὀκλάσαντας τοὺς προσθί-
ους καὶ ἐπὶ τοῦ τένοντος φέροντας ἢ παῖδα ἢ
κόρην. ὄψει δὲ ἄρα ταῦρον καὶ ἐπὶ τοῖς νώτοις
γυναῖκα ἄγοντα,² καὶ μετέωρον ἑστῶτα ἐπὶ τῶν
κατόπιν σκελῶν, καὶ τὸ πᾶν σῶμα ἐφ' ὅτου δὴ
κούφως ἐρείσαντα. εἶδον δὲ ἐπὶ ταύροις καὶ
ὀρχουμένους καὶ ἀκινήτους ἐκείνους καὶ ἀτρέ-
πτους ἑστῶτας.

5. Ἡ γῆ ἡ Λίβυσσα πολλῶν καὶ ποικίλων
θηρίων γόνιμός ἐστι, καὶ μέντοι καὶ τὸ κατώβλεπον
οὕτω καλούμενον καὶ αὐτὸ ἡ αὐτὴ ἔοικε τίκτειν.
καὶ ταύρῳ μέν ἐστι παραπλήσιον ὅσα ἰδεῖν, τὴν δὲ
ὄψιν δοκεῖ βλοσυρώτερον. ὑψηλαὶ μὲν γὰρ αἱ
ὀφρύες αὐτῷ καὶ δασεῖαι, οἱ δὲ ὀφθαλμοὶ ὑπόκεικ-
ται οὐ μάλα τι κατὰ τοὺς τῶν βοῶν μεγάλοι,
βραχύτεροι δὲ καὶ ὕφαιμοι· καὶ ὁρῶσιν οὐκ
εὐθύωρον, ἀλλὰ ἐς τὴν γῆν, ἔνθεν τοι καὶ κέκληται
κατώβλεπον. λόφος δὲ ἄρα ἄνωθεν ἐκ τῆς
κορυφῆς ἀρξάμενος αὐτῷ καὶ ἱππείᾳ τριχὶ παρα-
πλήσιος διὰ τοῦ μετώπου κάτεισι, καὶ τὸ πρόσω-
πόν οἱ καταλαμβάνει, καὶ ἐργάζεται φοβερώτερον
τῷ ἐντυχόντι. σιτεῖται δὲ ἄρα ῥίζας θανατηφόρους. ἐπειδὰν δὲ ὑποβλέψῃ ταυρηδόν, φρίττει μὲν
παραχρῆμα καὶ ἐγείρει τὴν λοφιάν· ὑπανισταμένης
δὲ ἄρα ταύτης καὶ ὀρθουμένης καὶ γυμνουμένων
τῶν περὶ τὸ στόμα χειλέων, ἐκπέμπει διὰ τῆς
φάρυγγος . . .³ ὀξοβαρὲς ⁴ καὶ βρωμῶδες, ὡς

¹ μεταβαλλομένου.
² ἄγοντα τὴν Εὐρώπην δή.

98

savage become gentle. At any rate Bulls remain quiet when harnessed to litters, or if you want them to lie still on their back or with their head on the ground or to sink down on their knees and carry a boy or a girl on their neck. And you will even see a Bull bearing a woman on its back or standing erect on its hind legs while it supports with ease the entire weight of its body on some object or other. And I have even seen men dancing on the backs of Bulls, and the same men motionless there also and standing undislodged.

5. Libya is the parent of a great number and a great variety of wild animals, and moreover it seems that the same country produces the animal called the *Katoblepon* (down-looking).[a] In appearance it is about the size of a bull, but it has a more grim expression, for its eyebrows are high and shaggy, and the eyes beneath are not large like those of oxen but narrower and bloodshot. And they do not look straight ahead but down on to the ground: that is why it is called 'down-looking.' And a mane that begins on the crown of its head and resembles horse-hair, falls over its forehead covering its face, which makes it more terrifying when one meets it. And it feeds upon poisonous roots. When it glares like a bull it immediately shudders and raises its mane, and when this has risen erect and the lips about its mouth are bared, it emits from its throat pungent[b] and foul-

The 'Catoblepon' or Gnu

---

[a] Generally considered to be the Gnu.
[b] Lobeck, *Path.* 476 ὀξοβαρές = *graveolens.*

AELIAN

καταλαμβάνεσθαι μὲν τὸν ὑπὲρ κεφαλῆς ἀέρα,
τῶν δὲ ζῴων τὰ πλησιάζοντα ἀναπνέοντα τοῦτον
κακοῦσθαι σφόδρα, καὶ ἀφωνίαις τε καὶ σπασμοῖς
θανατώδεσι περιπίπτειν.¹ συνίησί τε τῆς ἑαυτοῦ
δυνάμεως ὅδε ὁ θήρ· οἶδε δὲ αὐτὸν καὶ τὰ ζῷα,
καὶ ὡς ὅτι πορρωτάτω ἀποδιδράσκει.

6. Ἐλεφάντων θήρας ἐπιστήμονες ᾄδουσιν ἡμῖν,
ὅταν διώκωνται οἵδε οἱ θῆρες, ᾄττειν αὐτοὺς καὶ
φέρεσθαι ῥύμῃ ² ἀμάχῳ ³ καὶ ὁρμῇ ἀκατασχέτῳ,
καὶ ἀναστέλλεσθαι ὑπὸ μηδενός, καὶ μέντοι καὶ
διὰ τῶν μεγίστων ἵεσθαι δένδρων οἷον διὰ ληίων,
ὥσπερ οὖν στάχυς τινὰς κατακλῶντας τὰ δένδρα·
καὶ πῇ μὲν τὰ δένδρα αὐτῶν ὑπερέστηκε καὶ τὰς
κόμας ὑπερέχει, πῇ δὲ αὐτοὶ τῶν δένδρων εἰσὶν
ὑψηλότεροι. θέουσι μὲν οὖν ἀνὰ κράτος, καὶ
ὑποτέμνονται ταῖς ὁδοῖς τοὺς διώκοντας, καὶ
εἰκότως· εἰσὶ γὰρ τῶν χωρίων ἠθάδες. καὶ ὅταν
πολὺ ἀποσπάσωσι, καὶ πόρρω τῆς ἵππου τῆς
μετελθούσης ⁴ αὐτοὺς γένωνται, καὶ ἀναθαρρήσω-
σιν ὡς ἐν σκέπῃ τοῦ κινδύνου καὶ ἐλευθερίᾳ
γενόμενοι, ἑστᾶσί τε καὶ ἀναπαύονται, τὴν ἐκ τοῦ
δέους φροντίδα καὶ μάλα ἀσμένως ἐκβάλλοντες.
ἐνταῦθά τοι τοῦ χρόνου καὶ μνήμη τροφῆς αὐτοὺς
ἐσέρχεται· σιτοῦνται δὲ ἀκούω τήν τε σχῖνον
ἀμφιλαφῆ τοῖς δένδροις περιπεφυκυῖαν καὶ κιττὸν
ἄγριον τοῖς φυτοῖς ἐφέρποντα καὶ ὑπέρδασυν καὶ
φοινίκων μέντοι τὰς ἁπαλάς τε καὶ νεαρὰς κόμας
καὶ ἄλλων φυτῶν τοὺς ὄρπηκας καὶ τοὺς πτόρθους

¹ περιπίπτειν, καὶ ἄνθρωπος εἰ παραπέσοι.
² Jac: ῥώμῃ.
³ ἀμηχάνῳ.

100

smelling ⟨breath⟩, so that the whole air overhead is infected, and any animals that approach and inhale it are grievously afflicted, lose their voice, and are seized with fatal convulsions. This beast is conscious of its power; and other animals know it too and flee from it as far away as they can.

6. Those who are adept at hunting Elephants constantly tell us that when these beasts are pursued they dash forward and are carried along with irresistible force and an impetus that nothing can withstand; there is no stopping them; they even rush through the largest trees as though they were standing corn, smashing the trees like corn-stalks. In one place the trees overtop them and hold their leaves above them, in another they themselves are higher than the trees. Indeed they run with all their might and baffle their pursuers by the course they take; which is natural, for they are familiar with the country. And when they have got far away and are at a great distance ahead of the pursuing horsemen and have regained their courage through being secure from danger and feeling free, they pause and rest and are most glad to lay aside their anxious fears. And then at this time they bethink them of food. They feed, so I hear, on the bushy mastic that grows around *a* the trees and the wild ivy that creeps with its dense foliage over them, also upon the young and tender leaves of the date-palm and upon the more sappy shoots and twigs of other plants.

The Elephant when hunted

a It looks as if Ael. thought the mastic tree, *Pistacia lentiscus*, which may be anything up to 20 ft. high, was a parasite like ivy and *clung to* (περιπεφυκυῖαν) larger trees.

---

[4] μεταθεούσης Cobet.

AELIAN

τοὺς ὑγροτέρους. εἰ δὲ οἱ διώκοντες πάλιν προσ-
πελάζοιεν, οἱ δὲ ἐς φυγὴν ἐκτρέπονται αὖθις.[1]
οἵ γε μὴν διώκοντες αὐλίζονται ἑσπέρας καταλα-
βούσης, καὶ ἐμπρήσαντες τὴν ὕλην εἶτα μέντοι
τρόπον τινὰ τὴν ὁδὸν τὴν ὀπίσω διατειχίσαντες
αὑτοῖς[2] ἔστησαν. δεδοίκασι δὲ πῦρ οὐ μεῖον τῶν
λεόντων οἱ ἐλέφαντες.

7. Ἀριστοτέλους ἀκούω λέγοντος ὅτι ἄρα γέ-
ρανοι ἐκ τοῦ πελάγους ἐς τὴν γῆν πετόμεναι
χειμῶνος ἀπειλὴν ἰσχυροῦ[3] ὑποσημαίνουσι τῷ
συνιέντι. πετόμεναι δὲ ἄρα ἡσυχῇ αἱ αὐταὶ
ὑπισχνοῦνται εὐημερίαν τινὰ[4] καὶ εἰρήνην ἀέρος,
καὶ σιωπῶσαι δὲ ὅτι ἔσται[5] ὑπεύδια τοὺς οὐκ
ἀπείρως ἔχοντας τῇ σιωπῇ ὑπομιμνήσκουσιν αἱ
αὐταί. ἐὰν δὲ † καταπέτωνται †[6] καὶ βοῶσι καὶ
ταράττωσί τε καὶ ταράττωνται, ἀπειλοῦσι κἀν-
ταῦθα χειμῶνα ἰσχυρόν. ἐρῳδιὸς δὲ κνεφαῖος
βοῶν τὰ αὐτὰ ἔοικεν ὑποδηλοῦν.[7] πετόμενος δὲ
ἐρῳδιὸς τῆς θαλάττης εὐθὺ ὕδωρ ἐξ οὐρανοῦ
ῥαγήσεσθαι αἰνίττεται. εἰ δὲ εἴη χειμέρια, ᾄσασα
γλαῦξ εὐδίαν μαντεύεται καὶ ἡμέραν φαιδράν·
ἐὰν δὲ εὐδία μὲν ᾖ, ἡ δὲ ὑποφθέγγηται, χειμῶνα
δεῖ προσδέχεσθαι. κόραξ δὲ ἐπιτρόχως[8] φθεγ-
γόμενος καὶ κρούων τὰς πτέρυγας καὶ κροτῶν
αὐτάς, ὅτι χειμὼν ἔσται κατέγνω πρῶτος. κόραξ
δὲ αὖ καὶ κορώνη καὶ κολοιὸς δείλης ὀψίας εἰ

[1] αὖθις καὶ πολὺ ἀποστάντες ἀναπαύονται.
[2] αὐτούς Reiske.
[3] ἰσχυράν.
[4] εὐημερίας τινάς.
[5] ἐστίν.
[6] καταπέτωνται ‘corrupt.’ H, κάτω πέτ- Jac.

102

But if their pursuers again approach, the Elephants once more take to flight. And so when evening has overtaken them the pursuers bivouac, and by setting fire to the forest to some extent cut off the Elephants' retreat and so bring them to a standstill. For Elephants no less than lions have a horror of fire.

7. I learn from Aristotle [a] that cranes flying in to land from the sea indicate to the intelligent man that a violent storm is threatening. But if the same birds are flying tranquilly, that is a promise of fine weather and a calm atmosphere; and if they make no sound they are reminding those who have experience that it will be fairly calm. And if they ⟨fly in from the sea?⟩ uttering their cries and confusing their order in their agitation, there again they are threatening a heavy storm. And if a shearwater utters its cry at dusk, it apparently signifies the same; if it flies straight to the sea, it is giving a hint that a rainstorm will burst from the sky. If however the weather is stormy, the hooting of an owl portends fair weather and a bright day; whereas if the weather is fair and the owl hoots softly, you must expect storms. If a raven croaks volubly and pecks and shakes its wings, it is the first to observe that a storm is coming. Again, if the raven, the crow, and the jackdaw utter their cries in

Birds as weather-prophets

[a] The treatise *de Signis tempestatum*, on which this section appears to be based and which was formerly ascribed to Aristotle, is now counted among the writings of Theophrastus. See vol. 2 of Sir A. F. Hort's *Theophrastus* (Loeb Class. Lib.).

[7] ὑποδηλοῦν, ὡς αὐτὸς Ἀριστοτέλης φυλάξας λέγει.
[8] ταχέως καὶ ἐπιτρόχως.

φθέγγοιντο, χειμῶνος ἔσεσθαί τινα ἐπιδημίαν
διδάσκουσι. κολοιοὶ δὲ ἱερακίζοντες, ὡς ἐκεῖνος
λέγει, καὶ πετόμενοι πῆ μὲν ἀνωτέρω πῆ δὲ κατω-
τέρω, κρυμὸν καὶ ὑετὸν δηλοῦσι. κορώνη δὲ ἐπὶ
δείπνου [1] ὑποφθεγγομένη ἡσυχῇ, ἐς τὴν ὑστεραίαν
εὐδίαν παρακαλεῖ. φανέντες δὲ ὄρνιθες πολλοὶ μὲν
τὸν ἀριθμόν, λευκοὶ δὲ τὴν χρόαν, χειμὼν ὅτι ἔσται
πολὺς ἐκδιδάσκουσι. νῆτται δὲ καὶ αἴθυιαι πτερυ-
γίζουσαι πνεῦμα δηλοῦσιν ἰσχυρόν. ὄρνιθες δὲ ἐκ
τοῦ πελάγους ἐς τὴν γῆν σὺν ὁρμῇ πετόμενοι μαρ-
τύρονται χειμῶνα. ἐρίθακος δὲ ἐς τὰ αὔλια καὶ τὰ
οἰκούμενα παριὼν [2] δῆλός ἐστι χειμῶνος ἐπιδημίαν
ἀποδιδράσκων. ἀλεκτρυόνες γε μὴν καὶ ὄρνιθες οἱ
ἠθάδες πτερυσσόμενοι καὶ φρυαττόμενοι καὶ ὑπο-
τρύζοντες χειμῶνα δηλοῦσιν. ἀπειλοῦσι δὲ [3] πνεῦμα λούμεναί γε [4] ὄρνιθες, καὶ ἀνέμων τινὰς
ἐμβολὰς ὑποφαίνουσι. χειμῶνος δὲ ὄντος ἐς
ἀλλήλους ὄρνιθες πετόμενοι καὶ δι' ἀλλήλων
θέοντες σημαίνουσιν εὐδίαν. ὄρνιθες δὲ ἀθροι-
ζόμενοι περί τε λίμνας καὶ ποταμῶν ὄχθας χειμῶνα
ἐσόμενον οὐκ ἀγνοοῦσι. πάλιν τε ὄρνιθες οἱ μὲν
θαλάττιοι καὶ οἱ λιμναῖοι ἐς τὴν γῆν ἰόντες ὡς
ἔσται [5] χειμὼν πολὺς οὐκ ἀγνοοῦσιν, οἱ δὲ χερ-
σαῖοι σπεύδοντες ἐς τὰ νοτερὰ εὐδίας ἄγγελοί
εἰσιν, ἐὰν μέντοι σιωπῶσιν.

8. Αἰγυπτίων ἀκούω λεγόντων τὸν ὄρυγα συνιέναι τὴν τοῦ Σειρίου ἐπιτολὴν πρῶτον, καὶ [6]
μαρτύρεσθαι τῷ πταρμῷ αὐτήν.[7] νεανιεύονται δὲ
καὶ οἱ Λίβυες ἀνὰ κράτος φάσκοντες καὶ τὰς αἶγας

---

[1] δείπνῳ.   [2] *Abresch* : περιών.
[3] δὲ καί.   [4] τε.

the late afternoon, they teach us that we shall have
a visitation by a storm. And if jackdaws, as the same
writer says [Thphr. *Sig.* 16], scream like hawks [a] and fly
now high now low, they point to frost and rain. If a
crow caws softly at supper-time, it is inviting us to
expect fair weather next day. If birds appear in
great numbers and they are white, it is a certain
indication that there will be heavy storms. When
ducks and shearwaters flap their wings, they point to
violent winds. And when birds come speeding into
land from the sea, this is evidence of stormy weather.
If the robin comes to cattle-sheds and houses, he is
clearly trying to escape from a coming storm.
Cockerels too and domestic fowls, when they flap
their wings and step proudly and cluck, signify
stormy weather. When birds bathe, it is a sign that
wind is threatening, and it points to gusty weather.
If during a storm birds fly towards one another and
in and out, it is a sign of fine weather. When birds
congregate about meres and on river banks, they
know that a storm is coming. On the other hand
when birds of the sea and lake come in to land, they
know that there will be a heavy storm, whereas land
birds hastening to moist places are heralds of fine
weather, if, that is, they make no sound.

8. I have heard that the Egyptians assert that the *Animals as weather-prophets*
antelope is the first creature to know when the
Dog-star rises, and testifies to the fact by sneezing.
The Libyans are equally bold in stoutly maintaining

---

[a] Or ' hover like hawks ' ? (Hort *ad loc.*)

---

[5] ἔσοιτο or ἐσεῖται.   [6] καὶ τὴν ἐπιτολήν.
[7] αὐτόν.

AELIAN

παρ' αὐτοῖς προειδέναι τὸ αὐτὸ δήπου τοῦτο.
αὗται μὲν [1] καὶ μέλλοντα ὑετὸν προδηλοῦσιν.
ἐπειδὰν γὰρ προέλθωσι τῶν σηκῶν, δρόμῳ καὶ
μάλα γε ὤκιστα ὁρμῶσιν ἐπὶ τὸν χιλόν· εἶτα
ἐμπλησθεῖσαι αὐτὰς ἐπιστρέφουσιν ἐς τὰ οἰκεῖα,
καὶ ὁρῶσαι ἐκεῖσε ἀτρεμοῦσι,[2] τὸν ποιμένα
ἀναμένουσαι, ἵνα τὴν ταχίστην συνελάσῃ αὐτάς.
καὶ Ἵππαρχος μὲν ἐπὶ Ἱέρωνος [3] τοῦ τυράννου
καθήμενος ἐν θεάτρῳ καὶ φορῶν διφθέραν, ὅτι τὸν
μέλλοντα χειμῶνα ἐκ τῆς παρούσης αἰθρίας
προηπίστατο ἐξέπληξε· καὶ ἐθαύμαζεν Ἱέρων [3]
αὐτόν, καὶ Νικαεῦσι τοῖς Βιθυνοῖς συνήδετο ὅτι
Ἱππάρχου πολίτου ἔτυχον· ἐν Ὀλυμπίᾳ δὲ
θεώμενον Ἀναξαγόραν ἐν διφθέρᾳ καὶ αὐτὸν τὰ
Ὀλύμπια ἐπιρραγέντος ὑετοῦ τὸ Ἑλληνικὸν πᾶν
ᾖδεν, καὶ θειότερα νοεῖν ἢ κατὰ τὴν θνητὴν φύσιν
ἐκόμπαζεν. ὅτι δὲ βοῦς, ἐὰν μέλλῃ ὕειν ὁ Ζεύς,
ἐπὶ τὸ ἰσχίον τὸ δεξιὸν κατακλίνεται, ἐὰν δὲ εὐδία,
πάλιν ἐπὶ τὸ λαιόν, θαυμάζει ἤ τις ἢ οὐδείς. καὶ
ἐκεῖνα δὲ προσακήκοα ἐκπλῆξαι ἱκανά. βοῦς ἐὰν
βοᾷ καὶ ὀσφραίνεται,[4] ὕειν ἀνάγκη. ἄδην δὲ βόες
καὶ πέρα τοῦ ἔθους ἐσθίοντες δηλοῦσι χειμῶνα.
πρόβατα δὲ ὀρύττοντα ταῖς ὁπλαῖς τὴν γῆν ἔοικε
σημαίνειν χειμῶνα, ἀναβαινόμενα δὲ τὰ αὐτὰ
πρωὶ πρώιον [5] χειμῶνα ὁμολογεῖ. κοιμώμεναι δὲ

[1] Perh. δέ or μέντοι H.
[2] οὕτω νέμονται.
[3] Valesius : Νέρωνος . . . Νέρων.
[4] Schn : ὀσφραίνεται τῆς γῆς.
[5] πρωὶ πρώιον] Schn : πώεα.

[a] Hipparchus of Nicaea, famous astronomer, 2nd cent. B.C.

106

that in their country the goats also know in advance; they also give clear signs of impending rain. For when they emerge from their pens they rush at full speed to their fodder. Later, when satisfied, they turn towards home, and facing in that direction remain still and wait for the herdsman to gather them in as quickly as possible.

And Hipparchus [a] in the reign of Hiero the Tyrant [b] was sitting in the theatre wearing a leathern jerkin, and astonished people by knowing in advance out of the clear weather then prevalent that a storm was coming. And Hiero in his admiration of the man congratulated the people of Nicaea in Bithynia on having Hipparchus as a citizen. And when at Olympia Anaxagoras,[c] likewise clad in a leathern jerkin, was watching the Olympic Games and a storm of rain burst, all Hellas sang his praises, and claimed that his wisdom was more that of a god than of a man. And few if any are surprised that an ox, if rain threatens, lies down on his right side, contrariwise if fair weather is coming, on his left. And I have also heard the following facts which are calculated to astonish one. If an ox bellows and sniffs the air, rain is inevitable. And if oxen eat copiously and more than is their custom, it portends a storm. When sheep dig the ground with their hoofs, it is likely to mean a storm; and if the rams mount them early in the day, it promises an early storm; and the

[b] No 'Tyrant' of this name is known to have lived in the 2nd cent. B.C.
[c] Anaxagoras of Clazomenae, 5th cent. B.C., taught that physical phenomena were due to natural causes. His doctrines were regarded as impious and he was forced to quit Athens.

ἀθρόαι αἱ αἶγες τὰ αὐτὰ ὁμολογοῦσιν. ὕες δὲ ἐν
τοῖς ἀρώμασι[1] φαινόμεναι ὑετοῦ φυγὴν διδάσκουσιν. ἄρνες δὲ ἄρα καὶ ἔριφοι ἀλλήλοις ἐμπηδῶντές
τε καὶ ὑποσκιρτῶντες φαιδρὰν ἡμέραν ὁμολογοῦσιν. γαλαῖ δὲ ὑποτρίζουσαι καὶ μύες ἐκείναις
δρῶντες τὰ αὐτὰ χειμῶνα ἔσεσθαι συμβάλλονται
ἰσχυρόν. λύκοι δὲ φεύγοντες ἐρημίας καὶ εὐθὺ τῶν
οἰκουμένων ἰόντες χειμῶνος ἐμβολὴν μέλλοντος
ὅτι πεφρίκασι μαρτυροῦσι δι᾿ ὧν δρῶσι. λέοντος
δὲ ἐν τοῖς καρπίμοις χωρίοις ἐπιδημία αὐχμὸν
δηλοῖ. σκιρτῶντά ⟨γε⟩[2] μὴν τὰ ὑποζύγια καὶ
βοῶντα τοῦ ἔθους μᾶλλον νοτερὸν χειμῶνα ἐσόμενον δηλοῖ· εἰ δὲ καὶ ταῖς ὁπλαῖς κόνιν προσανα-
βάλλοι, ταὐτὰ ταῦτα δηλοῖ που. λαγὼ δὲ ἐν τοῖς
αὐτοῖς χωρίοις ὁρώμενοι πολλοὶ δηλοῦσιν εὐδίαν.
πάντων δὲ τούτων ἀπολείπονται οἱ ἄνθρωποι, καὶ
ἴσασιν αὐτὰ ὅταν γένηται.[3]

9. Ἱεράκων πέρι καὶ ταῦτα προσακήκοα. οἱ
τοῦ Ἀπόλλωνος ἐν τῇ Αἰγύπτῳ θεραπευταὶ
λέγουσι καλεῖσθαί τινας οὕτως ἱερακοβοσκούς,
οἵπερ οὖν εἰσι τῶν τοῦ θεοῦ ἱεράκων τροφεῖς τε
καὶ μελεδωνοὶ μέντοι οἱ αὐτοί. πᾶν μὲν οὖν τὸ
φῦλον[4] ἀνεῖται τῷ θεῷ τῷδε, ἤδη δέ τινες ἐκεῖθι
καὶ ἱεροὶ τρέφονται τροφῇ πεφροντισμένῃ, καὶ
δοκοῦσι τῶν ἀναθημάτων διαφέρειν οὐδὲ ἕν. οἱ
τοίνυν τὴν τούτων ἐγκεχειρισμένοι κομιδὴν πρὸς
τοὺς ἀγνοοῦντας λέγουσιν ἐν ταῖς νεοττιαῖς ἑκάσ-
τους (ἐν ἄλσει γὰρ ἱερῷ τρέφονται)[5] τίκτειν·

---

[1] ἀρόμασι.
[2] ⟨γε⟩ add. H.
[3] γένωνται.
[4] τὸ τῶν ὀρνίθων φῦλον.

same when goats lie huddled together. When pigs appear in cornland, they inform us that the rain is departing. Now when lambs and kids leap on one another and frisk about, they promise a bright day. But when martens squeak and mice likewise, they are conjecturing that there will be a violent storm. When wolves quit lonely places and make straight for inhabited districts, they show thereby that they dread the onslaught of a coming storm. If a lion visits cornlands, it presages a drought. And if beasts of burden gambol and low more than is their custom, it shows that storm and rain are on their way; and if besides, they toss up the dust with their hoofs, it signifies the same. If hares are seen in great numbers in the same places, it signifies fair weather. In all these matters men fall behind: they only know these changes when they occur.

9. Here are further facts which I have heard touching Hawks. The ministers of Apollo [a] in Egypt say that there are certain men called ' hawk-keepers ' for this reason: they feed and tend the Hawks belonging to the god. Now the whole race of Hawks is consecrated to this god, but there are certain sacred birds which are fed upon carefully pre-pared food and which seem in nowise to differ from offerings made to the god. Now the men who have been charged with the care of these birds tell the uninformed that each of them (they are tended in a

The Hawk in Egypt

----

[a] *I.e.* Horus; cp. *NA* 10. 14.

----

[5] ἐν ἄλσει ... τρέφονται *Jac would transpose to follow* διαφέρειν οὐδὲ ἕν.

AELIAN

†ὁμολογεῖσθαι δὲ τὴν ἄλλων μέν, ἐκ τούτων δὲ ἔτι
καὶ μᾶλλον†.[1] τοῖς δὲ ἀρτιγενέσι προβάλλουσιν
ὀρνίθων τεθηραμένων ἐξηρημένους τοὺς ἐγκεφά-
λους, τροφὴν ἁπαλὴν νεοττοῖς ὑγροῖς· τοῖς γε μὴν
τελείοις οὖσι παρατιθέασι σάρκας τε καὶ [2] ἶνας,
ὅσα ἰσχυρὰν τροφὴν ὄρνισιν ἁρπακτικοῖς ἐργάζε-
ται· τοῖς δὲ ἐν μεθορίῳ τῶν ἀρτιγενῶν καὶ τῶν
ἤδη τελείων καρδίαι παράκεινται,[3] καὶ τούτων
λείψανα ὁρᾶται. καὶ ἥ γε διαφορότης ἡ προειρημένη τῆς τροφῆς ὁμολογεῖ ὅτι τὸ ἁρμόττον
ἡλικίᾳ ἑκάστῃ καὶ πρόσφορον ἴσασιν οἱ ἱέρακες
καὶ μάλα γε ἀκριβοῦσι τοῦτο, καὶ τῆς παρ᾽
ἡλικίαν τροφῆς οὐκ ἂν ἅψαιντο. καθ᾽ ὥραν δὲ
ἄρα καὶ ὀρτύγων αὐτοῖς ἐπιδημίαι γίνονται, καὶ
τῶν ἄλλων ὀρνίθων ἐπιφοιτῶσιν ἀγέλαι, καὶ
ἔχουσί γε οἱ [4] ἱεροὶ ἐκεῖνοι καὶ ἐντεῦθεν θοίνην.

10. Κυνῶν ἐς τοὺς τρέφοντας αὐτοὺς ἄμαχον
εὔνοιαν ὁμολογεῖ καὶ ἐκεῖνο δήπου. ἔν τινι τῶν
ἐμφύλων πολέμων ἐν τῇ Ῥώμῃ Κάλβου τοῦ
Ῥωμαίου σφαγέντος, οὐδεὶς μέντοι τῶν ἐχθρῶν
τοῦ ἀνδρὸς ἠδυνήθη τὴν κεφαλὴν ἀποτεμεῖν,
καίτοι μυρίων ἀγώνισμα τιθεμένων σφίσι καὶ
καλλώπισμα τοῦτο, πρὶν ἢ τὸν παρεστῶτά οἱ
κύνα ἀποκτεῖναι ὑπ᾽ αὐτῷ [5] τραφέντα καὶ μέντοι

[1] ὁμολογεῖσθαι . . . μᾶλλον corrupt.
[2] Triller : καὶ κρέα καί.
[3] κεῖνται.
[4] ἔχουσιν οἵ γε.
[5] αὐτοῦ.

[a] The sentence appears pointless and perhaps there is a lacuna at the end.

110

sacred grove) lays eggs in its nest.[a] They have,
it is true, the care of all Hawks, but these sacred
ones are their special charge.[b] They take out the
brains of birds which have been caught and throw
them to the newly born Hawks: soft food for
tender chicks. But to those that are full-grown
the keepers serve flesh and sinews, which furnish
strengthening nourishment for birds of prey. Those
however that are in the intermediate stage between
chicks and full-grown birds are served with the
hearts,[c] and one may see the remains of them. So
the aforesaid difference of foods concedes the point
that Hawks know what is appropriate and agreeable
to each age; and they are particular about it and
would never touch food unsuited to their age. At
a certain season quails visit their country and other
birds arrive in flocks, and these sacred Hawks feast
on them also.

10. The following story, I think, also affords
evidence of the unbreakable affection which Dogs
have for those who keep them. In one of the civil
wars at Rome when Galba the Roman was murdered,[d] there was not one of the man's enemies that
was able to cut off his head, although countless
numbers competed for this trophy, until they had
killed the Hound at his side that had been reared
under his care and that maintained its affection with
the utmost loyalty and fought on behalf of its dead
master, as though it were a fellow soldier, sharer of

The Dog's
devotion to
its master;
Galba's dog

[b] The text is uncertain, and the translation provisional.

[c] But see *NA* 2. 42.

[d] This seems to be the Galba who was Roman Emperor for
six months, A.D. 68, and was murdered by his soldiers. Cp.
Suet. *Galba* 20. 2 and Mooney's note *ad loc.*

καὶ τὴν εὔνοιάν οἱ πιστότατα ἀποσῴζοντα καὶ
ὑπερμαχοῦντα τοῦ κειμένου, ὥσπερ οὖν συστρατιώτην τε καὶ σύσκηνον ἀγαθὸν καὶ ἐς τὰ ἔσχατα
φίλον. οἷον δ' αὖ καὶ τόδ' ἔρεξεν οὐκ ἀνὴρ μὰ
Δία, ἀλλ' ἀγαθὸς κύων καὶ τὴν γνώμην καρτερός,
μαθεῖν ἄξιον. ὁ Ἠπειρώτης Πύρρος ὡδοιπόρει,
εἶτα μέντοι περιτυγχάνει νεκρῷ πεφονευμένου,[1]
καὶ κυνὶ παρεστῶτι καὶ μέντοι καὶ φρουροῦντι τὸν
δεσπότην, ἵνα μὴ πρὸς τῷ φόνῳ καὶ τῷ νεκρῷ
λυμήνηταί τις. ἔτυχε δὲ ἄρα τρίτην ἔχων ὁ
κύων ἄπόσιτος τὴν ἡμέραν ἐπὶ τῇ φιλοπόνῳ καὶ
καρτερικωτάτῃ φρουρᾷ. ὅπερ οὖν διδαχθεὶς ὁ
Πύρρος τὸν μὲν ᾤκτειρε καὶ ταφῆς ἠξίωσε, τόν
γε μὴν κύνα προσέταξε τυχεῖν κηδεμονίας, καὶ
ἐδίδου ὅσα κυνὶ ὀρέγεται[2] ἐκ χειρός, καὶ μάλα
γε ἱκανὰ καὶ ἐφολκὰ ἐς τὴν ἑαυτοῦ φιλίαν τε καὶ
εὔνοιαν, κατὰ μικρὰ ὑπάγων τὸν κύνα ὁ Πύρρος.
καὶ ταῦτα μὲν ἐς τοσοῦτον. εἶτα μέντοι οὐ μετὰ
μακρὸν ἐξέτασις ὁπλιτῶν ἦν, καὶ ὁ βασιλεὺς ὃν
προεῖπον ἐθεᾶτο, καί οἱ παρῆν ἐκεῖνος ὁ κύων.
καὶ τὰ μὲν ἄλλα ἑαυτὸν σιγῇ κατεῖχε καὶ πρᾳό-
τατος ἦν· ἐπεὶ δὲ ἄρα τοὺς τοῦ δεσπότου φονέας
ἐν τῇ τῶν στρατιωτῶν εἶδεν ἐξετάσει, ὁ δὲ οὐκ
ἐκαρτέρησεν ἐνταῦθα ἀτρεμεῖν, ἀλλὰ ἐς αὐτοὺς
ἐπήδα καὶ ὑλάκτει ἀμύσσων τοῖς ὄνυξι, καὶ ἐς
τὸν Πύρρον θαμὰ[3] ἐπιστρεφόμενος ὡς οἷός τε ἦν
ἐπήγετο μάρτυρα ὅτι ἄρα τοὺς ἀνδροφόνους ἔχει.
οὐκοῦν ὑπόνοια ἐσέρχεται καὶ τὸν βασιλέα καὶ
τοὺς περιεστῶτας αὐτόν, καὶ ποιοῦνται ἐνθύμιον
τὴν τοῦ κυνὸς ὑλακὴν τὴν ἐς τοὺς προειρημένους.
καὶ συλληφθέντες στρεβλοῦνται, καὶ κατεῖπον ὅσα

___

[1] πεφονευμένῳ.　　　　[2] παρόντι.

the same tent, and friend to the very last. It is worth knowing ' what a deed was this, wrought ' not ' by a man ' [Hom. *Od.* 4. 242], I declare, but by a faithful Hound of valiant spirit.

Pyrrhus of Epirus was on a journey when he came upon the corpse of a man who had been killed, with his Dog standing beside and guarding its master to prevent anybody from adding outrage to murder. Now it happened that this was the third day for which the Dog was keeping its assiduous and most patient watch, unfed. And so when Pyrrhus learnt this he took pity on the dead man and ordered him to be buried; but as for the Dog, he directed that it should be cared for and gave it whatever one offers a dog with one's hand, in sufficient quantity and of a nature to induce it to be friendly and well-disposed towards him; and little by little Pyrrhus drew the Dog away. So much then for that. Now not so long after, there was a review of the hoplites, and the King whom I mentioned above was looking on, and that same Dog was at his side. For most of the time it remained silent and completely gentle. But directly it saw the murderers of its master in the review, it could not contain itself or remain where it was, but leaped upon them, barking and tearing them with its claws, and by frequently turning towards Pyrrhus did its best to make him see that it had caught the murderers. And so a suspicion dawned upon the King and those about him, and the way in which the Dog barked at the aforesaid men caused them to reflect. The men were seized and put on the rack and confessed their crime.

---

[3] *Jac*: ἅμα.

AELIAN

ἐτόλμησαν. καὶ δοκεῖ μὲν μῦθος ταῦτα τοῖς [1]
ὅσοι Διὸς ἑταιρείου καὶ φιλίου τοῦ αὐτοῦ θεσμὸν
πατήσαντες εἶτα μέντοι ζῶντας προὔδοσαν τοὺς
φίλους καὶ ἀποθανόντας· ἐγὼ δὲ οὐ πείθομαι τοῖς
νοοῦσι κακῶς τὰ τῆς φύσεως καλά, ἥπερ οὖν εἰ
τοῖς ἀλόγοις μετέδωκεν εὐνοίας τε καὶ στοργῆς,
πάντως που καὶ τῷδε τῷ ζῴῳ τῷ λογικῷ μετέ-
δωκε μᾶλλον. ἀλλὰ οὐ χρῶνται τῷ δώρῳ. καὶ
τί δεῖ τὰ λοιπὰ ἐπιλέγειν ὁπόσα ἄνθρωποι ὑπὲρ
τοῦ πονηροῦ κέρδους κακὰ τοὺς ἑαυτῶν φίλους
εἰργάσαντο, ἐπιβουλὰς ῥάπτοντες καὶ προδιδόντες;
ὡς ἐμέ γε ἀλγεῖν εἴπερ οὖν ἀνθρώπων πιστότερος
καὶ εὐνούστερος ἐλήλεγκται [2] κύων.[3]

11. Πολύποδος ἐς οὓς ἐμὸν καὶ ἐκεῖνο ἧκεν. ἦν
πέτρα προήκουσα μέν, οὐ μὴν ἄγαν ὑψηλή.
οὐκοῦν πολύπους ποτὲ ἀνερπύσας εἶτα ἥπλωσε
τὰς πλεκτάνας, καὶ μάλα γε ἀσμένως ὑπεθάλπετο
(καὶ γὰρ οὖν καὶ χειμέρια ἐδόκει πως), οὐ μὴν
ἑαυτὸν ἐς τὴν χρόαν τῆς πέτρας ἐκτρέψας ἤδη ἦν.
πεφύκασι δὲ ἄρα δρᾶν τοῦτο οἱ πολύποδες τὰς [4] ἐς
ἑαυτοὺς ἐπιβουλὰς φυλαττόμενοι καὶ μέντοι καὶ
αὐτοὶ τοὺς ἰχθῦς ἐλλοχῶντες. ἰδὼν οὖν ὀξὺ μέν,
ἑαυτῷ δὲ οὐκ ἀγαθὸν τὸ θήραμα ἀετός, ὡς [5]
ὁρμῆς τε ἅμα καὶ πτερῶν εἶχεν ἐμπηδᾷ τῷ
πολύποδι, καὶ μέντοι καὶ δεῖπνον ἕξειν ἕτοιμον
ἑαυτῷ τε καὶ τοῖς παισὶ τοῖς ἑαυτοῦ κατέγραφεν.
πλόκαμοι δὲ ἄρα ἐκείνου [6] περιβάλλουσι τῷ
ἀετῷ σφᾶς αὐτούς, καὶ ἀπρὶξ ἐχόμενοι εἶτα

114

To those who trample upon the ordinance of Zeus the god of fellowship and of affection and betray their friends in life and after death, all this seems a mere tale. But for my part I do not follow those who fail to appreciate the excellence of Nature which, if she has given brutes a share of kindliness and affection, has certainly given a larger share to us rational beings. But they make no use of her gift. And what need is there to add to my story all the other crimes which men have committed against their friends for the sake of base gain, hatching plots and acting the traitor? It fills me with pain that a Dog should be shown to have more loyalty, more kindly feeling than man.

11. Here is another story which has come to my ears: it is about the Octopus. There was a rock rising from the sea, though not to a great height. Now once upon a time an Octopus crawled up it and spread out its tentacles and was glad to warm itself (the weather was inclined to be stormy), though it did not at once assume the colour of the rock. Octopuses do this naturally, to protect themselves against those who have designs upon them, and also that they themselves may ambush fishes. Now an Eagle, quick to mark its prey (though it got no good thereby), swooped with all the force of its wings upon the Octopus, reckoning to secure a ready meal for itself and its young. But the creature's tentacles wreathed themselves round the Eagle, and clinging fast to its hated enemy dragged it down, and it was

Octopus and Eagle

---

¹ τοῖς ἄλλοις.   ² ἐλήλεγκται καί.
³ *Reiske*: κύων ὤν.   ⁴ καὶ τάς.
⁵ ὥσπερ οὖν.   ⁶ ἰχθύος ἐκείνου.

AELIAN

ἕλκουσι κάτω τὸν ἔχθιστον, καὶ χανὼν λύκος ὡς
ἂν εἴποις εἶτα μέντοι νεκρὸς ἐπενήχετο τῇ θαλάττῃ
ὁ ἀετὸς ὑπὲρ τοῦ δείπνου. μυρία μὲν δὴ τοιαῦτα
πάσχουσιν ὄρνιθες, πλείω δὲ ἄνθρωποι· ἐν δὲ
τοῖς ᾀδομένοις ὑφ᾽ Ἡροδότου Μασσαγέταις ὁ
Καμβύσου Κῦρος ὁ ἕτερος καὶ μέντοι καὶ Πολυ-
κράτης ἐς Ὀροίτου σπεύσας ὡς τὸν χρυσὸν
ἁρπασόμενος καὶ ἄλλος

τεύχων ὡς ἑτέρῳ τις ἑῷ κακὸν ἥπατι τεύχει.

καὶ ταῦτα μὲν οὐκ οἶδε τὰ ἄλογα, ἄνθρωποι δὲ εἰδότες [1] οὐ φυλάττονται. καὶ τί δεῖ γλώττης καὶ
λόγων καὶ διδασκάλων καὶ πληγῶν, ὦ Κῦρε καὶ
Πολύκρατες; τοὺς δὲ ἄλλους ἐῶ· τί γάρ μοι
κωφοῖς καὶ ἀνοήτοις συμβουλεύειν τὰ λυσιτελέ-
στατα;

12. Μέγα φρονείτωσαν [2] αἱ Παιονίδες γυναῖκες
καὶ τὸ φρύαγμα αἱρέτωσαν δρῶσαι τὰ ὑμνούμενα.
ἔστι δὲ τοιαῦτα.[3] τῇ μὲν κεφαλῇ φέρουσιν
ὑδρίαν μεστὴν ὕδατος, καὶ τὸν αὐχένα ἀνέστησαν,
ὥστε αὐταῖς βαδιζούσαις ἄτρεπτόν τε καὶ ἀκλινῆ
διαμένειν τὴν ὑδρίαν· ἐξαρτήσασαι δὲ τοῦ κόλπου
θηλάζουσι τὰ βρέφη, καὶ ἐς τὸν βραχίονα τὸν
ῥυτῆρα ἐνάψασαι τὸν τοῦ γήμαντος ἵππον ἐς
ἀρδείαν ἄγουσι, καὶ ταῖς χερσὶ νῶσι λίνον.

---

[1] οἱ εἰδότες.
[2] Jac: νῦν (or μὴ νῦν) φρονείτωσαν.
[3] ταῦτα.

---

[a] The proverb took its origin from Æsop's fable (223, ed.
Chambry) of the hungry wolf who overhears a mother

a case of ' The hungry wolf,' *a* as you might say.
And presently the Eagle was floating dead upon the
sea for the sake of its meal. Birds in fact suffer
countless misadventures of this kind, and men even
more: for example, Cyrus the Second, the son of
Cambyses,*b* among the Massagetae celebrated by
Herodotus [1. 214]; Polycrates *c* also [*id.* 3. 125]
who hastened to Oroetes with the intention of laying
hands on his gold, and any who

'working for another's ill, wreaks ill for his own
heart.' [Anon.]

Brute beasts do not realise these dangers; human
beings do, but fail to guard against them. What use
to you, Cyrus and Polycrates, were a tongue, speech,
teachers, beatings? I say nothing of the others,
for why should I give the most profitable advice to
men who are deaf and senseless?

12. Let the women of Paeonia be proud: let
them assume arrogant airs, since their conduct is
celebrated. This is what they do: on their head
they carry a vessel full of water, their neck held
straight so that as they walk the vessel shall remain
erect without upsetting. They attach their children
to their breast before suckling them; and fastening
the rein of their husband's horse to one arm lead it to
drink, while they use their hands to spin thread. It

The Women of Paeonia

threatening to give her child to the wolf unless it stops crying.
Later she says to the child, ' If the wolf comes we will kill it.'
See Leutsch, *Paroemiog. Gr.* 1. 273; 2. 121, 510; Babrius 16.
   b Cyrus the *First* was the son of Cambyses.
   c Polycrates, Tyrant of Samos, fell victim to a plot by the
Persian satrap Oroetes, *c.* 522 B.C.

AELIAN

ταῦτά τοι καὶ Δαρεῖος ἐθαύμασεν, ὅτε Παίονες
νεανίαι τὴν ἑαυτῶν ἀδελφὴν οὕτω σκευάσαντες,
δικάζοντος αὐτοῦ, παρήγαγον αὐτήν, ἵνα ἐς ἔρωτα
ἐμπεσὼν τῆς οὕτως ἀθρόας αὐτουργίας ἐλεῇ [1]
Παίονας. ἀλλὰ ἡ φύσις πόσῳ Παιονίδων [2] σοβα-
ρωτέρα. κύων θηράσασα (λαγὼς δὲ ἦν τὸ ἄγρευ-
μα αὐτῇ, καὶ ἐκύει ἡ κύων) ἐπεὶ [3] τῆς σπουδῆς
τῆς προκειμένης ἐτετυχήκει,[4] τῷ μὲν δεσπότῃ τοῦ
θηράματος ἀπέστη, ἀναχωρήσασα δὲ ἐννέα φασὶ
σκύλακας ἀποκυήσασα εἶτα ἐξέθρεψεν αὐτούς. εἰ
δὲ Λιγυστίνων [5] αἱ γυναῖκες μέγα φρονοῦσιν ὅτι
κἀκεῖναι τὴν ὠδῖνα ἀπολύσασαι καὶ ἐξαναστᾶσαι
τῶν ἔργων ἔχονται τῶν κατὰ τὴν οἰκίαν, ἀκούσα-
σαι τὸ τῆς κυνὸς ἔργον τῆς προειρημένης τοῦ
φυσήματος ἀποστᾶσαι πάντως ἐγκαλύψονται.

13. Τὸ μὲν τοῦ ἡμιόνου τοῦ φιλοπόνου [6] Ἀριστοτέλης εἶπε καὶ ἡμεῖς ἄνω που,[7] τὸ δὲ τοῦ κυνὸς
καὶ τοῦτο ἐν ταῖς Ἀθήναις γενόμενον εἰπεῖν
οὐδὲν ἄτοπον. ἐς Ἀσκληπιοῦ παρῆλθε θεοσύλης
τό τε μεσαίτατον τῆς νυκτὸς παραφυλάξας καὶ
τῶν καθευδόντων τὸν βαθύτατον [8] ὕπνον ἐπιτηρή-
σας, εἶτα ὑφείλετο τῶν ἀναθημάτων πολλά, καὶ
ὥς γε ᾤετο ἐλελήθει. ἦν δὲ ἄρα σκοπὸς ἀγαθὸς
ἔνδον κύων καὶ τῶν ζακόρων ἀμείνων ἐς ἀγρυ-
πνίαν, ὅσπερ οὖν εἵπετό οἱ διώκων, καὶ ὑλακτῶν

[1] (or ἐλεήσῃ) Grasberger : ἔλη mss, H.
[2] πόσων Παιόνων.
[3] ἐπεὶ δέ.
[4] τετύχηκε.
[5] Λιγυστίων Jac, Αἰγυπτίων mss.
[6] φιλοπόνου καὶ τῆς δημοσίας αὐτῷ δοθείσης τροφῆς ἐξ Ἀθη-
ναίων ὑπὲρ τοῦ ἐθελουργοῦ καθ᾽ ἡλικίαν.

was this that moved Darius to admiration when some young Paeonians, having equipped their sister in the manner described, brought her before him as he sat in judgment, in order that he might be attracted by such a concentration of self-help and show mercy to their country.

And yet how far more impressive is Nature than the Paeonian women. A bitch was hunting; the quarry was a hare and the bitch was pregnant. As soon as she had attained the object of her pursuit, she left it to her master and drawing aside, dropped (so they say) nine puppies, which she then reared. And if the women of Liguria pride themselves that they also after giving birth rise up and devote themselves to their household duties, they will, on hearing what the aforesaid bitch did, forgo their pride and hide their heads in shame.

A pregnant Hound

13. Aristotle has told the story of the labour-loving Mule, and so have we earlier on,[a] but the episode of the Dog, which also occurred in Athens, is not irrelevant.

Dog reveals sacrilege

A temple-thief who had waited for the midmost hour of night and had watched till men were deep asleep, came to the shrine of Asclepius and stole a number of offerings without, as he supposed, being seen. There was however in the temple an excellent watcher, a Dog, more awake than the attendants, and it gave chase to the thief and never stopped

[a] See 6. 49.

---

[7] που καλῶς δρῶντες. [8] *Bernard* : βαρύτατον.

οὐκ ἀνίει, ᾗπερ¹ οὖν ἔσθενε δυνάμει τὸ πραχθὲν
μαρτυρόμενος. τὰ μὲν οὖν πρῶτα ἔβαλλεν αὐτὸν
λίθοις αὐτός τε καὶ οἱ τῆς κακῆς ἐκείνης πράξεως
κοινωνοί, τὰ δὲ τελευταῖα προύσειεν² ἄρτους τε
καὶ μάζας. ἐπήγετο δὲ ἄρα ταῦτα³ δέλεαρ
κυνῶν προμηθῶς, ὥς γε ὑπελάμβανεν. ἐπεὶ δὲ
καὶ παρελθόντος ἐς τὴν οἰκίαν οὗ κατήγετο
ὑλάκτει καὶ πάλιν προϊόντος, ἐγνώσθη μὲν ὁ
κύων ἔνθεν ἦν, τὰ λείποντα δὲ τῶν ἀναθημάτων
ἐπόθουν αἱ γραφαί τε καὶ αἱ χῶραι ἔνθα ἀνέκειντο.
συνέβαλον οὖν τοῦτον ἐκεῖνον εἶναι οἱ Ἀθηναῖοι,
καὶ στρεβλώσαντες τὸ πᾶν κατέγνωσαν. καὶ ὁ
μὲν ἐδικαιώθη τὰ ἐκ τοῦ νόμου, ὁ δὲ κύων
ἐτιμήθη δημοσίᾳ τροφῇ καὶ κηδεμονίᾳ, οἷα δήπου
φύλαξ πιστὸς καὶ τῶν νεωκόρων οὐδενὸς μείων
τὴν ἐπιμέλειαν.

14. Ἀγαθὴ δὲ ἄρα ἦν αἲξ καὶ τὴν τῶν ὀφθαλμῶν
ἀχλὺν ἥνπερ οὖν παῖδες Ἀσκληπιαδῶν ὑπόχυσιν
καλοῦσιν ἀκέσασθαι, καὶ λέγονταί γε οἱ ἄνθρωποι
παρ' ἐκείνης μαθεῖν τόδε τὸ ἴαμα. τὸ δὲ ἄρα
τοιοῦτόν ἐστιν. ὅταν αἲξ νοήσῃ τὸν ὀφθαλμὸν
ἐπιθολωθέντα αὐτῇ, πρόσεισι βάτῳ, καὶ παραβάλλει τῇ ἀκάνθῃ τὸ ὄμμα.⁴ καὶ ἡ μὲν ἐκέντησε,
τὸ δὲ ὑγρὸν ἐξεχώρησε, μένει δὲ ἀπαθὴς ἡ κόρη,
καὶ ὁρᾷ αὖθις, καὶ δεῖται σοφίας καὶ χειρουργίας
ἀνθρωπικῆς οὐδὲ ἕν.

15. Ποταμὸν ἐλέφαντες διέρχονται οἱ μὲν ἔτι
νέοι διανηχόμενοι· οἱ δὲ ἤδη τέλειοι, καὶ εἰ

barking, as with all its might it summoned others to witness what had been done. And so at first the thief and his companions in that crime pelted the Dog with stones; finally he dangled bread and cakes in front of it. He had been careful to bring these things with him as an attraction to Dogs, as he supposed. Since however the Dog continued to bark when the thief came to the house where he lodged and when he came out again, it was discovered where the Dog belonged, while the inscriptions and the places where the offerings were set up lacked the missing objects. The Athenians therefore concluded that this man was the thief, and by putting him on the rack discovered the whole affair. And the man was sentenced in accordance with the law, while the Dog was rewarded by being fed and cared for at the public expense for being a faithful watcher and second to none of the attendants in vigilance.

14. The Goat, it seems, is in fact skilful at curing that mist of the eyes which doctors call 'cataract,' and it is even said that men have learnt this cure from the Goat. The method is as follows. When the Goat perceives that its sight has become clouded it goes to a bramble and applies its eye to a thorn. The thorn pricks it and the fluid is discharged, but the pupil remains unharmed and the Goat regains its sight without any need of man's skill and manipulation. *The Goat cures cataract*

15. Young Elephants cross a river by swimming, but the full-grown ones, if covered by the stream, *Elephants, their mutual devotion*

---

³ καὶ ταῦτα.   ⁴ ὄμμα νύξαι αὐτό.

AELIAN

καλύπτοιντο ὑπὸ τοῦ ῥεύματος, ἀνέχουσι μέντοι
τὰς προβοσκίδας ὑπὲρ τὸ ὕδωρ, τὰ δὲ ἀρτιγενῆ
πώλια ἐπὶ τῶν ὀδόντων[1] φέρουσιν αἱ μητέρες.
τῶν μὲν οὖν κινδύνων καὶ πόνων οἱ νέοι κατάρχονται,
ποτοῦ δὲ ἄρα καὶ τροφῆς ἀφίστανται τοῖς πρεσβυτέροις αἰδῶ νέμοντες, καὶ τῶν Λυκούργου νόμων[2]
δέονται οἶδε οὐδὲ ἕν. γήρᾳ δὲ παρειμένον ἐλέφαντα
ἢ νόσῳ κατειλημμένον οὐκ ἄν ποτε οἱ συναγελα-
ζόμενοι καταλίποιεν, ἀλλὰ πιστῶς παραμένουσι,
καὶ ἀναρρῶσαι σπεύδουσι τῇ τε ἄλλῃ καὶ ἐὰν
διώκωνται, καὶ ὑπὲρ αὐτοῦ μάχονται, καὶ τιτρώ-
σκονται παραμένοντες,[3] φυγεῖν δυνάμενοι. καὶ τὰ
βρέφη δὲ τὰ νεαρὰ οὐκ ἄν ποτε αἱ τεκοῦσαι
προδοῖεν, ἀλλὰ καὶ ἐκείνοις ἐκεῖναι πιστῶς παραμένουσι, καίτοι τῶν θηρώντων ἐγκειμένων, καὶ
πρότερόν γε τὴν ψυχὴν ἀπολίποιεν ἂν ἢ τὰ τέκνα.

Ἐγὼ δὲ ᾔδειν μειράκιον ὢν ἄνθρωπον πρεσβῦτιν
Λαινίλλαν ὄνομα, καὶ ἐδείκνυτο ὑπὸ πάντων, καὶ
ἐπ᾽ αὐτῇ μῦθος ἐλέγετο, καὶ ὅ γε μῦθος τοιόσδε
ἦν. οἱ πρεσβύτεροι πρός με ἔφασκον ἐκείνην τὴν
ἄνθρωπον ἐρασθεῖσαν οἰκέτου δριμέως αὐτῷ μὲν
συγκαθεύδειν, κηλῖδα δὲ ἄρα παισὶ τοῖς ἑαυτῆς
περιάπτειν. οἱ δὲ εὐγενεῖς ἦσαν, καὶ ἐς τὴν
βουλὴν τὴν Ῥωμαίων ἐτέλουν ἐκ πατέρων τε καὶ
τῶν ἄνω τοῦ γένους.[4] οἱ τοίνυν ⟨παῖδες⟩[5] αἰδού-
μενοι τῇ μητρὶ ἤχθοντο τοιαῦτα δρώσῃ, καὶ
πράως ὑπενουθέτουν, καὶ τοῦ πραττομένου τὴν
αἰσχύνην ἐπέλεγον ἡσυχῇ· ἡ δὲ κυμαίνουσα ἐκ
τῆς ἐπιθυμίας, καὶ τὸν ἔρωτα ἐπίπροσθεν τῶν
υἱέων ποιησαμένη, καταγορεύει αὐτῶν πρὸς τὸν

---

[1] ὀδόντων ἢ κεράτων.
[2] Jac: τιμῶντες τὸ γῆρας νόμων.

122

raise their trunks above the water, while the mother-
elephants carry their newly born young upon their
tusks.   It is the young who take the lead in danger
and hardship; out of respect for their elders they give
way to them in drinking and feeding, and they have
no need at all of the laws of Lycurgus.   An Elephant
old and weak or stricken with disease would never be
abandoned by his fellows in the herd, but they stay
beside him loyally and hasten to lend him strength
on all occasions, especially when they are being
pursued; and they fight on his behalf and through
staying by him receive wounds, when they could
escape.   The females would never desert the young
they have borne, but they too remain loyally at their
side even though hunters press hard upon them, and
they would sooner relinquish their life than their
offspring.

When I was a boy I knew an aged woman, Laenilla   Laenilla and
by name, and everybody used to point at her, and   her sons
a story was told of her to this effect.   My elders used
to tell me that she had passionately loved a servant
and used to sleep with him, thereby bringing a slur
upon her own children.   They were well-born and
belonged to the Senatorial order in Rome by descent
from their fathers and remoter ancestors.   Now the
children for very shame were angry with their
mother for her behaviour and admonished her gently
and spoke to her in private of the shamefulness of her
conduct.   But she, seething with lust and putting her
love above her sons, accused them before the magis-

---

[3] *Reiske* : παρόντες.   [4] γένους ἀρξάμενοι.
[5] ⟨παῖδες⟩ *add. H.*

ἄρχοντα,¹ καὶ λέγει ὡς ἐπιβουλεύοιεν αὐτῷ. ὁ
δὲ ἔχων ἐς διαβολὰς τὸ οὖς ῥᾴδιον, καὶ ὕποπτης
ὢν καὶ δειλὸς (πάθη δὲ ταῦτα ἀγεννοῦς διανοίας)
ἐπίστευσε. καὶ οἱ μὲν οὐδὲν ἀδικοῦντες ἀπέθνη-
σκον, ἡ δὲ ἆθλον τοῦ κατειπεῖν ἠνέγκατο δούλῳ
συγκαθεύδειν ἀνέδην. ὦ πατρῷοι θεοὶ καὶ Ἄρτεμι
λοχεία Εἰλείθυιαί τε θυγατέρες Ἥρας, τί ἂν ²
ἔτι Μήδειαν εἴποιμεν τὴν Κόλχον ³ ἢ Πρόκνην
τὴν Ἀτθίδα, τῶν ἔναγχός τε καὶ καθ' ἡμᾶς
παθῶν μνημονεύσαντες;

16. Τὰς χερσαίας χελώνας οἱ ἀετοὶ συλλαβόντες
εἶτα ἄνωθεν προσήραξαν ταῖς πέτραις, καὶ τὸ
χελώνιον συντρίψαντες οὕτως ἐξαιροῦσι τὴν σάρκα
καὶ ἐσθίουσι. ταύτῃ τοι καὶ Αἰσχύλον τὸν
Ἐλευσίνιον τὸν τῆς τραγῳδίας ποιητὴν τὸν βίον
ἀκούω καταστρέψαι. ὁ μὲν Αἰσχύλος ἐπί τινος
πέτρας καθῆστο, τὰ εἰθισμένα δήπου φιλοσοφῶν
καὶ γράφων· ἄθριξ δὲ ἦν τὴν κεφαλὴν καὶ ψιλός.
οἰηθεὶς οὖν ἀετὸς ⁴ πέτραν εἶναι τὴν κεφαλὴν
εἶτα μέντοι κατ' αὐτῆς ἀφῆκεν ἣν κατεῖχε χελώνην,
καὶ ἔτυχε τοῦ προειρημένου τὸ βέλος, καὶ ἀπέκτεινε τὸν ἄνδρα.

17. Κηρύλος δὲ καὶ ἀλκυὼν ὁμόνομοι καὶ
σύμβιοι. . . .⁵ καὶ γήρᾳ γε παρειμένους αὐτοὺς
ἐπιθέμεναι αἱ ἀλκυόνες περιάγουσιν ἐπὶ τῶν
καλουμένων μεσοπτερυγίων. ἄνθρωποι δὲ καὶ
τῶν ἀνδρῶν ὑπογηρώντων καταφρονοῦσι καὶ πρὸς
τὰ μειράκια ἀφορῶσι· καὶ οἱ γήμαντες περὶ τὰς

¹ ἄρχοντα ὃς ἦν τότε.   ² ἂν οὖν or οὖν ἄν.

trate, alleging that they were plotting against him.
The magistrate having a ready ear for calumny, and
being of a suspicious and cowardly nature (those are
attributes of an ignoble character), believed her.
So her sons who had done no wrong were put to
death, while the woman reaped the reward of her
informing and slept freely with the slave.

O gods of our fathers, O Artemis of the child-bed,
and ye goddesses of birth, daughters of Hera, why,
when we recall calamities that befell recently and in
our own day, should we speak any more of Colchian
Medea or Attic Procne?

16. Eagles seize Tortoises and then dash them on
rocks from a height, and having smashed the Tortoise's shell they extract and eat the flesh. It was
in this way, I am told, that Aeschylus of Eleusis, the
tragic poet, met his end. Aeschylus was seated
upon a rock, meditating, I suppose, and writing as
usual. He had no hair on his head and was bald.
Now an Eagle supposing his head to be a rock, let
the Tortoise which it was holding fall upon it. And
the missile struck the aforesaid poet and killed him.

17. The Ceryl and the Halcyon feed side by side
and live together. . . . And when the Ceryls are
feeble with age the Halcyons place them on their
back and carry them about upon their middle wing-
feathers, as they are called. Women however look
down upon those who are ageing, and cast their eyes
on youths. And husbands are eager after girls and

[3] Κόλχιν.   [4] ὁ ἀετός.
[5] Lacuna.

νέας ἠνέμωνται, τῶν ἀφηλικεστέρων γαμετῶν
ὥραν μὴ τιθέμενοι, καὶ οὐκ αἰδοῦνται οἱ ἔμφωνοι
τῶν ἀλόγων ζώων βιοῦντες ἀλογώτερον.

18. Λέγουσι δὲ οἱ ⟨Αἰγύπτιοι⟩[1] περὶ τὴν
καλουμένην Κοπτὸν δύο μόνους ὁρᾶσθαι κόρακας.
ἀλλὰ καὶ τῶν Ῥωμαίων οἱ τὴν ὄρειον παραφυλάττοντες διὰ τὸ τῆς σμαράγδου μέταλλον διισχυρί-
ζονται καὶ οἵδε τοσούτους ὄρνιθας τοῦ γένους
τοῦδε οἰκεῖν ἐκεῖθι.[2] νεὼς δὲ Ἀπόλλωνι τιμᾶται
ἐν τῷ χωρίῳ ἐκείνῳ, οὗπερ οὖν ἱεροὺς εἶναί
φασιν αὐτούς.

19. Καὶ ταύτῃ δὲ τὰ ἴδια τῶν ζώων εἰπεῖν οὐ
χεῖρόν ἐστι. νωθέστερά πως δοκεῖ πρόβατον καὶ
ὄνος, ἄτολμα δὲ νεβροὶ καὶ πρόκες καὶ ζόρκες τε
καὶ πύγαργοι[3] καὶ οἱ λαγῴ, οὓς δὴ καὶ πτῶκας
οἱ ποιηταὶ καλοῦσιν.[4] ἀλλὰ καὶ[5] τῶν πετεινῶν
ἐστιν ἄτολμα οἱ στρουθοὶ[6] καὶ τῶν ἐνύδρων οἱ
κεστρεῖς. ἀκόλαστα δὲ κυνοκέφαλοί τε καὶ τράγοι,
καὶ μέντοι[7] καὶ ὁμιλεῖν γυναικί φασιν αὐτούς,
καὶ ἔοικεν αὐτὸ θαυμάζειν Πίνδαρος. καὶ κύνες
δὲ γυναιξὶν ἐπιτολμᾶν ἐλέχθησαν, καὶ μέντοι καὶ
κριθῆναι λέγεται γυνὴ ἐν τῇ Ῥώμῃ μοιχείας ὑπὸ
τοῦ γήμαντος, καὶ ὁ[8] μοιχὸς ἐν τῇ δίκῃ κύων
εἶναι ἐλέγετο. ἤκουσα δὲ κυνοκεφάλους καὶ παρθέ-
νοις ἐπιμανῆναι καὶ μέντοι καὶ βιάσασθαι ὑπὲρ

---

[1] ⟨Αἰγύπτιοι⟩ add. H.
[2] Reiske: καὶ ἐκεῖθι.
[3] Ges: πυλαργοί.
[4] καλοῦσιν ἐκ τοῦ πτώσσειν δηλονότι.
[5] καὶ ἄλλα καί.
[6] ἄτολμα ὥσπερ οὖν οἱ σ.

take no notice of their elderly legal wives: creatures gifted with speech are not ashamed to live more unreasonably than unreasoning animals.

18. The Egyptians who live about the region called Coptus assert that no more than a pair of Ravens is seen there. And even those Romans who guard the mountain district because of the Emerald Mine,[a] they also maintain that the same number of this species live there. And in that place there is a temple in honour of Apollo to whom, they say, the birds are sacred.

The Raven

19. Here again I may as well speak of the peculiarities of animals. The sheep and the ass seem inclined to be sluggish; fawns, roe-deer, gazelles, antelopes, hares (which poets style ' cowerers ') are timorous creatures. Timorous also are sparrows among birds, and the mullet among fishes. Baboons and goats are lecherous, and it is even said that the latter have intercourse with women—a fact which Pindar [fr. 201 S] appears to marvel at. And even hounds are said to have assaulted women, and indeed it is reported that a woman in Rome was accused by her husband of adultery, and the adulterer in the case was stated to be a hound. And I have heard that baboons have fallen madly in love with girls and have even raped them, being more wanton than the

Animal peculiarities

[a] Smaragdus, the Egyptian Emerald Mine, lay E of the Nile near the Red Sea, between Berenice and the mountain range of Lepte. See *Geogr. Jl* 16 (1900) 537.

[7] *Reiske*: οὗτοι μέν. [8] ὁ μέν.

τὰ μικρὰ μειράκια τὰ τοῦ Μενάνδρου ἐν ταῖς
παννυχίσιν ἀκόλαστα. λαγνίστατον δὲ καὶ ὁ
πέρδιξ καὶ μοιχικόν. λάθρᾳ γοῦν ἐπὶ τὰς θηλείας
καί πως ἀψοφητὶ λέγονται φοιτᾶν. τροφῆς δὲ τὴν
κοινωνίαν ἥκιστα ἐνδέχονται κύνες. πολλάκις
γοῦν καὶ ὑπὲρ ὀστοῦ[1] ἀλλήλους σπαράττουσιν,
ὥσπερ οὖν ὁ Μενέλεως καὶ ὁ Πάρις ὑπὲρ τῆς
Ἑλένης. μόνους δὲ ἀκούω τοὺς Μεμφίτας κύνας
ἐς μέσον τὰς ἁρπαγὰς κατατίθεσθαι καὶ ἐσθίειν
κοινῇ. ἄσπονδον δὲ καὶ ἔκδικον ὁ σῦς. ἀλλήλων
γοῦν οὗτοι νεκρῶν ἐσθίουσι. καὶ οἱ ἰχθῦς δὲ οἱ
πλεῖστοι δρῶσιν αὐτό. ἀσεβέστατον δὲ ὁ ποτάμιος ἵππος· γεύεται γὰρ καὶ τοῦ πατρός. ἀναιδῆ
δὲ καὶ μὴ ῥᾳδίως ὑποστελλόμενα μυῖαι καὶ
κύνες.

20. Ἀγριώτατον δὲ λύκοι. λέγουσι δὲ οἱ Αἰγύπτιοι ὅτι καὶ ἀλλήλους ἐσθίουσι, καὶ τὸν τρόπον
τῆς ἐπιβουλῆς ἐκεῖνόν φασιν. ἐς κύκλον ἑαυτοὺς
περιαγαγόντες[2] εἶτα μέντοι θέουσιν. ὅταν δέ τις
αὐτῶν ὑπὸ τοῦ κατὰ τὸν δρόμον ἰλίγγου σκοτο-
δινιάσῃ καὶ περιτραπῇ, οἱ λοιποὶ κειμένῳ προσπεσόντες σπαράττουσιν αὐτὸν καὶ ἐσθίουσι. δρῶσι
δὲ ἄρα τοῦτο ἐπὰν ἀθηρίᾳ περιπέσωσι. πρὸς γὰρ
τὸ μὴ πεινῆν πάντα λῆρον ἥγηνται ὥσπερ οὖν οἱ
τῶν ἀνθρώπων κακοὶ πρὸς τὸ ἀργύριον.

21. Κακοηθέστατον δὲ ἄρα τῶν ζῴων ὁ πίθηκος
ἦν, καὶ ἔτι πλέον ἐν οἷς πειρᾶται μιμεῖσθαι τὸν
ἄνθρωπον. αὐτίκα γοῦν ἰδὼν ἐξ ἀπόπτου τροφὸν

---

[1] ὀστέου.   [2] περιάγοντες.

little boys in the all-night revels of Menander.[a] The partridge is extremely lecherous and given to adultery; at any rate these birds are said to go after the hens stealthily and with hardly a sound. Dogs do not admit others to share their food on any account; at any rate they often tear one another over a bone, just like Menelaus and Paris over Helen. I am told that the dogs of Memphis are the only ones that pool their prey and share their food. The hog is implacable and devoid of justice; at any rate these creatures eat one another's dead bodies. And the majority of fishes do the same. But the most impious of all is the hippopotamus, for it even eats its own father. Flies and dogs are without shame and are not easily checked.

20. Wolves are exceedingly fierce, and the Egyptians assert that they even eat one another, and that the way in which they plot against each other is, they say, as follows. They gather round in a circle and then start to run. And when any of their number is overcome with dizziness from running round and round and collapses, the rest fall upon him as he lies, tear him to pieces, and eat him. They do this whenever their hunting is unsuccessful. For with them, provided they do not go hungry, nothing else counts; just as with evil men nothing counts but money. *Hungry Wolves*

21. It seems that the Monkey is the most mischievous of animals; and even worse when it attempts to copy man. For example, a Monkey observed from *Monkey and baby*

---

[a] No comedy of Menander of the name of Παννυχίς(-ίδες) is known; the reference is presumably general.

AELIAN

λούουσαν παιδίον ἐν σκάφῃ, καὶ πρῶτον μὲν
ὑπολύουσαν τὰ σπάργανα, εἶτα ¹ ἐκ τοῦ λουτροῦ
κατειλοῦσαν αὐτό, παραφυλάξας ἔνθα ἀνέπαυσε
τὸ βρέφος, ὡς εἶδεν ἐρημίαν, ἐσέθορε διά τινος
ἀνεῳγμένης θυρίδος, ἐξ ἧς οἱ πάντα ² σύνοπτα ἦν,
καὶ ἄρας ἐκ τῆς εὐνῆς τὸ παιδίον, καὶ γυμνώσας
ὡς ἔτυχεν ἰδών, καὶ κομίσας ἐς μέσον τὴν σκάφην,
ζέον ὕδωρ (καὶ γὰρ ἦν ἐπί τινων ἀνθράκων
θερμαινόμενον) τοῦ δυστυχοῦς παιδίου κατέχεε,
καὶ μέντοι καὶ ἀπέκτεινεν αὐτὸ οἴκτιστα.

22. Κακόηθες δὲ ἄρα καὶ ὕαινα ἦν καὶ ὅν φασι
κοροκότταν. ἡ γοῦν ὕαινα πρὸς τὰ αὔλια νύκτωρ
φοιτᾷ, καὶ μιμεῖται τοὺς ἐμοῦντας. ἀκούοντες δὲ
οἱ κύνες προσίασιν ὡς ἐπ' ἄνθρωπον· ἡ δὲ αὐτοὺς
συλλαμβάνει καὶ ἐσθίει. πανουργίαν δὲ κοροκόττα,
ἣν ἤκουσα καὶ αὐτήν,³ ἔοικα λέξειν νῦν. ἐς τοὺς
δρυμοὺς ἑαυτὸν ἐγκρύψας εἶτα μέντοι τῶν ὑλουρ-
γούντων ἀκούει καλούντων ⁴ ἀλλήλους ἐξ ὀνόματος καὶ μέντοι ⟨καὶ⟩⁵ λαλούντων ἄττα.⁶ εἶτα
μέντοι μιμεῖται τὰς φωνάς, καὶ φθέγγεται, εἰ καὶ
μυθῶδες τὸ εἰρημένον, ἀνθρωπίνῃ γοῦν φωνῇ,
καὶ καλεῖ τὸ ὄνομα ὃ ἤκουσε. καὶ ὁ κληθεὶς
πρόσεισιν, ὁ δὲ ἀναχωρεῖ καὶ πάλιν καλεῖ· ὁ δὲ
καὶ μᾶλλον κατὰ τὴν φωνὴν ἔρχεται. ὅταν δὲ
αὐτὸν τῶν συμπονούντων ἀπαγάγῃ καὶ ἔρημον
ἀποφήνῃ, συλλαβὼν ἀπέκτεινε καὶ ποιεῖται τροφὴν τὸ ἐντεῦθεν φωνῇ δελεάσας.

¹ εἶτα δέ. ² καὶ πάντα.
³ αὐτός. ⁴ καὶ καλούντων.
⁵ ⟨καὶ⟩ add. H.
⁶ Jac : αὐτά.

a distance a nurse washing a baby in a tub, observed
how first of all she took off its swaddling clothes and
then after the bath wrapped it up; it marked where
she laid it to rest, and when it saw the place unguarded,
sprang in through an open window, from which it
had a view of everything; took the baby from its cot;
stripped it as it had chanced to see the nurse do;
brought the tub out, and (there was water heating on
some embers) poured boiling water over the wretched
baby and even caused it to die most miserably.

22. It seems that the Hyena also and the *Coro-* The Hyena
cottas,[a] as they call it, are viciously clever animals.
At any rate the Hyena prowls about cattle-folds by
night and imitates men vomiting. And at the
sound dogs come up, thinking it is a man. Whereupon it seizes and devours them. I shall now relate The 'Coro-
the villainy of the Corocottas, of which I have cottas'
actually heard. It conceals itself in thickets and
then listens to woodcutters calling one another by
name, and even to anything they say. And then it
imitates their voices and speaks (though the story
may be fabulous) with a voice that sounds human at
any rate, calling out the name which it has heard.
And the man who has been called approaches: the
animal withdraws and calls again: the man follows
the voice all the more. But when it has drawn him
away from his fellow-workers and has got him alone,
it seizes him and kills him and then makes a meal off
him after luring him on with its call.

---

[a] Κορόκοττας: 'perh. *hyena*' (L-S⁹); O. Keller (*Antike
Tierwelt* 1. 152) says that the word is of Libyan origin and
denotes the speckled Libyan hyena, *Hyaena crocuta*, as distinct
from the common striped species.

AELIAN

23. Ἀμύνεσθαι δὲ τὸν προαδικήσαντα ὁ λέων
οἶδε, καὶ εἰ μὴ παραχρῆμα αὐτῷ τιμωρήσειεν,
ἀλλά γε καὶ μετόπισθεν ἔχει κότον, ὄφρα τελέσσῃ,
ἐν στήθεσσιν ἑοῖσιν.

καὶ τούτου μαρτύριον Ἰόβας ὁ Μαυρούσιος ὁ τοῦ
παρὰ Ῥωμαίοις ὁμηρεύσαντος πατήρ. ἤλαυνέ
ποτε διὰ τῆς ἐρήμης ἐπί τινα ἔθνη τῶν ἀποστάν-
των, καί τις αὐτῷ τῶν παραθεόντων μειρακίσκος
εὐγενὴς μὲν καὶ ὡραῖος ἤδη δὲ ⟨καὶ⟩[1] θηρατικὸς
λέοντά πως παρὰ τὴν ὁδὸν ἐκφανέντα ἀκοντίῳ
βάλλει, καὶ σκοποῦ μὲν ἔτυχε καὶ ἔτρωσεν, οὐ
μὴν ἀπέκτεινε. κατὰ σπουδὴν δὲ τῆς ἐλάσεως
οὔσης, τὸ μὲν θηρίον ἀνεχώρησε, παρέδραμε δὲ
καὶ ὁ τρώσας καὶ οἱ λοιποί. ἐνιαυτοῦ γε μὴν
διελθόντος ὁλοκλήρου ὁ μὲν Ἰόβας κατορθώσας
ἐφ' ἃ ἐστάλη, τὴν αὐτὴν ὑποστρέφων ἔρχεται κατὰ
τὸν τόπον, ἔνθα ἔτυχεν ὁ λέων τρωθείς. καὶ
ὄντος πλήθους παμπόλλου πρόσεισι τὸ θηρίον
ἐκεῖνο, καὶ τῶν μὲν ἄλλων ἀπέχεται, συλλαμβάνει
δὲ τὸν τρώσαντα πρὸ ἐνιαυτοῦ, καὶ τὸν θυμόν,
ὅνπερ οὖν παρὰ τὸν χρόνον τὸν προειρημένον
ἐφύλαττεν, ἀθρόον ἐκχεῖ καὶ διασπᾷ τὸ μειράκιον γνωρίσας. ἐτιμώρησε δὲ οὐδείς, φοβηθέντες ὀργὴν λέοντος ἰσχυρὰν καὶ δεινῶς ἐκπληκτικήν·
ἄλλως τε καὶ ἡ πορεία ἤπειγεν.

24. Καρκίνων γένη διάφορα καὶ φῦλα ποικίλα
ἀκούω εἶναι. καὶ γὰρ οὖν καὶ πετραῖοί εἰσιν·

[1] ⟨καὶ⟩ add. Reiske.

132

23. The Lion knows how to take vengeance on one who has previously done him an injury, and even though the vengeance be not immediate,

' yet doth he keep his anger thereafter in his bosom, until he accomplish it ' [Hom. *Il.* 1. 82].

And Juba of Mauretania,[a] the father of the boy who was a hostage at Rome, bears witness to this. He was marching once through the desert against some tribes who had revolted, when one of the youths who ran beside him, well-born, handsome, and already fond of the chase, struck with a javelin a Lion that chanced to appear by the roadside : he hit the mark and wounded the beast, but failed to kill it. But the expedition was in haste ; the animal drew off, and the boy who had wounded it hurried by with the rest. Now when a whole year had passed and Juba had accomplished his purpose, returning by the same way he arrived at the spot where the Lion had happened to be wounded. And in spite of the multitude of men that same Lion came forward and, without touching anyone else, seized him who a year ago had wounded it, and pouring forth the gathered anger which it had been nursing all that while, tore to pieces the boy whom it had recognised. But not a soul took vengeance : they were afraid of the fierce and absolutely terrifying anger of the Lion. And besides, their journey made them hasten.

24. I have heard that there are different species and various tribes of Crabs, for there are some that

---

[a] Juba I, King of Numidia (not Mauretania), 1st cent. B.C., took the side of Pompey in the Civil War ; after the battle of Thapsus he committed suicide.

ἀλλὰ καὶ πηλοὶ τίκτουσι καρκίνους, καὶ φυκία
καὶ ψάμμος. ἰδέαι τε αὐτῶν καὶ ἐπωνυμίαι
πολλαί. πλανῶνται δὲ δεῦρο καὶ ἐκεῖσε ⟨οἱ⟩[1]
καλούμενοι δρομίαι (ὧδε γὰρ καλεῖν[2] αὐτοὺς
πρεπωδέστατον)· ἀτρεμεῖν γὰρ καὶ ἡσυχάζειν
ἐπὶ τῆς αὐτῆς χώρας οὔτε ἐθέλουσιν οὔτε πεφύκασιν, ἀλλὰ περὶ τοὺς αἰγιαλοὺς ἀλῶνται, ὅθενπερ
καὶ ἐξέφυσαν· ἤδη δὲ καὶ στέλλονται πορρωτέρω,
ὥσπερ οὖν οἱ τῶν ἀνθρώπων φιλαπόδημοι.[3] ὑπόθεσις δὲ αὐτοῖς τῆς τοσαύτης ἄλης τὸ ἐθέλειν
πλείονός τινος ἀπολαῦσαι. ἐν δὲ τῷ Θρακίῳ
Βοσπόρῳ ἐπειδὰν τὸ ῥεῦμα βίαιον ἐκ τοῦ Πόντου
καταφέρηται, βούλονται μὲν οἱ καρκίνοι πρὸς
ἐναντίον ὠθούμενοι τὸν ῥοῦν βαδίζειν, ταῖς δὲ
ἄκραις ὡς τὸ εἰκὸς βιαιότερον τὸ ῥεῦμα περιρ-
ρήγνυται. τοὺς οὖν καρκίνους ὠθήσει τε πάντως
καὶ ἀνατρέψει, εἰ μέλλοιεν[4] ἰέναι τῷ ῥῷ ὁμόσε.
οἱ δὲ ταῦτα προΐσασι, καὶ ἐπειδὰν ἀφίκωνται τῆς
ἄκρας πλησίον, ἕκαστος ἔν τινι κολπώδει χωρίῳ
ἐπέχει,[5] καὶ τοὺς λοιποὺς ἀναμένει. εἶτα ἀθροι-
σθέντες ἐν ταὐτῷ προσανέρπουσιν ἐς τὴν γῆν, καὶ
ἀναρριχῶνται ἐπὶ τοὺς κρημνούς, καὶ τὸ μάλιστα
ῥοῶδες καὶ βίαιον τοῦ πελάγους πεζοὶ διέρχονται.
εἶτα ἔξω τῆς ἄκρας γενόμενοι καὶ παραμείψαντες
αὐτὴν ἐς τὴν θάλατταν κατίασιν αὖθις. φείδονται
δὲ αὐτῶν οἱ ἁλιεῖς, ὅτι ἑκόντες ἐπὶ τὴν γῆν
προσέρπουσι, σωθῆναι δεόμενοι ὁμοῦ αὐτοῖς.
οὔκουν[6] ὑπομένουσι τοῦ κλύδωνος ἀγριώτεροι
δοκεῖν οἱ ἄνθρωποι.

live on rocks, but there are others besides, which mud,
seaweed, and sand generate. And they have many
shapes and many names. And the Runner-crabs
as they are called (and most appropriately) roam
hither and thither, for it is neither their wish nor
their nature to remain quiet and at rest in the same
place, but they wander about the beaches where they
were born; and they do in fact go further afield, just
as human beings who are fond of travel. The
occasion of their wandering so far is their desire for
more food of some kind. Now in the Thracian
Bosphorus whenever the current comes down
strongly from the Euxine, the Crabs wish to force
their way upstream, but, as is natural, the stream
breaks with too great violence round the headlands,
so that if they should want to go against it, it will
altogether thrust them back and defeat them. Now
the Crabs are already aware of this, and whenever
they come near a headland each one halts in some
bay-like spot and waits for the others. Then when
they have congregated in one spot, they crawl up on
to the land and scramble up on to the cliffs and so
pass by on foot that part of the sea where the current
is strongest. Then having surmounted and passed
the promontory, they descend once more to the sea.
But the fishermen spare them because it is of their
own free will that the Crabs crawl out on to the land:
the men wish also to be spared themselves: they
cannot bear to appear more cruel than the waves.

[1] ⟨οἱ⟩ add. Schn.   [2] ἐπαινεῖν.
[3] φιλαπόδημοι. ἡ δὲ πρόφασις τῆς πλάνης εἰς τὰ πετρώδη
χωρία ἐλθεῖν καὶ εἰς τὰ πηλώδη πολλάκις.
[4] μέλλουσιν.
[5] εἶτα ἑαυτὸν ἐπέχει.   [6] οὔκουν αὐτοῖς.

AELIAN

25. Ζηλοτυπίαν ζῴου φρονιμωτάτου καὶ μέντοι καὶ σωφρονεστάτου ἄνω που οἶδα εἰπών (πορφυρίων δὲ ἄρα τὸ ζῷον ἦν, εἴ τι παρ' ἡμῖν μνήμης ὑγιές ἐστιν[1]), ἤδη δὲ καὶ κυνίδιον μοιχοῖς πολέμιον καὶ ἔχθιστον τῷδε τῷ φύλῳ πέπυσμαι Σικελικόν. ὁ μὲν μοιχὸς ἔνδον ὑπεκέκρυπτο,[2] τῆς γυναικὸς τῆς μάχλου πυθομένης ἥκειν τὸν ἄνδρα ἐξ[3] ἀποδημίας, καὶ ὥς γε ᾤετο σκέπης ἐν καλῷ ἦν·[4] οἱ γάρ τοι τῶν οἰκετῶν δεκασθέντες, ὅσοι γοῦν τῇ δεσποίνῃ τὸ κακὸν συναπέκρυπτον (ἦσαν δὲ ἄρα ὅσοι κατόπτρων καὶ μύρων ἐπιστάται, φησὶν Εὐριπίδης), καὶ οἱ θυρωροὶ δέ εἶτα μέντοι θαρρεῖν ἐποίουν τὸν τῆς εὐνῆς κλῶπα. οὐ μὴν ἀπήντησε ταύτῃ ταῦτα, ἐπεὶ καὶ πολλοῦ δεῖ· τὸ γάρ τοι κυνίδιον[5] ὑλακτεῖ τε ἅμα καὶ μέντοι καὶ ταῖς θύραις τοὺς πόδας προσαράττει, ὡς ἐκπλῆξαι τὸν δεσπότην καὶ συμβαλεῖν ἐκ τοῦ δρωμένου κακὸν εἶναί τι ὑπολανθάνον. καὶ οἷα εἰκὸς[6] τὰς θύρας ἐκβαλὼν τὸν μοιχὸν καταλαμβάνει. καὶ εἶχε ξίφος ἐκεῖνος, καὶ νύκτα ἀνέμενεν, ἵνα ἀποκτείνῃ τῆς οἰκίας τὸν δεσπότην καὶ τὴν προειρημένην ὑπογήμῃ γυναῖκα.

26. Σοφὰ δὲ αἰγῶν ἐστι καὶ ἐκεῖνα. πτύελον ἀνθρώπου θανατηφόρον εἶναι ζῴῳ ἑτέρῳ καλῶς ἴσασι καὶ φυλάττονται, ὥσπερ οὖν καὶ ἡμεῖς πειρώμεθα ἀποδιδράσκειν ὅσα ἀνθρώπῳ κακόν ἐστιν, εἴπερ οὖν ἀπογεύσαιτο αὐτῶν. ἤδη μέντοι

[1] εἴ ... ἐστιν] εἴ τι μεντοι καὶ παρ' ἡμῖν μνήμης ὑγιοῦς ἀγαθόν ἐστιν.
[2] ὑπεκρύπτετο.
[3] Jac: ὡς ἐξ.   [4] εἶναι.

25. I know that I have somewhere earlier on [a] Lap-dog and
spoken of jealousy on the part of an animal not only adulterer
extremely prudent but also extremely continent:
it was, if my memory is sound, the Purple Coot.
And I have now heard of a Lap-dog in Sicily that was
the enemy of adulterers and a bitter foe to all of
that class. The adulterer had concealed himself
indoors, the lecherous woman having heard that her
husband was returning from a journey; and the man
was, as he supposed, well-situated for a hiding-place:
for the servants, or those who were in league with
their mistress to conceal the crime (there were ' such
as were stewards of mirrors and of perfumes,' as
Euripides says [Or. 1112]),[b] and the doorkeepers too
had been bribed, and this made the adulterer bold.
However matters did not turn out as intended; far
from it. For the Lap-dog kept barking and even
scratching with its paws at the door in such a way as
to alarm the master and to cause him by its action to
guess that there was some mischief lurking. So
naturally enough he threw open the door and caught
the adulterer. The man had a sword and was waiting
till night fell so that he might kill the master of the
house and thereupon marry the aforesaid woman.

26. Here is another example of the cleverness of The Goat
Goats. They know full well that human spittle is and human
deadly to other animals and they keep away from it, spittle
just as we also try to avoid anything that would
injure a man were he to taste of it. Indeed it has

[a] See 3. 42.
[b] Our texts of Euripides have οἴους ἐνόπτρων . . . ἐπιστάτας.

---

[5] κυνίδιον ἔνθα ὁ μοιχὸς ἦν.   [6] εἰκὸς δείσας.

AELIAN

τις καὶ ἄνθρωπος ἀγνοῶν καὶ λαθών ⟨τι κακὸν⟩[1]
κατέπιεν, αἱ δὲ αἶγες, οὐκ ἂν αὐτὰς λάθοι τὸ[2]
προειρημένον. ἀποκτείνειν δὲ καὶ τὰς θαλαττίας
σκολοπένδρας τὸ αὐτὸ δήπου πτύελον δεινότατόν
ἐστι. μέλλουσα δὲ ἡ αἲξ ἀποσφάττεσθαι σαφῶς
οἶδε. καὶ τὸ μαρτύριον, οὐκ ἂν ἔτι τροφῆς
προσάψαιτο. οὐκ ἀξιοῖ δὲ προβάτων οὐραγεῖν,
ἀλλὰ ἡγεῖσθαι[3] αὐτὴν δεῖν καὶ ἐκ τῆς βαδίσεως
ὁμολογεῖ. προθεῖ γοῦν ἐκείνων, καὶ μέντοι καὶ
αὐτῶν τῶν αἰγῶν[4] ὁ τράγος, τῷ γενείῳ θαρρῶν
καὶ κατά τινα φύσιν θαυμαστὴν τοῦ θήλεος
προκρίνων τὸ ἄρρεν.

27. Εὐπειθέστατα[5] δὲ ἄρα τῶν ζῴων τὰ
πρόβατα ἦν καὶ ἄρχεσθαι φύσει πεπαιδευμένα.
ὑπακούει γοῦν καὶ τῷ νομεῖ καὶ τοῖς κυσί, καὶ
μέντοι καὶ ἔπεται[6] ταῖς αἰξί. φιλεῖ δὲ καὶ
ἄλληλα ἰσχυρῶς, καὶ ὑπό γε τῶν λύκων ἐπιβου-
λεύεται ἧττον· οὐ γὰρ πλανᾶται ἰδίᾳ ἕκαστον,
οὐδὲ μὴν ἀπὸ τοῦ συννόμου σχίζεται, ὥσπερ οὖν
αἱ αἶγες. λέγουσι δὲ Ἄραβες ὅτι ἄρα τὰ παρ'
αὑτοῖς ποίμνια πιαίνεται ὑπὸ μουσικῆς μᾶλλον
ἢ ὑπὸ τοῦ χιλοῦ. τῶν δὲ ἁλμυρῶν ἐσθίει ἥδιον·
ποιεῖται γὰρ τὴν τοιαύτην τροφὴν ποτοῦ ὄψον.
τά γε μὴν πρόβατα κἀκεῖνο οἶδεν, ὅτι αὐτοῖς ὁ
βορρᾶς καὶ ὁ νότος συμμάχονται πρὸς τὸ τίκτειν
οὐ μεῖον τῶν ἀναβαινόντων αὐτὰ κριῶν· οἶδε δὲ
καὶ τοῦτο, ὅτι ἄρα ὁ μὲν βορρᾶς ἀρρενοποιός
ἐστιν, ὁ δὲ νότος θηλυγόνος εἶναι πέφυκε· καὶ
ἐὰν δέηται τοῦδε τοῦ ἐκγόνου ἢ τοῦδε ὀχευομένη

---

[1] ⟨τι κακὸν⟩ add. H. [2] Reiske: τὸ ἤδη.
[3] ἡγεῖσθαι οὖν. [4] αὐτὰς τὰς αἶγας.

138

happened before now that a man has in his ignorance and unconsciously swallowed some poison; but as to Goats, the aforesaid spittle would never take them unawares. And doubtless the same spittle is most effective at killing even sea-scolopendras. A Goat that is destined for slaughter is well aware of it: witness the fact that it will no longer touch food. And a Goat disdains to bring up the rear of a flock of sheep, but must take the lead, and proclaims it by its gait. At any rate she walks ahead of them, and the He-goat of the She-goats as well: his beard gives him confidence, and by some mysterious natural instinct he sets the male above the female.

27. It seems that Sheep are in fact the most The Sheep readily obedient of animals and have been taught by Nature to submit to rule. At all events they give heed to the shepherd and his dogs, and they even follow goats. Also they are devoted to one another and consequently less exposed to the attacks of wolves. For a Sheep does not wander away by itself, nor yet does it separate itself from its fellow, as goats do. The Arabians maintain that their flocks grow fat upon music rather than upon fodder. They like eating saline things, because they add a flavour to their drink. Moreover Sheep know this too, viz that the north wind and the south wind, no Winds promote fertility less than the rams which mount them, are their allies in promoting fertility. And this also they know, that whereas the north wind tends to produce males, the south wind produces females. And a Sheep that is being covered faces in this direction or in that

---

[5] εὐπειθέστατα A, εὐπειθέστερα L.   [6] ἕπονται.

AELIAN

ἢ οἷς, πρὸς τὸν ἀπέβλεψεν ἢ πρὸς τόν. Ἀχιλλεὺς
μὲν οὖν ἵνα ὁ φίλος αὐτῷ κείμενος ἐπὶ τῆς πυρᾶς
καυθῇ, καὶ εὐχῆς ἐδεῖτο, καὶ ἡ Ἶρις παρεκάλει
τοὺς ἀνέμους αὐτῷ, ὦ καλὲ Ὅμηρε, καὶ ὑπισχ-
νεῖτο[1] ἥκουσιν ἱερουργίαν οἱονεὶ μισθόν, καὶ ὁ
τοῦ Νεοκλέους δὲ Ἀθηναίους ἐδίδασκε θύειν τοῖς
πνεύμασιν· αἱ δὲ οἷς ἀπραγμόνως τοὺς ἀνέμους
ἐς ὠδῖνα τὴν σφετέραν ὑπηρέτας ἑτοίμους καὶ
ἀκλήτους ἔχουσι. σκοποὶ δὲ ἄρα τούτων εἰσὶ καὶ
οἱ ποιμένες ἀγαθοί. ὅταν γοῦν ὁ νότος πνέῃ,
τότε τοὺς κριοὺς ἐπὶ τὰς οἷς ἄγουσιν, ἵνα ἡ γονὴ[2]
θηλυγόνος ᾖ αὐτοῖς μᾶλλον.

28. Ὅτε τὸν Ἰκάριον ἀπέκτειναν οἱ προσήκοντες
τοῖς πρῶτον πιοῦσιν οἶνον καὶ ἐς ὕπνον ἐμπεσοῦσιν,
οὐκ εἰδότες πω μὴ θάνατον εἶναι τὸ πραχθὲν ἀλλὰ
οἰνηρὸν κάρον, ἐνόσησαν οἱ κατὰ τὴν Ἀττικήν,
ἐμοὶ δοκεῖν τοῦ Διονύσου τιμωροῦντος τῷ πρώτῳ
γεωργῷ τῶν ἑαυτοῦ φυτῶν καὶ πρεσβυτάτῳ. ὁ
γοῦν Πύθιος ἔχρησεν, εἰ βούλονται τυχεῖν σωτηρίας, Ἰκαρίῳ θύειν καὶ Ἠριγόνῃ τῇ τούτου παιδὶ
καὶ τῷ κυνὶ τῷ ᾀδομένῳ, ὅτι ἄρα δι' ὑπερβολὴν
εὐνοίας τῆς πρὸς τὴν δέσποιναν βιῶναι μετ'
αὐτὴν οὐκ ἔγνω. παίζει δὲ Εὐριπίδης λέγων

χρηστοῖσι δούλοις συμφορὰ τὰ δεσποτῶν
κακῶς πίτνοντα καὶ φρενῶν ἀνθάπτεται.

[1] ὑπισχνεῖται.   [2] ἐπιγονή.

140

according as it wants a male or a female offspring.
So Achilles needed to pray in order that his friend
lying on the pyre might be burned, and Iris summoned the winds for him, O noble Homer [*Il.* 23.
194 ff.], promising them, if they came, a sacrifice by
way of reward. And the son of Neocles [a] taught the
Athenians to sacrifice to the Winds. But Sheep
without any trouble have them ready and un-
summoned to help them to pregnancy. And so
shepherds also are good at looking out for them.
At any rate when the south wind blows they put
the rams to the Sheep, in order that their offspring
may preferably be female.

28. When Icarius was slain by the relatives of
those who, after drinking wine for the first time, fell
asleep (for as yet they did not know that what had
happened was not death but a drunken stupor), the
people of Attica suffered from a disease, Dionysus
thereby (as I think) avenging the first and the most
elderly man who cultivated his plants.[b] At any rate
the Pythian oracle declared that if they wanted to
be restored to health they must offer sacrifice to
Icarius and to Erigone his daughter and to her hound
which was celebrated for having in its excessive love
for its mistress declined to outlive her. Euripides
is not serious when he says [*Med.* 54]

' Good slaves are grieved and their hearts are
gripped when things go ill with their masters,'

Icarius and
the Hound
of Erigone

[a] Themistocles. Cp. Hdt. 7. 179.
[b] Icarius was instructed by Dionysus in the cultivation
of the vine. Wine and its possible effects were till then
unknown.

ποῦ γὰρ ἄνθρωπος ἐπὶ τῷ δεσπότῃ τέθνηκε,
κυνὸς δούλου δράσαντος αὐτό;

29. Ἰδίαν δὲ ἄρα κυνῶν ἐς τοὺς τρέφοντας
εὔνοιαν [1] καὶ ἐκεῖνο μαρτυρεῖ. Κολοφώνιος ἀνὴρ
παραγίνεται ἐς τὴν Τέων συνωνησόμενός τινα·
καὶ γὰρ ἦν ἐμπορικός, καὶ τὴν ἐκ τῶν ὠνίων
καπηλείαν τε καὶ μεταβολὴν πρόσοδον εἶχεν.
ἀργύριον δὲ ἐπήγετο καὶ οἰκέτην καὶ κύνα, ἔφερε
δὲ τὸ ἀργύριον ὁ δοῦλος. ἐπεὶ δὲ πρὸ ὁδοῦ ἦσαν,
ὁ οἰκέτης ἐξετράπετο· ἤπειγε γάρ τι αὐτὸν τῶν
κατὰ φύσιν, ἠκολούθησε δὲ καὶ ὁ κύων. τὸ
τοίνυν φασκώλιον ἀνέπαυσεν ὁ νεανίας, καὶ
ἀνελέσθαι πάλιν οὐκ ἐνενόησεν, ἀλλὰ ᾤχετο
ἀπιών· ὁ δὲ κύων ἑαυτὸν κατακλίνας ἐπὶ τῷ
ἀργυρίῳ ἔμενεν ἥσυχος. ἐλθόντες δὲ ἐς τὴν
Τέων ὅ τε δεσπότης καὶ ὁ οἰκέτης εἶτα μέντοι
ἄπρακτοι ἐπανῆλθον, ὅτου ὠνήσωνται [2] οὐκ ἔχον-
τες· τὴν αὐτήν γε μὴν ἐκτρέπονται πάλιν ἔνθα ὁ
οἰκέτης ἀπέλιπε τὸ βαλάντιον, καὶ καταλαμ-
βάνουσι τὸν σφέτερον κύνα ἐπικείμενον αὐτῷ καὶ
μόλις ἐμπνέοντα ὑπὸ τοῦ λιμοῦ. ὁ δὲ ὡς εἶδε τὸν
δεσπότην καὶ τὸν ὁμόδουλον, ἑαυτὸν ἀποκλίνας
τοῦ φασκωλίου, κατὰ τὸν αὐτὸν χρόνον καὶ τὴν
φρουρὰν καὶ τὴν ψυχὴν ἀφῆκεν. οὔκουν οὐδὲ
Ἄργος ὁ κύων μυθοποίημα ἦν, ὦ θεῖε Ὅμηρε,
σόν, οὐδὲ κόμπος ποιητικός, εἴπερ οὖν καὶ τῷ
Τηΐῳ ταῦτα ἀπήντησεν ὅσα [3] προεῖπον.

---

[1] ἴδιον . . . τῆς τῶν κ. εἰς τοὺς τ. εὐνοίας.
[2] Ward: ὀνήσονται.
[3] ὅσα καί.

for where is the man who died in consequence of his master's death,[a] although this is what a dog—a slave —did?

29. Now here is a further testimony to the peculiar goodwill which Dogs bear towards those who keep them. A man of Colophon arrived at Teos with the intention of buying up certain articles, for he was a merchant and made his profits by retailing and exchanging his purchases. And he brought with him money, a servant, and a Dog; and the slave carried the money. But on the journey the servant stepped aside—he had a pressing call of nature—and the Dog followed him. Now the young man put down the money-bag and forgot to pick it up again and went on his way. But the Dog lay down on the money and remained quietly there. And when the master and his servant arrived at Teos they returned without doing any business, not having the means to make purchases. They turned aside however along the same road where the servant left the purse and found their own Dog lying upon it and hardly breathing from starvation. But directly the Dog saw its master and its fellow-slave it moved off the money-bag and in the same instant gave up its post of guardian and its life.

A faithful Hound

So then even the dog Argus,[b] O divine Homer, was no fiction of yours, no poetical exaggeration, if indeed the events which I have narrated really befell the man of Teos.[c]

[a] See 6. 25 fin.
[b] Homer *Od.* 17. 291; cp. Ael. *NA* 4. 40.
[c] This is a slip; the man came from Colophon.

AELIAN

30. Γένος καρκίνων ἔστι, καὶ πετηλίαι [1] ὄνομα αὐτοῖς. τῶν μὲν οὖν ἄλλων ἰδεῖν εἰσι λευκότεροι, τίκτονται δὲ ἐν τῷ πηλῷ. δείσαντες δὲ οὗτοι καὶ πέτονται· ἔχουσι γὰρ πτερύγια μικρά, ἅπερ οὖν αὐτοὺς ἡσυχῇ μετεωρίζει τε καὶ ἐλαφρίζει. βαδίζοντες μὲν οὖν ἥκιστα τούτων δέονται, φοβηθέντες δὲ ἔχουσί τινα ἐπικουρίαν οὐ πάνυ τι [2] καρτεράν· ἁλίσκονται γὰρ οὔτε ὑψηλοὶ πετόμενοι, οὔτε μὴν μετεωροπορεῖν οἷοί τε ὄντες. τούτους τοι [3] τοὺς καρκίνους ἐσθίουσί τινες· φασὶ δὲ καὶ ἰσχίου πόνῳ ἀγαθὸν εἶναι, εἴ τις φάγοι ἀλγῶν.

31. Αἱ δὲ καρκινάδες τίκτονται μὲν γυμναί, τὸ δὲ ὄστρακον ἑαυταῖς αἱροῦνται ὡς οἰκίαν οἰκῆσαι τὴν ἀρίστην. ὑποδύονται δὲ καὶ πορφύρας ὄστρακον κενῷ περιτυχοῦσαι καὶ στρόμβου. καὶ ἐς ὅσον μὲν αὐτὴν στέγει, χαίρει τῇ καταγωγῇ· ἐὰν δὲ αὐξήσῃ τὴν σάρκα, ἐς ἄλλον μετοικίζεται οἶκον. περιτυγχάνει δὲ τοῖς προειρημένοις πολλοῖς.[4]

32. Οἱ δὲ στρόμβοι καὶ βασιλέα ἔχουσι, καὶ μάλα γε εὐπειθῶς ἄρχονται. καὶ ὁ μὲν βασιλεὺς οὗτος μεγέθει μέν ἐστι μέγιστος, κάλλιστος δὲ κάλλος.[5] καὶ εἰ μὲν εἴη οἱ καταδῦναι λῷον, ἔδρασε τοῦτο πρώτιστος· εἰ δὲ ἀναδῦναι, καὶ τοῦδε ἄρχει· μετακινουμένῳ δὲ ἕπονται καὶ οἱ λοιποί. ὅστις δ' ἂν ἕλῃ τόνδε τὸν βασιλέα, ὅτι

[1] πηλαῖοι H.   [2] πάντῃ.
[3] μέντοι.
[4] πολλοῖς καὶ κενοῖς ὀστράκοις.

30. There is a species of Crab called *Peteliae* The 'Flying (flyers).[a] They are paler in appearance than other Crab' crabs and are generated in the mud. And when scared they actually fly, for they possess tiny wings which give them a slight lift and lessen their weight. When walking however they have no need of them, but when frightened these wings afford them a certain not very considerable assistance, for as they do not fly high and are unable to travel through the air, they are caught; and some people eat these crabs. And they do say that they are good for sciatica if eaten during an attack.

31. Hermit-crabs are born without a shell and The Hermit select for themselves the shell that makes the best crab house for them to live in. They even enter the shell of the purple-shellfish if they can find one empty, and the shell of the whelk. And so long as it is large enough to cover them they are satisfied with their lodging. But if their body grows they migrate to another dwelling, and they find quantities of such shells.

32. Whelks even have a King and submit most The King obediently to his rule. And this King exceeds Whelk all others in size and beauty. And if it is expedient for him to sink, he is the first to do so; if to come up again, he leads the way; and when he moves to another place the rest follow him. The man who succeeds in catching this King knows well that his affairs will prosper. Moreover if a man

[a] Thompson, *Gk. fishes*, s.v. καρκίνος, ' they suggest the little sand-hoppers,' which leap about but cannot fly.

---

[5] κάλλει Ges.

AELIAN

ἄμεινον πράξει καλῶς οἶδε. καὶ μέντοι καὶ εἴ τις
ἴδοι θηρώμενον, εὐθυμότερος ἀπῆλθεν, ὥς τι
χρηστὸν καὶ ἐκεῖνος ἕξων. ἐν Βυζαντίῳ δὲ καὶ
ἆθλον πρόκειται τῷ θηράσαντι τὸν προειρημένον·
διδόασι δὲ οἱ συνθηραταὶ δραχμὴν Ἀττικὴν
ἕκαστος τῷ ἑλόντι, καὶ τό γε ἆθλον τοῦτό ἐστιν.

33. Τοὺς ἐχίνους ὁ κλύδων κυλίων ἐς τὰ ἔξω
καὶ προσαράττων τῷ ξηρῷ τῆς θαλάττης βιαιότατα
ἐκβάλλει. τοῦτο τοίνυν ἐκεῖνοι δεδιότες, ὅταν
αἴσθωνται φρίττον τὸ κῦμα καὶ μέλλον[1] ἁδρό-
τερον ὑπανίστασθαι, ταῖς ἀκάνθαις ἀναιροῦνται
λιθίδια, ὅσα εὔκολά ἐστι φέρειν αὐτοῖς, καὶ
ἔχουσιν ἕρμα, καὶ οὐ ῥᾳδίως κυλίονται, οὐδὲ
πάσχουσιν ὃ δεδοίκασιν.

34. Ἡ πορφύρα λίχνον ἐστὶν ἰσχυρῶς, καὶ
ἔχει γλῶτταν προμηκεστέραν, καὶ διείρει διὰ
παντὸς οὗπερ ἂν καὶ δύνηται, καὶ διὰ ταύτης
ἕλκει ὅσα ἐσθίει, καὶ διὰ ταύτης δὲ ἁλίσκεται.
καὶ ὁ τρόπος τῆς θήρας ἐκεῖνός ἐστι. διαπλέκεται
κυρτὶς μικρὰ μέν, πυκνὴ δέ· καὶ ἐντὸς ἔχει
στρόμβον,[2] καὶ διείρται οὗτος ἐν τῇ κυρτίδι μέσῃ.
ἀγώνισμα οὖν τῇ πορφύρᾳ διατεῖναι τὴν γλῶτταν
ἐστι καὶ ἐφικέσθαι αὐτοῦ· καὶ ἀνάγκη πᾶσαν
αὐτὴν προβάλλειν, εἰ μέλλοι μὴ ἁμαρτήσεσθαι
οὗ γλίχεται.[3] καὶ ἐμβαλοῦσα τὴν γλῶτταν ἐκμυζᾷ,
εἶτα διῴδησεν αὐτῇ[4] ἡ γλῶττα ὑπὸ πλησμονῆς,
καὶ ἐξελκύσαι ἀδύνατός ἐστιν αὖθις. μένει τοίνυν

[1] μέλλον πνεῦμα.
[2] στρόμβον τῇ πορφύρᾳ τὸ δέλεαρ.

146

sees a King Whelk being caught, he goes away in
more cheerful spirits, imagining that he too will have
some good fortune. And at Byzantium a prize is
offered for the man who catches the aforesaid fish:
each of his fellow-anglers contributes an Attic
drachma to the one who catches it, and that is the
prize.

33. Waves roll Sea-urchins out of their haunts, The Sea-
dash them on to the dry land, and hurl them with the urchin
utmost violence out of the sea. So for fear of this,
whenever these creatures perceive the waves rippling
and beginning to swell to greater violence, they pick
up with their prickles as many pebbles as they can
carry and have some ballast, so that they are not
easily rolled about and do not undergo what they
dread.

34. The Purple Shellfish is exceedingly gluttonous The Purple
and possesses an unusually long tongue which it Shellfish
thrusts through everything that it can. By this
means it draws in whatever it eats, and by this means
it is caught. And the way in which it is hunted is
this : men weave a weel, small and of close texture,
and inside there is a whelk and this has been inserted in the centre of the weel. Now the Purple
Shellfish struggles to extend its tongue to the utmost
and to reach its prey. And it is forced to project
the whole length if it is not to miss what it longs for.
And when it has inserted its tongue it sucks until the
tongue is so swollen with surfeiting that the creature
cannot withdraw it again. So there it remains

---

[3] γλίχεται λαβεῖν.　　　[4] ἑαυτήν.

ἁλοῦσα, καὶ ὁ πορφυρεὺς αἰσθόμενος ἐθήρασε
δεύτερος τὴν ὑπὸ τῆς λιχνείας προῃρημένην.[1]

35. Σκολόπενδρα θαλάττιον θηρίον, καὶ τῷ
χερσαίῳ ⟨ὅσα⟩[2] ἰδεῖν[3] ὁμοιότατόν ἐστιν· εἰ δὲ
αὐτῆς προσάψαιτο[4] ἀνθρωπεία σάρξ, ὀδαξᾶταί τε
παραχρῆμα καὶ κνησιᾷ,[5] καὶ πάσχει τοιαῦτα,
ὁποῖα καὶ ὑπὸ τῆς βοτάνης, ἣν καλοῦσι κνίδην.
ποιοῦσι δὲ καὶ ἀκαλῆφαι κνησμονήν, ἀλλὰ οὔπω
τοσοῦτον. εἰσὶ δὲ ἐδώδιμοι μᾶλλον διελθούσης
ἰσημερίας αἱ ἀκαλῆφαι.

36. Ὅταν ὑπὸ τῶν θηρατῶν ὡς ἐν πολέμῳ
στρατιῶται τραπῶσιν οἱ ἐλέφαντες καὶ ἐς φυγὴν
ὁρμήσωσιν, οὐ φεύγουσι διῃρημένοι οὐδὲ καθ᾽
ἕνα, ἀλλὰ κοινῇ, καὶ πιέζουσιν ἀλλήλους τῶν
συννόμων ἐχόμενοι. καὶ κύκλῳ μὲν οἱ νέοι, ὡς
εἰπεῖν, τὸ μαχιμώτατον, ἐν μέσῳ δὲ οἱ γεγηρα-
κότες καὶ αἱ μητέρες, ὑπὸ ταύταις δὲ τὰ πώλια,
ἑκάστη τὸ ἴδιον ἀποκρύπτουσα· καὶ ὁρῶνταί γε
οἵδε οἱ μικροὶ σπανιώτατα. ἀθρόους δὲ αὐτοὺς
ἐὰν θεάσωνται καὶ λέοντες, ἢ φεύγουσι προ-
τροπάδην ἢ ἄλλος ἄλλῃ κατέπηξαν ὡς νεβροί,
τοὺς ἐλέφαντας οἱ τέως φοβεροὶ καὶ ἐκπληκτικοὶ
καταδείσαντες. οὐκ ἀνθίσταται δὲ τοῖς διώκουσιν
ἐλέφας, εἰ μή ποτε ἄρα ὑπὲρ τῶν τέκνων καὶ τῶν
νοσούντων. ἐνταῦθα δὲ ἄμαχός ἐστιν.

---

[1] Jac: ὑπὸ τῆς λ. τὴν προειρημένην.
[2] ⟨ὅσα⟩ add. H.
[3] Ges: εἰπεῖν.
[4] προσάψεται.
[5] Reiske: κνησιᾷ or κνησίει.

caught, and the fisherman observing this, catches
for the second time what has already been caught by
its own gluttony.

35. The Scolopendra [a] is a creature of the sea and
looks exactly like the land-scolopendra (centipede).
And if a man's skin come in contact with it, he at
once feels a stinging and irritation, and has the same
kind of pain as from the plant they call the nettle.
And Sea-anemones also produce an itching, but not
so violent; and they are better to eat when the
equinox is past.

The Sea-scolopendra

36. Whenever Elephants are routed by hunters
and begin to stampede like soldiers in war, they do
not scatter and take to flight singly but in a herd,
and they press against one another as they cling to
their fellows. Round the outside are the young
animals, the most pugnacious, you might say; in the
middle the old elephants and the mothers, and
beneath them the baby elephants, each mother
hiding her own. And these little ones are very
seldom to be seen. And even lions, if they catch
sight of them herded together, lions which up to
that moment have inspired fear and consternation,
either flee at full speed or cower down one here
and another there, like fawns, in terror of the
Elephants.

A stampede of Elephants

The Elephant does not turn and face its pursuers,
unless it be to protect its young or sick ones: then it
is irresistible.

[a] Not certainly identified; thought by some to be an
annelid worm, e.g. *Nereis*, but for the fact that this does not
sting.

37. Πώρου τοῦ τῶν Ἰνδῶν βασιλέως ὁ ἐλέφας
ἐν τῇ πρὸς Ἀλέξανδρον μάχῃ τετρωμένου πολλὰ
ἡσυχῇ καὶ μετὰ φειδοῦς τῇ προβοσκίδι ἐξῄρει τὰ
ἀκόντια, καὶ μέντοι καὶ αὐτὸς τετρωμένος πολλὰ
οὐ πρότερον εἶξε πρὶν ἢ συνεῖναι ὅτι ἄρα ὁ δεσπότης
αὐτῷ διὰ τὴν ῥοὴν τοῦ αἵματος τὴν πολλὴν
παρεῖται καὶ ἐκθνῄσκει. οὐκοῦν ἑαυτὸν ὑπέκλινε,
καὶ ὀκλάσας ἔμεινεν, ἵνα μὴ ἄνωθεν πεσὼν ὁ
Πῶρος εἶτα μέντοι κακωθῇ τὸ σῶμα ἐπὶ μᾶλλον.

38. Ὑρκανοῖς καὶ Μάγνησιν οἱ κύνες συνεστρα-
τεύοντο, καὶ ἦν καὶ τοῦτο συμμαχικὸν ἀγαθὸν
αὐτοῖς καὶ ἐπικουρικόν. συστρατιώτην δέ τις
Ἀθηναῖος ἐν τῇ μάχῃ τῇ ἐν Μαραθῶνι ἐπήγετο
κύνα, καὶ γραφῇ εἴκασται ἐν τῇ Ποικίλῃ ἑκάτερος,
μὴ ἀτιμασθέντος τοῦ κυνός, ἀλλὰ ὑπὲρ τοῦ
κινδύνου μισθὸν εἰληφότος ὁρᾶσθαι σὺν τοῖς ἀμφὶ
τὸν Κυνέγειρον καὶ Ἐπίζηλόν τε καὶ Καλλίμαχον.
ἔστι δὲ [1] καὶ οὗτοι καὶ ὁ κύων Μίκωνος [2] γράμμα.
οἱ δὲ οὐ τούτου, ἀλλὰ τοῦ Θασίου Πολυγνώτου
φασίν.

---

[1] εἰσὶ δέ *Schn.*   [2] *Meursius* : Νίκωνος.

---

[a] At the crossing of the Hydaspes (mod. Jhelum), 327 B.C.
[b] Porus survived to become the ally of Alexander.

37. When Porus the King of the Indians had Porus and his
Elephant received many wounds in the battle ^a against Alexander, his Elephant proceeded with its trunk to pick out the javelins gently and cautiously; and in spite of its own numerous wounds it did not pause until it knew that its master was collapsing through copious loss of blood and was swooning.^b And so it lay down beneath him and remained crouching to prevent Porus from falling from a height and damaging his body even more.

38. Their hounds used to accompany the people The Dog as
companion of Hyrcania and Magnesia to war, and in fact these allies were an advantage and a help to them. An Athenian took with him a Dog as fellow-soldier to the battle of Marathon, and both are figured in a painting in the *Stoa Poecile*,^c nor was the Dog denied honour but received the reward of the danger it had undergone in being seen among the companions of Cynegirus,^d Epizelus, and Callimachus. They and the Dog were painted by Micon,^e though some say it was not his work but that of Polygnotus ^f of Thasos.

^c ' Painted Porch ' : a series of colonnades surrounding the Agora at Athens, decorated with paintings of episodes from the Persian wars.

^d Brother of the poet Aeschylus, famed for his bravery at Marathon, 490 B.C.—Epizelus (or Polyzelus) blinded at Marathon by a remarkable vision; see Hdt. 6. 117.—Callimachus, Athenian Polemarch, distinguished himself at Marathon and died there in a heroic attack on the Persian fleet.

^e Athenian painter and sculptor, 5th cent. B.C., contemporary of Polygnotus; both artists painted frescoes in the Stoa Poecile.

^f Polygnotus of Thasos, lived and worked at Athens, second half of 5th cent. B.C. One of the foremost of Greek painters.

AELIAN

39. Ὅσοι λέγουσι θῆλυν ἔλαφον κέρατα¹ οὐ
φύειν, οὐκ αἰδοῦνται τοὺς τοῦ ἐναντίου μάρτυρας,
Σοφοκλέα μὲν εἰπόντα

> νομὰς τέ τις κερούσσ᾽ ἀπ᾽ ὀρθίων πάγων
> καθεῖρπεν ἔλαφος·

καὶ πάλιν

> ἄρασα μύξας . . . καὶ κερασφόρους
> στόρθυγγας εἷρφ᾽ ² ἔκηλος.

καὶ ταῦτα μὲν ὁ τοῦ Σοφίλλου ἐν τοῖς Ἀλεάδαις·
ὁ δὲ Εὐριπίδης ἐν τῇ Ἰφιγενείᾳ

> ἔλαφον δ᾽ Ἀχαιῶν χερσὶν ἐνθήσω φίλαις
> κερούσσαν, ἣν σφάζοντες αὐχήσουσι σὴν
> σφάζειν θυγατέρα.

ἐν δὲ τοῖς Τημενίδαις τὸν Ἡράκλειον ἆθλον
κέρατα ἔχειν ὁ αὐτὸς Εὐριπίδης φησί, τὸν τρόπον
τόνδε ᾄδων

> ἦλθεν δ᾽
> ἐπὶ χρυσόκερων ἔλαφον, μεγάλων
> ἄθλων ἕνα δεινὸν ὑποστάς,
> κατ᾽ ἔναυλ᾽ ³ ὀρέων ἀβάτους ἐπί τε
> λειμῶνας ποίμνιά τ᾽ ἄλση.

ὁ δὲ Θηβαῖος μουσοποιὸς ἔν τινι τῶν ἐπινικίων
ὑμνεῖ λέγων

> Εὐρυσθέος ἔντυ᾽ ἀνάγκα πατρόθεν
> χρυσόκερων ἔλαφον θήλειαν ἄξονθ᾽.

¹ τὰ κέρατα.   ² Jac : εἷρπεν.   ³ Nauck : ἐναύλων.

---

[a] In neither of the extant plays on Iphigenia.
[b] The third ' Labour ' was to capture the Arcadian stag.
[c] Pindar.

152

39. Those who maintain that Hinds do not grow horns have no regard for witnesses to the contrary, none for Sophocles who says A horned Hind

' And down from the steep crags came roaming an antlered hind ' [*fr.* 89 P];

and again

' Lifting its nostrils . . . and the tynes of its antlers ⟨the hind⟩ moved on in peace ' [*ib.*].

This is what the son of Sophillus wrote in his *Aleadae*. And Euripides in his *Iphigenia* [a] says

' But I will place in the very hands of the Achaeans an antlered hind, which they will slay and boast they have slain thy daughter ' [*fr.* 857 N].

And the same Euripides says in his *Temenidae* that the ' Labour ' of Heracles [b] had horns, in the following verses:

' And he came in quest of the golden-horned deer, braving one fearful task in his mighty labours, over mountain haunts to meadows untrodden, and to groves where flocks graze ' [*fr.* 740 N].

And the Theban minstrel [c] in one of his *Epinician* odes sings thus:

' Necessity laid upon him by Eurystheus through his father urged him on to fetch the hind with the golden horns ' [Pind. *O.* 3. 28].[d]

[d] In consequence of an oath of his father Zeus, Heracles was forced to submit to the will of Eurystheus; see Hom. *Il.* 19. 95 ff. Ridgeway (*Early age of Greece*, 1. 360) considered the ' horned doe ' to be the reindeer of N Asia and Europe; it is the only kind of deer in which the female possesses horns.

καὶ Ἀνακρέων ἐπὶ θηλείας φησίν

οἷά τε νεβρὸν νεοθηλέα
γαλαθηνόν, ὅς τ' ἐν ὕλῃ κεροέσσης
ὑπολειφθεὶς ὑπὸ μητρὸς ἐπτοήθη.

πρὸς δὲ τοὺς μοιχῶντας τὸ λεχθὲν καὶ μέντοι καὶ
φάσκοντας δεῖν ἐροέσσης γράφειν ἀντιλέγει κατὰ
κράτος Ἀριστοφάνης ὁ Βυζάντιος, καὶ ⟨ἐμέ⟩[1] γε
αἱρεῖ τῇ ἀντιλογίᾳ.

40. Ἴδια δὲ ἄρα κυνῶν καὶ ἐκεῖνα εὐνοίας
ὑπερβολὴν πᾶσαν ἐκνενικηκότα. Πώλῳ μὲν[2] τῷ
τῆς τραγῳδίας ὑποκριτῇ ὁ κύων ὁ τρόφιμος
αὐτοῦ τεθνεῶτι καὶ καομένῳ ἑαυτὸν συγκατέπρησε
τῇ πυρᾷ ἐμπηδήσας. καομένῳ δὲ καὶ Μέντορι
σκύλακες Ἐρετρικαὶ ἑαυτὰς συγκατέπρησαν ἑκοῦ-
σαι κοινωνήσασαι τοῦ τέλους. Θεόδωρον δὲ
ἄνδρα ψαλτικὴν ἀγαθόν, τὸν μὲν ἐς τὴν σορὸν
ἐνέθεσαν οἱ προσήκοντες, κυνίδιον δὲ Μελιταῖον
ἑαυτὸ ἐνέβαλεν ἐς τὴν θήκην τοῦ νεκροῦ καὶ
συνετάφη. πέπυσμαι δὲ καὶ Αἰθιόπων εἶναι ἔθνος,
ἐν ᾧ βασιλεύει κύων, καὶ τῇ ἐκείνου ὁρμῇ πείθονται, κνυζωμένου τε ἴσασιν ὅτι μὴ θυμοῦται, καὶ
ὑλακτοῦντος τὴν ὀργὴν συνιᾶσι. τοῦτο εἴ τῳ
ἱκανὸς Ἕρμιππος τεκμηριῶσαι, μάρτυρά οἱ τοῦ

---

[1] ⟨ἐμέ⟩ add. H.   [2] μὲν οὖν.

And Anacreon says of the Hind

' Even as a new-born fawn unweaned, which,
when forsaken by its horned mother in the forest,
is affrighted ' [*fr.* 39 D].

Those who falsify the reading and go so far as to say
that we should write ἑροέσσης (for κεροέσσης) are
soundly refuted by Aristophanes of Byzantium ; and
I am convinced by his refutation.[a]

40. Now here are further instances afforded by
Dogs of loyalty unsurpassable. When Polus [b] the
tragic actor died and his body was burning,
the Dog which he had kept sprang on to the
pyre and was burned to death along with him.
When the body of Mentor [c] was burning, his Eretrian
Hounds of their own accord were burned to death
and shared his end. Theodorus,[d] an excellent
harp-player, was placed in the coffin by his relatives,
and his Maltese Lap-dog threw itself into the
receptacle and was buried along with him.   And I
have heard that there is a race of beings in Ethiopia
among whom a Dog is king, and they obey his
wishes : when he whimpers they know that he is in
a good temper, but when he barks they understand
that he is angry.   If Hermippus is in anyone's view
a competent authority, he should carry conviction

The Dog's devotion to its master

Dog as King

---

[a] In all the examples except that from Anacreon the
feminine can, as often, be taken as sexless = a deer.

[b] Polus, of Athens, 5th cent. B.C.   He excelled in Sopho-
clean parts.

[c] Perh. Mentor of Rhodes, 4th cent. B.C., mercenary soldier,
later general, in the Persian army.

[d] No harpist of this name is known.   The ' Theodorus '
mentioned in Ael. *VH* 12. 17 was a piper, *c.* 300 B.C.

λόγου ἐπαγόμενος Ἀριστοκρέωνα[1] πειθέτω· ἐμὲ
δὲ μὴ λαθὼν εἶτα ἐν καλῷ τῆς μνήμης ἀφίκετο.

41. Λακύδῃ τῷ περιπατητικῷ κτῆμα ἦν χηνός
τι χρῆμα θαυμάσιον. ἐφίλει γοῦν τὸν τροφέα
ἰσχυρῶς, καὶ βαδίζοντι μὲν συνεβάδιζε, καθημένου
δὲ ἀνεπαύετο, οὐκ ἀπελείπετο δὲ αὐτοῦ ἔμβραχυ.
ὅπερ καὶ ἀποθανόντα ὁ Λακύδης ἔθαψε καὶ
πάνυ φιλοτίμως, ὥσπερ οὖν ἢ υἱὸν ἢ ἀδελφὸν
ἐκεῖνος θάπτων. Πύρρῳ δὲ τῷ Ἠπειρώτῃ ἦν
ἐλέφας, ὅσπερ οὖν τὸν ἑαυτοῦ πωλευτὴν οὕτως
ἠγάπησεν, ὥστε[2] ἀποθανόντος ἐν Ἄργει τοῦ
Πύρρου, ἐκπεσόντος δὲ τοῦ ἐλαύνοντος, οὐ πρότερον ὑπέμεινεν ἀτρεμῆσαι καὶ ἡσυχάσαι πρὶν ἢ
ἀνασώσασθαι αὐτὸν[3] ἐκ τῶν πολεμίων καὶ ἐς τὸ
φίλιον μεταγαγεῖν.

42. Κακουργίαν δὲ ὀρέως Θαλῆς ὁ Μιλήσιος
ἠμύνατο, καταφωράσας πάνυ ἀπορρήτως. ἅλας
ἡμίονος ἦγε φόρτον, καί ποτε διὰ ποταμοῦ ἰὼν
κατὰ τύχην κατώλισθε καὶ περιετράπη. βραχέντες
οὖν οἱ ἅλες κατετάκησαν, καὶ κοῦφος ὁ ὀρεὺς
γενόμενος ἥσθη· καὶ συνιδὼν ὁπόσον τὸ μεταξὺ
ἦν τοῦ μόχθου καὶ τῆς ῥᾳστώνης, τοῦ λοιποῦ τὴν
τύχην διδάσκαλον ποιησάμενος, ὃ πρότερον ἄκων
ἔπαθεν, εἶτα μέντοι τοῦτο εἰργάζετο ἑκών. ἄλλην
δὲ τῷ ὀρεωκόμῳ[4] ἐλαύνειν καὶ ἔξω τοῦ ποταμοῦ

---

[1] *Schn* : Ἀριστοκλέωνα.　　　　[2] ὡς.
[3] τοῦτον αὐτόν.　　　　[4] ὀρεοκόμῳ.

---

[a] Head of the 'Middle Academy,' *c.* 240–215 B.C.; his
copious writings have perished. The above story may well
be spiteful gossip.

from having cited Aristocreon as a witness to his story. This has not escaped my notice and it was opportune that I remembered it.

41. Lacydes the peripatetic philosopher [a] possessed a remarkable goose. At any rate it was deeply devoted to its keeper: when he went for a walk, it went too; when he sat down, it would remain still and would not leave him for a moment. And when it died Lacydes gave it a most costly funeral as though he were burying a son or a brother. And Pyrrhus of Epirus had an elephant which was so fond of its master that when Pyrrhus was killed at Argos,[b] though its driver had fallen off, it would not halt and remain still until it had rescued him from the hands of the enemy and had brought him back to his friends.

Lacydes and his Goose

Pyrrhus and his Elephant

42. Thales of Miletus [c] repaid the malice of a Mule which he detected with great subtlety. A Mule was carrying a load of salt and once, when crossing a river, by accident stumbled and was upset. Consequently the salt was soaked and melted, and the Mule was delighted to be eased of its burden. So the Mule realising the difference between labour and relaxation took a lesson for the future from its accident and deliberately contrived what before it had unwillingly undergone. It was impossible for the muleteer to drive it by any other road away from

Thales and his Mule

[b] Pyrrhus, King of Epirus, struck on the head by a tile and killed while fighting at Argos, 272 B.C.
[c] Thales, 7th/6th cent. B.C., one of the Seven Sages of Greece, philosopher and mathematician.

AELIAN

ἄπορον ἦν. τοῦτό τοι διηγουμένου ὁ Θαλῆς ὡς ἐπύθετο, σοφίᾳ ἀμύνασθαι τῆς κακουργίας τὸν ὀρέα ᾠήθη δεῖν, καὶ προστάττει ὑπὲρ τῶν ἁλῶν σπογγιαῖς καὶ ἐρίοις ἐπισάξαι αὐτόν. ὁ δὲ τὴν ἐπιβουλὴν οὐκ εἰδὼς κατὰ τὸ σύνηθες ὤλισθε, καὶ ἀναπλήσας ὕδατος τὰ ἐπικείμενα,[1] ᾔσθετο ὅπως[2] οἱ τὸ σόφισμα ἐτράπη ἐπὶ κακόν,[3] καὶ ἐξ ἐκείνου ἡσυχῇ διερχόμενος καὶ κρατῶν τῶν σκελῶν ἀπαθεῖς τοὺς ἅλας διεφύλαττεν.

43. Πυνθάνομαι δὲ ἐν τῇ Ἀντιόχου πόλει τῇ Σύρων πρᾶον γενέσθαι ἐλέφαντα, ἰόντα τε αὐτὸν ἐπὶ τὰς νομὰς στεφανόπωλιν γυναῖκα ὁρᾶν ἡδέως, καὶ προσεστάναι αὐτῇ καὶ τῇ προβοσκίδι τὸ πρόσωπον τῆς ἀνθρώπου καθαίρειν. δέλεαρ δὲ ἄρα ἐκείνη καθίει τοῦ φίλτρου αὐτῷ στέφανον ἐκ τῶν καθ᾽ ὥραν πλεκόμενον,[4] καὶ τῷ μὲν ὁσημέραι λαμβάνειν ἔργον ἦν, τῇ δὲ διδόναι. χρόνῳ δὲ ὕστερον ἡ μὲν ἄνθρωπος τὸν βίον κατέστρεψεν, ὁ δὲ ἐλέφας τῆς συνηθείας διαμαρτάνων καὶ οὐχ ὁρῶν ἣν ἐπόθει γυναῖκα, ὥσπερ οὖν ἐραστὴς ἐρωμένης ἀτυχῶν ἐξηγριώθη· καὶ ὁ τέως πρᾷότατος ὢν ἐς θυμὸν ἐξήφθη ὥσπερ οὖν καὶ τῶν ἀνθρώπων οἱ ἄγαν ὑπὸ τῆς λύπης ἐπικλυσθέντες καὶ ἔκφρονες γεγενημένοι.

44. Τὸν ἥλιον ἀνίσχοντα προσκυνοῦσιν ἐλέφαντες, τὰς προβοσκίδας εὐθὺ τῆς ἀκτῖνος ὡς χεῖρας ἀνατείνοντες, ἔνθεν τοι καὶ τῷ θεῷ φιλοῦνται.

---

[1] ἐπικείμενα τοῦ ἄχθους.   [2] καὶ ὅπως.
[3] Reiske: κακῷ.

the river. So when Thales heard the man's explanation, he thought that he must contrive to punish the Mule for its malice and ordered the man to load it with sponges and wool on top of the salt. But the Mule all unaware of the plot stumbled as usual, and having saturated its burden with water, realised that its trick was turned to its own undoing; so after that it made the crossing without disturbance and kept control of its legs and preserved the salt undamaged.

43. I learn that at Antioch in Syria there was a tame Elephant and that as it went to its feeding-grounds it used to take great pleasure in the sight of a woman who sold garlands, and would stand close by her and clean her face with its trunk. Accordingly the woman used to hang out as a bait to charm it a garland woven of the season's flowers, and every day it was the Elephant's practice to accept, and hers to offer it. In course of time the woman departed this life, and the Elephant, missing its customary fare and not seeing the woman of its desire, grew savage like a lover who has lost his loved one. And the creature that till then had been of the gentlest was inflamed with passion like men who are overwhelmed with excess of grief and driven out of their senses.

Elephant and flower-seller

44. Elephants do obeisance to the rising sun by lifting their trunks like hands to face its beams, and that, you see, is why they are beloved of the god.

The Elephant, a sun-worshipper

---

[4] ἐκ . . . πλεκόμενον] ἀεὶ τῶν ἐκ τῆς τέχνης τὸν καθ' ὥ. π.

μάρτυς ἀγαθὸς ὁ Φιλοπάτωρ ἡμῖν Πτολεμαῖος
ἔστω. ἡ μὲν κατὰ Ἀντιόχου νίκη σὺν αὐτῷ
ἐγένετο, θύων δὲ ἐπινίκια καὶ ἱλεούμενος τὸν
Ἥλιον ὁ Πτολεμαῖος τῇ τε ἄλλῃ μεγαλοπρεπῶς
ἔθυσεν καὶ οὖν καὶ τέτταρας ἐλέφαντας μεγέθει
μεγίστους [1] παρέστησεν ἱερεῖα, ὥς γε ᾤετο, καὶ
ταύτῃ τῇ θυσίᾳ γεραίρων ἐκεῖνος τὸ θεῖον.
ἐνύπνιον δὲ αὐτὸν διετάραξεν, ὡς ἀπειλοῦντος
τοῦ θεοῦ ἐπὶ τῇ ἀήθει τε καὶ ξένῃ θυσίᾳ· καὶ
δείσας ἐκεῖνος χαλκοῦς τέτταρας ποιησάμενος
ἀνῆψε τῷ θεῷ ὑπὲρ τῶν ἀνῃρημένων ἱλεούμενός [2]
τε καὶ εὐμενιζόμενος αὐτόν. ἐλέφαντες μὲν οὖν
θεοὺς προσκυνοῦσιν, οἱ δὲ ἄνθρωποι ἆρά γε [3]
εἰσὶ θεοὶ καὶ ὄντες εἰ φροντίζουσιν ἡμῶν διαπο-
ροῦσιν.

45. Οἱ ἐν τῇ Αἰγύπτῳ ἱερεῖς ἑαυτοὺς περιρ-
ραίνουσιν οὐ παντὶ ὕδατι, οὐ μὴν οὐδὲ τῷ παρατυ-
χόντι, ἀλλὰ ἐκείνῳ ἐξ οὗ πεπιστεύκασιν ὅτι ἄρα [4]
ἶβις πέπωκεν. ἴσασι γὰρ κάλλιστα ὅτι μήποτ᾽
ἂν πίοι [5] ὕδατος ἐκείνη ῥυπαροῦ καὶ λελυμασμένου
ἔκ τινων φαρμάκων· ἔχειν γάρ τι πιστεύουσιν ἐν
ἑαυτῷ τὸ ζῷον καὶ μαντικῆς, ἅτε ἱερόν.

Ἐλέφαντας δὲ ἀκούω τῶν τετρωμένων τοὺς
ἀτρώτους πεφεισμένως ἐξαιρεῖν καὶ ξυστὰ καὶ
ἀκόντια, ὥσπερ οὖν χειρουργίας ἐπιστήμονας καὶ
μαθόντας τὴν ἐν τοῖσδε σοφίαν.

Οὕτω δὲ ἄρα ἦν διὰ σπουδῆς τοῖς ἄνω τοῦ
χρόνου καὶ τὰ ἄλογα. ἔχαιρε μὲν ἀκούων Ἀετὸς
ὁ Ἠπειρώτης Πύρρος, ὁ δὲ τὸ δὴ λεγόμενον

---

[1] μεγάλους.   [2] δεόμενος.   [3] εἰ ἄρα γε.
[4] ἄρα καί.   [5] ὅτι ἂν μ. πίῃ.

Let Ptolemy Philopator be a trustworthy witness to the fact. With the aid of the god he overcame Antiochus,[a] and in sacrificing for his victory and to propitiate the Sun he not only offered sacrifices on a magnificent scale but even went so far as to offer four of the very largest elephants as victims, paying homage, as he supposed, to the god by this very sacrifice. But a vision in his sleep troubled him: the god seemed to threaten him for this unusual and strange offering. And he in his fear caused four elephants to be made of bronze and offered them to the god in place of those he had slaughtered, hoping to placate him and to ensure his favour. Elephants for their part worship the gods, whereas mankind is in doubt whether in fact there are gods, and, if there are, whether they take thought for us.

45 (i). The Priests of Egypt do not purify themselves with water of every kind, nor even with such water as they may chance upon, but only with that from which they believe an Ibis has drunk. For they know full well that this bird would never drink water that was dirty or that had been tainted with any drugs; for they believe that the bird possesses a certain prophetic faculty, seeing that it is sacred. *Egyptian Priests and their ablutions*

(ii). I learn that unwounded Elephants pick spears and javelins out of those that have been wounded, with caution, just as though they understood the practice of surgery and had acquired skill in these matters. *The Elephant as surgeon*

(iii). It seems that people in olden times paid regard even to brute beasts in the following way. Pyrrhus of Epirus delighted to be called ' the *Nicknames*

[a] At the battle of Raphia, 217 B.C.

Ἱέραξ ὁ Ἀντίοχος. διάφορα μὲν δὴ ταῦτα καὶ
ἀθρόα εἴρηται,¹ τῷ συνιέντι μαθεῖν ἄξια.

46. Μιθριδάτης ὁ Ποντικὸς τὴν ἑαυτοῦ φρουρὰν
καθεύδων ἐπίστευεν ἧττον καὶ τοῖς ὅπλοις καὶ
τοῖς δορυφόροις, καὶ διὰ τοῦτο ἡμερωθέντας εἶχε
φύλακας ταῦρον καὶ ἵππον καὶ ἔλαφον. καθεύδοντα οὖν ἐφρούρουν αὐτὸν οἵδε οἱ θῆρες, εἴ τις
προσίοι τάχιστα ἐκ τῆς ἀναπνοῆς αἰσθανόμενοι.
καὶ ὁ μὲν τῷ μυκήματι, ὁ δὲ τῷ χρεμετίσματι, ὁ
δὲ τῇ μηκῇ διύπνιζον αὐτόν.

47. Τῶν ἀγρίων ζῴων τὰ ἔκγονα τὰ νέα διαφόρως ὀνομάζεται, καὶ τά γε πλείω διπλῆν τὴν
ἐπωνυμίαν ἔχει. λεόντων γοῦν σκύμνοι καὶ λεοντιδεῖς ὀνομάζονται, ὡς Ἀριστοφάνης ὁ Βυζάντιος
μαρτυρεῖ, παρδάλεων δὲ σκύμνοι τε καὶ ἄρκηλοι·
εἰσὶ δὲ οἳ φασι γένος ἕτερον τῶν παρδάλεων τοὺς
ἀρκήλους εἶναι. θώων δὲ μόνον ² σκύμνοι φιλοῦσι
καλεῖσθαι, καὶ τίγρεων ὁμοίως, καὶ μυρμήκων ³
δὲ καὶ πανθήρων. ἔοικε δὲ καὶ τὰ τῶν λυγκῶν ⁴
ἔκγονα ὁμοίως ὀνομάζεσθαι. ἐν γοῦν τοῖς Λάσου
λεγομένοις Διθυράμβοις οὕτως εὑρίσκεται ⁵ εἰρημένον τὸ βρέφος τὸ τῆς λυγκός.⁶ πιθήκων δὲ

¹ Jac : εἰρήσεται.
² Gow : μόνοι mss, H.
³ μυρμήκων corrupt ? Ges.
⁴ Jac : λυγγῶν.
⁵ εὑρίσκεται σκύμνος.
⁶ Jac : λυγγός.

ᵃ Younger son of Antiochus I, whom he succeeded 245 B.C.;
driven out of Asia Minor and killed in Egypt 227 B.C. Justin.
27. 2 'Hierax' est cognominatus, quia non hominis sed
accipitris ritu in alienis diripiendis vitam sectaretur.

Eagle,' and Antiochus, so it is said, to be called 'the Hawk.' [a] I have mentioned these cases together, different though they are; an intelligent man will find them worth knowing.

46. Mithridates of Pontus [b] when asleep was unwilling to entrust his own safety to weapons and spearmen, and for that reason he kept as bodyguard a bull, a horse, and a stag that had been tamed. Accordingly these animals guarded him while he slept, and if ever anyone approached they at once perceived it by his breathing. And they would wake the King, the bull by bellowing, the horse by neighing, and the stag by bleating.

Mithridates, his bodyguard

47. The young offspring of wild animals have different appellations, and the majority at any rate have two names. The young of Lions, for instance, are called σκύμνοι and λεοντιδεῖς, as Aristophanes of Byzantium testifies; and of Leopards, σκύμνοι and ἄρκηλοι, although there are those who assert that ἄρκηλοι are a different kind of leopard. But the young of Jackals are habitually called σκύμνοι only; and the same with Tigers and Ants [c] and Panthers. And it appears that the young of Lynxes are similarly spoken of; at any rate in the *Dithyrambs*, as they are called, of Lasus we find the young of a lynx spoken of in this way. We hear too of the σκύμνοι and also of the πιθηκιδεῖς of Monkeys, and of

Names for the young of Animals

[b] Mithridates VI, Eupator, 2nd/1st cent. B.C., Rome's most formidable adversary in the East; defeated at length by Pompey, 65 B.C.

[c] Perhaps μύρμηξ is here to be interpreted as 'marmot'; see Hdt. 3. 102 with the commentators *ad loc.*

ἀκούομεν σκύμνους τε καὶ πιθηκιδεῖς τοὺς αὐτούς, βουβαλίδων δὲ πώλους· 'εἰ δὲ καὶ ὀρύγων, οὐκ ἂν θαυμάσαιμι,' ὁ αὐτὸς Ἀριστοφάνης φησί. 'κυνῶν δὲ καὶ λύκων σκύλακες καλοῖντο ἄν,' ἦ δ' ὅς· ἤδη δὲ καὶ λυκιδεῖς καλοῦνται οἱ τῶν λύκων, ὁ δὲ τέλειος καὶ μέγιστος καλοῖτο ἂν μονόλυκος. τῶν γε μὴν [1] λαγῶν λαγιδεῖς· ὁ δὲ τέλειος, πτῶκα [2] αὐτὸν φιλοῦσιν ὀνομάζειν οἱ ποιηταί, ταχίναν δὲ Λακεδαιμόνιοι. ἀλωπέκων δὲ τὰ ἔκγονα ἀλωπεκιδεῖς κέκληνται· αὐτὴ δὲ ἡ μήτηρ καὶ κερδὼ καὶ σκαφώρη καὶ σκινδαφός. τῶν δὲ ἀγρίων ὑῶν τὰ τέκνα μολόβρια ὀνομάζουσιν· ἀκούσειας δ' ἂν τοῦ Ἱππώνακτος καὶ αὐτὸν τὸν ὗν μολοβρίτην που λέγοντος. καλοῦνται δὲ καὶ τῶν ὑῶν μονίαι τινές. τάς γε μὴν δορκάδας καὶ ζόρκας καὶ πρόκας εἰώθασιν ὀνομάζειν. τῶν δὲ ὑστρίχων καὶ τῶν τοιούτων [3] τὰ ἔκγονα ὄβρια [4] καλεῖται· καὶ μέμνηταί γε Εὐριπίδης ἐν Πελιάσι τοῦ ὀνόματος καὶ Αἰσχύλος ἐν Ἀγαμέμνονι καὶ Δικτυουλκοῖς. τὰ δὲ τῶν ὀρνίθων καὶ τὰ τῶν ὄφεων καὶ τὰ τῶν κροκοδίλων ἔνιοι καὶ ψακάλους καλοῦσιν,[5] ὧν εἰσι καὶ Θετταλοί. τὰ δὲ πρόσφατα ὀρνύφια ὀρταλίχους, ἀλεκτρυόνων τε νεοττοὺς [6] ἀλεκτοριδεῖς [7] λέγουσι,[8] καὶ αὖ πάλιν χηνιδεῖς καὶ χηναλωπεκιδεῖς καὶ τὰ τούτοις ὅμοια κατὰ τὰ αὐτὰ σχηματίζουσιν. Ἀχαῖος δὲ ὁ τῆς

[1] τῶν μὲν δή.
[2] καὶ πτῶκα.
[3] τοιούτων ἀγρίων.
[4] ὀβρίκαλα Valck.
[5] τὰ δὲ τῶν ὀρνίθων . . . καλοῦσιν] τὰ δὲ ἔτι ἐν τῇ γαστρὶ ἔμβρυα καλοῦσι, τὰ δὲ τῶν ὀ. . . . κροκοδείλων ἔνιοι ἔμβρυα καὶ ψ. κ.

the πῶλοι of Antelopes, ' And I should not be surprised if we heard of the πῶλοι of Gazelles also ' says
the same Aristophanes. ' But the young of Dogs
and Wolves would be called σκύλακες ' he says. And
young wolves are in fact also called λυκιδεῖς, whereas a full-grown wolf of the largest size would be called
μονόλυκος. The young of Hares are λαγιδεῖς, but a
fully grown Hare poets like to call πτώξ; the Spartans,
ταχίνας. The young of Foxes are called ἀλωπεκιδεῖς,
while their mother is called κερδώ and σκαφώρη and
σκινδαφός. Men call the young of Wild Swine
μολόβρια, and you may hear Hipponax in some
passage [fr. 68 D] speaking of an actual Boar as
μολοβρίτης. And there are certain Pigs that are
called μονίαι. People habitually call Gazelles
ζόρκες and πρόκες. And the young of Porcupines
and similar creatures are called ὄβρια; the word is
mentioned by Euripides in his Peliades [fr. 616 N]
and by Aeschylus in his Agamemnon[a] and his
Dictyulci [fr. 48 N]. But the young of Birds and of
Snakes and of Crocodiles are called ψάκαλοι by some,
among whom are the people of Thessaly. And
people call little new-hatched birds ὀρτάλιχοι, and
the young of chickens ἀλεκτοριδεῖς; and again they
speak of χηνιδεῖς and χηναλωπεκιδεῖς and form words
like them on the same principle. But Achaeus the

[a] At Ag. 143 Aesch. wrote ὀβρικάλοισι, it was therefore in
the Dict. that he must have written ὄβρια.

6 ὀρταλίχους . . . νεοττούς] νεοττοὺς καὶ ὀρταλίχους ἀλεκτρυόνων τε ἀλεκτορίδας.
7 Valck: ἀλεκτορίδας.
8 λέγουσι· καὶ τόν γε πέρυσιν ὀνομάζουσιν ὡς καὶ τὸν οἶνον.

τραγῳδίας ποιητὴς τὸν νεοττὸν τῆς χελιδόνος
μόσχον ὠνόμασεν.

48. Μνήμην δὲ παρακολουθεῖν καὶ τοῖς ζῴοις,
καὶ ἴδιον αὐτῶν καὶ τοῦτο εἶναι χωρὶς τῆς ἐς
αὐτὴν τέχνης τε καὶ σοφίας, ἣν τερατευόμενοί
τινες ἐπινοῆσαι κομπάζουσι, τεκμηριοῖ καὶ ἐκεῖνα.
τὸν δεσπότην ὄντα τῶν ἐκ τῆς Ῥωμαίων βουλῆς
ἀπέδρα Ἀνδροκλῆς ὄνομα, οἰκέτης τὴν τύχην, ὅ
τι κακουργήσας καὶ ἡλίκον οὐκ οἶδα εἰπεῖν. ἧκε
δ᾽ οὖν ἐς τὴν Λιβύην, καὶ τὰς μὲν πόλεις ἀπελίμ-
πανε, καὶ τοῦτο δὴ τὸ λεγόμενον ἄστροις αὐτὰς
ἐσημαίνετο, προῄει δὲ ἐς τὴν ἐρήμην. φρυγόμενος
δὲ ὑπὸ πολλῆς ⟨καὶ⟩[1] διαπύρου τῆς ἀκτῖνος,
ἀσμένως ὕπαντρόν τινα πέτραν ὑποδραμὼν ἡσύ-
χαζε· λέοντος δὲ ἄρα κοίτη ἥδε ἡ πέτρα ἦν.
ἐπάνεισι τοίνυν ἐκ θήρας ὁ λέων, σκόλοπι βιαίῳ
περιπαρεὶς καὶ κολαζόμενος, καὶ ἐντυχὼν τῷ
νεανίᾳ εἶδεν αὐτὸν πράως, καὶ σαίνειν ἤρξατο, καὶ
προύτεινε τὸν πόδα, καὶ ἐδεῖτο ὡς ἠδύνατο
ἐξαιρεθῆναι τὸν σκόλοπα. ὁ δὲ τὰ μὲν πρῶτα[2]
κατέπτηξεν· ἐπεὶ δὲ πρᾶον τὸν θῆρα ἐθεάσατο
καὶ τὸ κατὰ τὸν πόδα συνεῖδε πάθος, ἐξεῖλε τὸ
λυποῦν τοῦ ποδός, καὶ τῆς ὀδύνης ἀπήλλαξεν.
ἡσθεὶς οὖν τῇ θεραπείᾳ ὁ λέων ἰατρεῖά οἱ ἐκτίνων
ξένον τε ἐνόμιζε καὶ φίλον, καὶ ὧν ἐθήρα ἐκοινώ-
νει. καὶ ὁ μὲν ἐσιτεῖτο ὠμὰ ᾗ λεόντων νόμος, ὁ
δὲ ἑαυτῷ ὤπτα· καὶ κοινῆς ἀπέλαυον τραπέζης

---

[1] ⟨καὶ⟩ add. H.   [2] πρῶτα καίτοι θανάτου δεόμενος.

tragic poet called the young of the Swallow μόσχος
[*fr.* 47 N].

48. That memory is an attribute even of animals, Androcles
and the Lion
and that this is a characteristic acquired without
the system and science of mnemonics which certain
wonder-workers claim to have invented, the following facts demonstrate.   One Androcles by name,
who happened to be a slave in the household of
a Roman senator, ran away from his master after
committing some offence, the nature and extent
of which I am unable to state.   Well, he arrived
in Libya and was for avoiding towns and, as the
saying is, 'marked their places only by the stars'[a]
and went on into the desert.   And being parched
by the excessive and fiery heat of the sun, he was
glad to take refuge and to rest under a caverned
rock.   This rock, it seems, was the lair of a Lion.
Now the Lion returned from hunting, injured from
having been pierced with a sharp stake, and when
it encountered the young man it looked at him
in a gentle manner and began to fawn upon him,
extending its paw and imploring him as best it could
to have the stake plucked out.   Androcles at first
shrank back.   But when he saw that the beast was
in gentle mood, and realised what was the matter
with its paw, he extracted what was hurting it and
rid the Lion of its pain.   The Lion therefore in joy
at being healed paid him his fee by treating him as
its guest and friend, and shared with him the spoils
of its chase.   And while the Lion ate its food raw,
as is the custom of lions, Androcles used to cook his
for himself.   And they enjoyed a common table each

[a] Cp. 2. 7.

κατὰ φύσιν τὴν αὑτοῦ ἑκάτερος. καὶ τριῶν μὲν
ἐτῶν τὸν τρόπον τοῦτον διεβίωσεν ὁ Ἀνδροκλῆς·
εἶτα ὑπεράγαν κουριῶν καὶ ὀδαξησμῷ βιαίῳ
κατειλημμένος τὸν μὲν λέοντα ἀπολιμπάνει, ἑαυτὸν
δὲ μεθίησι τῇ τύχῃ. εἶτα ἀλώμενον αὐτὸν συλλαβόντες καὶ ὅτου εἴη πυθόμενοι ἐς τὴν Ῥώμην τῷ
δεσπότῃ δήσαντες ἀποπέμπουσιν. ὁ δὲ ἐφ᾽ οἷς
ἠδικήθη εὐθύνει τὸν οἰκέτην, καὶ κατεγνώσθη
ἐκεῖνος θηρίοις βορὰ παραδοθῆναι. ἐθηράθη δέ
πως καὶ ὁ Λίβυς ἐκεῖνος λέων καὶ ἀφείθη ἐν τῷ
θεάτρῳ, καὶ ὁ νεανίας δὲ ὡς ἀπολούμενος ὅ ποτε
αὐτῷ ἐκείνῳ τῷ λέοντι σύνοικός τε καὶ σύσκηνος
γεγενημένος. καὶ ὁ μὲν ἄνθρωπος οὐκ ἐγνώρισε
τὸν θῆρα, ἐκεῖνος δὲ παραχρῆμα ἀνέγνω τὸν
ἄνθρωπον, καὶ ἔσαινεν αὐτόν, καὶ ὑποκλίνας τὸ
πᾶν σῶμα ἔρριπτό οἱ παρὰ τοῖς ποσίν. ὀψὲ δὲ
καὶ ὁ Ἀνδροκλῆς ἐγνώρισε τὸν ἑαυτοῦ ξένον,
καὶ περιλαβὼν τὸν λέοντα ὡς ἥκοντα ἑταῖρον ἐξ
ἀποδημίας κατησπάζετο. ἐπεὶ δὲ ἐδόκει γόης,
ἐφείθη[1] οἱ καὶ πάρδαλις. ὁρμώσης δὲ αὐτῆς ἐπὶ
τὸν Ἀνδροκλέα, ὁ λέων ἀμύνων τῷ ποτε ἰασα-
μένῳ, καὶ κοινῆς τραπέζης μεμνημένος διασπᾷ τὴν
πάρδαλιν. οἷα τοίνυν εἰκὸς οἱ θεώμενοι ἐκπλήτ-
τονται, καὶ ὁ διδοὺς τὰς θέας καλεῖ τὸν Ἀνδροκλέα,
καὶ τὸ πᾶν μανθάνει. καὶ θροῦς ἐς τὸ πλῆθος
διαρρεῖ, καὶ τὸ σαφὲς ὁ δῆμος μαθόντες ἐλευθέρους
ἐκβοῶσιν ἀφεῖσθαι καὶ τὸν ἄνδρα καὶ τὸν λέοντα.
ἴδιον δὴ τῶν ζῴων καὶ ἡ μνήμη. καὶ συνῳδὸν
τοῖς προειρημένοις καὶ ἐς τὸ αὐτὸ δέ ἐστιν

as was his nature. And this was how Androcles
lived for the space of three years. After a time, as
his hair grew to an excessive length and he was
troubled with a violent itching, he forsook the Lion
and trusted himself to fortune. Then as he was
wandering about he was caught, questioned as to
whom he belonged to, and sent bound to his master
in Rome. The master punished his servant for the
injury he had done him and he was condemned to be
given to the wild beasts to eat. It chanced that the
same Libyan lion had also been caught and was let
loose in the arena together with the young man
destined for death, him who had shared that very
Lion's home and dwelling. The man for his part
did not know the Lion again, but the animal immediately recognised the man, fawned upon him, and
letting its whole body sink down, threw itself at
his feet. And at last Androcles recognised his host
and throwing his arms round it, greeted it like a
comrade returned after absence. But as he was
thought to be a magician; a leopard also was let
loose upon him. And when it rushed at Androcles
the Lion came to the rescue of its former healer
and remembering how they used to feed together,
tore the leopard to pieces. The spectators, as was
natural, were dumbfounded, and the man who was
giving the shows summoned Androcles and learnt
the whole story. And the report spread through
the multitude, and the populace on learning the
truth shouted aloud that both man and Lion must
be set free. Memory is indeed one of the attributes
of animals.

And there is a corresponding story to the same

---

[1] ἀφείθη.

† εὕδοντος ἐν τῇ Σάμῳ ἐπὶ τοῦ κεχηνότος Διονύσου † [1] νομίζοιτο ἂν καὶ τὸ φωλιὸν εἰδέναι. καὶ
τοῦτο ἀκουέτω Ἐρατοσθένους τε καὶ Εὐφορίωνος
καὶ ἄλλων περιηγουμένων αὐτό.

---

[1] εὕδοντος . . . Διονύσου corrupt.

effect as the above . . . in Samos in front of Dionysus
of the Open Mouth . . .,^a might be thought to
know the lair also. For this too he must consult
Eratosthenes, Euphorion, and others who narrate it.

^a The passage is corrupt. The reference is to the story
recorded by Pliny (*HN* 8. 57) of one Elpis of Samos who
relieved a suffering lion, of how it showed its gratitude, and
how Elpis dedicated a temple to Dionysus who had saved him.

BOOK VIII

Η

1. Ἰνδικοὶ λόγοι διδάσκουσιν ἡμᾶς καὶ ἐκεῖνα. τὰς κύνας ἄγουσιν ἐς τὰ ἔνθηρα χωρία οἱ θηρατικοὶ τὰς εὐγενεῖς τε καὶ ἴχνη καταγνῶναι θηρίων ἀγαθὰς καὶ ὠκίστας ἐς δρόμον, καὶ τοῖς δένδροις προσδήσαντες εἶτα μέντοι ἀπαλλάττονται, τοῦτο δήπου τὸ λεγόμενον ἀτεχνῶς κύβον ἀναρρίψαντες. οἱ δὲ τίγρεις ἐντυχόντες αὐταῖς, ἀθηρίᾳ μὲν καὶ λιμῷ συμπεσόντες διασπῶσιν αὐτάς· ἐὰν δὲ ὀργῶντες ἀφίκωνται καὶ κεκορεσμένοι, συμπλέκονταί τε αὐταῖς καὶ τῆς ἀφροδίτης ἐν πλησμονῇ καὶ ἐκεῖνοι μέμνηνται. ἐκ δὲ τῆς ὁμιλίας ταύτης οὐ κύων φασὶν ἀλλὰ τίγρις τίκτεται. ἐκ δὲ τούτου καὶ κυνὸς θηλείας ἔτι τίγρις τεχθείη ἄν, ὁ δὲ ἐκ τούτου καὶ κυνὸς ἐς τὴν μητέρα ἀποκρίνεται, καὶ κατώλισθεν ἡ σπορὰ πρὸς τὸ χεῖρον, καὶ κύων τίκτεται. πρὸς ταῦτα Ἀριστοτέλης οὐκ ἀντιφήσει. οὗτοι δὲ ἄρα οἱ κύνες, οἷς πάρεστι πατέρα αὐχεῖν τίγριν,[1] ἔλαφον μὲν θηρᾶσαι ἢ συὶ συμπεσεῖν ἀτιμάζουσι, χαίρουσι δὲ ἐπὶ τοὺς λέοντας ἄττοντες καὶ τοὺς ἄνω τοῦ γένους ἀποδεικνύντες ἐντεῦθεν. Ἀλεξάνδρῳ γοῦν τῷ Φιλίππου πεῖραν ἔδοσαν οἱ Ἰνδοὶ τῆς τῶν κυνῶν τῶνδε ἀλκῆς τὸν τρόπον τοῦτον. ἀφῆκαν ἔλαφον, καὶ ὁ κύων ἡσύχαζεν· εἶτα σῦν, ὁ δὲ ἀτρεμῶν κατέμενεν· καὶ ἄρκτον ἐπὶ τούτοις, καὶ ἔκνιζεν αὐτὸν ⟨ἡ⟩[2] ἄρκτος οὐδὲ ἕν· λέοντος δὲ ἀφεθέντος, ὁ δὲ ὡς εἶδ᾽[3] ὥς μιν μᾶλλον ἔδυ χόλος, καὶ οἷα δήπου

---

[1] τὸν τίγριν.   [2] ⟨ἡ⟩ add. H.   [3] εἶδε.

BOOK VIII

1. Indian histories teach us the following facts Indian Hounds bred from tigers also. Huntsmen take thoroughbred bitches which are good at tracking wild animals and are very swift of foot to places infested by these animals; they tie them to trees and then go away, simply, as the saying is, trying a throw of the dice. And if tigers find them when they have caught nothing and are famished, they tear them to pieces. If however they arrive on heat and full-fed they couple with the bitches, for tigers too when gorged turn their thoughts to sexual intercourse. From this union, so it is said, a tiger is born, not a hound. And from this tiger and a bitch again a tiger would be born, although the offspring of this last and of a bitch takes after its dam, and the seed degenerates and a hound is born. Nor will Aristotle contradict this [*HA* 607 a 4, *GA* 746 a 34]. Now these hounds which can boast a tiger for sire scorn to pursue a stag or to face a boar, but are glad to rush at lions and thereby to give proof of their pedigree. At any rate the Indians gave Alexander the son of Philip a test of the strength of these hounds in the following manner. They let loose a stag, and the Hound stayed quiet; then a boar, and it never moved; after that a bear, but the bear caused it no excitement whatever. But when a lion was let loose, and

' when ' the Hound ' beheld it, then came wrath upon him the fiercer ' [Hom. *Il.* 19. 16],

175

AELIAN

θεασάμενος τὸν ὄντως ἀντίπαλον οὔτε ἤμελλεν
οὔτε ἠτρέμει, ἀλλ᾽ ἄξας ἐπ᾽ αὐτὸν εἶτα μέντοι
καρτερᾷ τῇ λαβῇ εἴχετο πιέζων καὶ ἄγχων. ὁ
τοίνυν Ἰνδὸς ὁ τὴν θέαν τῷ βασιλεῖ τήνδε παρέχων
κάλλιστα εἰδὼς τοῦ κυνὸς τὸ καρτερικόν, προσέ-
ταξέν οἱ τὴν οὐρὰν ἀποκοπῆναι. καὶ ἡ μὲν
ἀπεκόπτετο, ὁ δὲ οὐκ ἐφρόντιζε. προσέταξεν οὖν
ὁ Ἰνδὸς καὶ τῶν σκελῶν ἓν ἀποκόψαι, καὶ ἀπεκόπη·
ὁ δὲ ὡς ἐξ ἀρχῆς ἐνέφυ εἴχετο, καὶ οὐκ ἀνίει,
ὥσπερ οὖν ἀλλοτρίου κοπτομένου σκέλους καὶ
ὀθνείου. καὶ ἄλλο ἀπεκόπτετο, καὶ τὸ δῆγμα ὁ
κύων οὐ κατελίμπανε· καὶ τρίτον ἕτερον, ὁ δὲ
εἴχετο· καὶ τὸ τέταρτον ἐπ᾽ ἐκείνοις, καὶ ἦν
ἐγκρατὴς τοῦ δήγματος ἔτι. καὶ τελευτῶντες
τῆς κεφαλῆς τὸ λοιπὸν σῶμα ἀφεῖλον· ὀδόντες δὲ
ἐκείνῳ [1] ἤρτηντο τῆς ἐξ ἀρχῆς ἀντιλαβῆς, καὶ ἡ
κεφαλὴ ᾐωρεῖτο μετέωρος ἐκ τοῦ λέοντος, αὐτοῦ
μέντοι τοῦ δάκοντος [2] οὐκέτι ὄντος. Ἀλέξανδρος
οὖν ἐνταῦθα ἠνιᾶτο, τὸν κύνα ἐκπλαγεὶς ὅτι ἄρα
πεῖραν ἑαυτοῦ δοὺς [3] εἶτα ἀπωλώλει,[4] τὸ ἐναντίον
τοῖς δειλοῖς παθών, θάνατον δὲ ὑπὲρ τῆς ἀνδρείας
ἠλλάξατο. ἰδὼν οὖν ὁ Ἰνδὸς αὐτὸν ἀνιώμενον,
τέτταρας ὁμοίους ἐκείνῳ κύνας ἔδωκέν οἱ. ὁ δὲ
ἥσθη λαβὼν καὶ ἀντέδωκεν ὁποῖα ἦν εἰκός, καὶ
τῆς γε ἐπὶ τῷ πρώτῳ λύπης ἔλαβε λήθην ὁ τοῦ
Φιλίππου παῖς λαβὼν τοὺς τέτταρας.

2. Κύων ἀγρευτικὸς ἅπας αὐτὸς μὲν λαβὼν
θηρίον ᾔδεται, καὶ κέχρηται τῇ ἄγρᾳ ὡς ἄθλῳ,
ἐὰν αὐτῷ συγχωρήσῃ ὁ δεσπότης· εἰ δὲ μή,

[1] ἐκείνη A, ἐκεῖνοι L, Shorey.
[2] δάκοντος ἐξ ἀρχῆς.

176

and as though it had seen its real adversary, it
neither hesitated nor remained still but leapt upon
the lion and clung to it with a vigorous grip, pressing
and throttling it. So then the Indian who was
giving the King this exhibition, knowing full well the
Hound's power of endurance, ordered the men to cut
off its tail. The tail was cut off, but the Hound paid
no heed. So the Indian ordered one of its legs to
be cut off, and cut off it was. But the Hound clung
as fast as ever, and would not let go, as though the
leg of some other creature unconnected with it were
being cut off. Then another leg was cut off and still
the Hound would not relax its bite; then a third,
and it continued to cling; and after these the fourth,
and still it was capable of biting. And finally they
severed the rest of its body from its head. But the
Hound's fangs maintained their original grip, while
the head hung aloft on the lion, although the biter
himself was no more. At this Alexander was
grieved and amazed that the Hound in giving proof
of its mettle had perished, a fate the reverse of a
coward's, and had met its death by reason of its
courage. Accordingly the Indian seeing Alexander's
grief, presented him with four hounds of the same
breed. And he was delighted to receive them and
gave the Indian a suitable gift in return. And when
the son of Philip received the four he forgot his
grief over the first.

2. Every Hound that is good at hunting delights
to catch unaided a wild animal and regards the
catch as its prize, provided its master consents to

The Hound's
delight in
hunting

---

³ διδούς.   ⁴ ἀπολώλει.

φυλάττει ζῶντα ἔστ' ἂν ὁ θηρατὴς ἀφίκηται καὶ
κρίνῃ γε ὑπὲρ τοῦ ληφθέντος ὅ τι καὶ ἐθέλει.[1]
νεκρῷ δὲ ἐντυχὼν ἢ λαγῷ[2] ἢ συὶ οὐκ ἂν ἅψαιτο,
τοῖς ἀλλοτρίοις ἑαυτὸν πόνοις οὐκ ἐπιγράφων,
οὐδὲ ἀξιῶν σφετερίσασθαι τὰ προσήκοντά οἱ
ἥκιστα. ἔοικε δὲ ἐκ τούτων ἔχειν τι καὶ φιλοτιμίας
ἐν ἑαυτῷ φυσικῆς· μὴ γὰρ δεῖσθαι κρεῶν, ἀλλὰ
νίκης ἐρᾶν. ἀκοῦσαι δὲ ἄξιον ὅ τι καὶ δρᾷ παρὰ
τὸν τῆς θήρας καιρὸν ὁ κύων ὁ θηρατικός. προη-
γεῖται τοῦ κυνηγέτου ἱμάντι μακρῷ προσημμένος,
καὶ ῥινηλατεῖ τῆς φωνῆς ἔχων ἐγκρατῶς.[3] καὶ
ἐς ὅσον μὲν ἀθηρία ἀπαντᾷ αὐτῷ καὶ οὐδενὶ
ἐντυγχάνει, πρόεισιν ὅσα[4] ἰδεῖν καὶ τεκμήρασθαι
κατηφέστερος, καὶ μέντοι καὶ ἐς τὸ πρόσω ἰὼν
ἐπάγεται τὸν θηρατὴν προθύμως τε καὶ καρτερικῶς
εὖ μάλα ὁ κύων· εἰ δὲ ἰχνεύσειε[5] καὶ ὀσμῇ
τινι προσπέσοι[6] θηρίου, ἐνταῦθα ἕστηκεν. ὁ δὲ
κυνηγέτης ἔρχεται πλησίον, καὶ ὁ κύων περιχαρὴς
τῇ εὐερμίᾳ ὢν αἰκάλλει τὸν δεσπότην καὶ φιλεῖ
τὼ πόδε, καὶ πάλιν τῆς ἐξ ἀρχῆς ἰχνεύσεως
ἔχεται, καὶ πρόεισι βάδην ἔστ' ἂν ἀφίκηται πρὸς
τὴν κοίτην, καὶ περαιτέρω οὐ πρόεισι. συνῆκεν
οὖν ὁ θηρατής, καὶ ὑποθωΰξας σημαίνει τοῖς
ἀρκυωροῖς· οἱ δὲ περιβάλλουσι τὰς ἄρκυς. καὶ
ἐνταῦθα τοῦ καιροῦ ὑλάκτησεν ὁ κύων· νοεῖ δὲ
αὐτῷ τηνικαῦτα ἡ βοὴ ἐς ἀνάστασιν τὸν σῦν
ὑποθῆξαι, ἵνα ἐκπέσῃ[7] φεύγων καὶ τοῖς δικτύοις
καταληφθῇ. ἁλόντος δὲ τοῦ θηρός, ὁ δὲ ἐπινίκιόν
τινα οἱονεὶ παιᾶνα ἐκβοᾷ, καὶ γέγηθε καὶ σκιρτᾷ,
178

this.[1] Otherwise it preserves the animal alive until the huntsman comes up and decides what he wants to do with the capture. But if it comes upon a dead hare or boar it will not touch it, refusing to claim credit for another's labours and declining to appropriate what does not belong to it. From these facts it appears to have a certain natural love of distinction: it is not meat that it wants; it is victory that it loves. And it is worth hearing how the Hound behaves when it is hunting. It goes ahead of the huntsman, to whom it is attached by a long leash, and controlling its bark,[3] tracks the game by scent. And so long as no game comes its way and it finds nothing, it goes forward rather despondently to judge from its looks; for all that, it goes ahead and leads the huntsman on with the utmost keenness and pertinacity. But if it tracks out some beast and comes upon some scent,[5] then it halts. And the huntsman approaches while the Hound overjoyed at its good luck fawns upon its master, licks his feet, and resumes its original quest, advancing step by step until it comes upon the lair; further it does not go. So then the huntsman understands and with a low call gives the signal to the men with the nets. And they set the nets in a ring. Thereupon the Hound barks. The intention of its baying just then is to provoke the boar to rise in order that he may emerge and as he flees may be caught in the nets.[7] And when the beast is captured, the Hound raises a loud cry of victory, as[4] it were a hymn of praise, and is delighted and leaps about,

---

¹ θέλει.
² λαγῷ τινι.
³ ἐγκρατῶς καὶ σιωπῶν.
⁴ ὡς.
⁵ ἰχνεύσειε τυχόν.
⁶ προσπέσοι που.
⁷ ἐμπέσῃ.

ὥσπερ οὖν ἐχθροὺς ¹ ὁπλῖται νενικηκότες. ταῦτα
ἐπὶ συῶν καὶ ἐλάφων δρῶσιν οἱ κύνες.

3. Χάριν δὲ ἄρα καὶ δελφῖνες ἀποδοῦναι τῶν
ἀνθρώπων ἦσαν δικαιότεροι, καὶ τῷ νόμῳ τῶν
Περσῶν ὃν ἐπαινεῖ καὶ Ξενοφῶν οὐκ ἐνέχονται.
ὃ δὲ λέγω τοιοῦτόν ἐστι. Κοίρανος ὄνομα, τὸ
γένος ἐκ Πάρου, δελφίνων τινῶν ἐν Βυζαντίῳ
βόλῳ περιπεσόντων καὶ ἑαλωκότων, δοὺς ἀργύριον οἱονεὶ λύτρα τοῖς ἠγρευκόσιν ἀφῆκεν αὐτοὺς
ἐλευθέρους, ἀνθ' ὧν τὴν χάριν ἀπείληφεν. ἔπλει
γοῦν ποτε πεντηκόντορον ἔχων, ὡς λόγος, Μιλη-
σίους τινὰς ἄγουσαν ἄνδρας, ἐν δὲ τῷ μεταξὺ
⟨Νάξου καὶ⟩² Πάρου πορθμῷ τῆς νεὼς ἀνατραπεί-
σης καὶ τῶν ἄλλων διαφθαρέντων, τὸν Κοίρανον
ἔσωσαν δελφῖνες, ὑπὲρ ἧς φθάσαντες εἶχον εὐεργεσίας τὴν ἴσην ἀντιδιδόντες. καὶ ἔνθα ἐξενήξαντο
ὀχοῦντες αὐτὸν ἄκρα δείκνυται καὶ ὕπαντρος
πέτρα, καὶ καλεῖται ὁ χῶρος Κοιράνειος. χρόνῳ
δὲ ὕστερον τεθνεῶτα τόνδε τὸν Κοίρανον θαλάττης
πλησίον ἔκαον. εἶτα μέντοι αἰσθόμενοί ποθεν οἱ
δελφῖνες ἠθροίσθησαν, ὥσπερ οὖν ἐπὶ τὸ κῆδος
ἥκοντες, καὶ ἐς ὅσον ἡ πυρὰ ἐνήκμαζε ³ καομένη,
παρέμειναν ὡς φίλῳ φίλος πιστός· εἶτα μέντοι
κατασβεσθείσης οἱ δὲ ἀπενήξαντο. ἄνθρωποί γε
μὴν ζῶντάς τε καὶ πλουτοῦντας καὶ εὖ πράττειν
δοκοῦντας θεραπεύουσι, νεκροὺς δὲ ἀποστρέφονται

¹ οἱ τοὺς ἐχθρούς.
² ⟨Νάξου καί⟩ add. Wesseling.   ³ ἤκμαζε.

ON ANIMALS, VIII. 2–3

like soldiers who have overcome their enemies. This
is what Hounds do in dealing with boars and stags.

3. It seems that even Dolphins are more scrupulous   The Dolphin,
than men in showing their gratitude and are not con-   its gratitude
trolled by the Persian custom applauded by Xeno-
phon [*Cyr.* 1. 2. 7].[a]   And what I have to tell is as
follows.   One Coeranus by name, a native of Paros,
when some Dolphins fell into the net and were
captured at Byzantium, gave their captors money,
as it were a ransom, and set them at liberty ; and
for this he earned their gratitude.   At any rate he
was sailing once (so the story goes) in a fifty-oar
ship with a crew of Milesians, when the ship capsized in the strait between Naxos and Paros, and
though all the rest were drowned, Coeranus was
rescued by Dolphins which repaid the good deed
that he had first done them by a similar deed.   And
the headland and caverned rock to which they swam
with him on their backs are pointed out, and the spot
is called Coeraneus.   Later when this same Coeranus
died they burnt his body by the sea-shore.   Whereupon the Dolphins, observing this from some point,
assembled as though they were attending his funeral,
and all the while that the pyre was ablaze they remained at hand, as one trusty friend might remain
by another.   When at length the fire was quenched
they swam away.

Men however are subservient to the wealthy and
the seemingly prosperous while they are alive, but
when dead or in misfortune they turn their backs

[a] The Persians punish those who could, but do not, show
their gratitude ; want of gratitude they regard as the parent
of other vices.

181

AELIAN

ἢ καὶ δυστυχοῦντας, ἵνα μή τινα ἐκτίσωσιν εὖ
παθόντες χάριν.

4. Ἦσαν δὲ ἄρα καὶ ἰχθύες πρᾶοί τε ἅμα καὶ
χειροήθεις καὶ οἷοι καλούμενοί τε ὑπακούειν καὶ
διδόντων τροφὰς ἑτοίμως δέχεσθαι, ὥσπερ οὖν ἡ
ἐν Ἀρεθούσῃ ἱερὰ ἔγχελυς. τὴν Κράσσου τε τοῦ
Ῥωμαίου μύραιναν ᾄδουσιν, ἥπερ οὖν καὶ ἐνωτίοις
καὶ ὁρμίσκοις διαλίθοις ἐκεκόσμητο,[1] οἷα δήπου
ὡραία κόρη, καὶ καλοῦντος τοῦ Κράσσου τὸ
φώνημα ἐγνώριζε, καὶ ἀνενήχετο, καὶ ὀρέγοντος
ὅ τι οὖν ἡ δὲ ἤσθιε προθύμως καὶ ἑτοίμως λαμ-
βάνουσα. ταύτην τοι καὶ ἔκλαυσεν ὁ Κράσσος,
ὡς ἀκούω, τὸν βίον καταστρέψασαν, καὶ ἔθαψε.
καί ποτε Δομετίου πρὸς αὐτὸν εἰπόντος 'ὦ μωρέ,
μύραιναν ἔκλαυσας τεθνεῶσαν', ὁ δὲ ὑπολαβὼν
'ἐγὼ θηρίον' ἔφατο, 'σὺ δὲ τρεῖς γυναῖκας θάψας
οὐκ ἔκλαυσας'.

Αἰγυπτίων δὲ ἀκούω λεγόντων τοὺς ἱεροὺς
κροκοδίλους εἶναι πράους, καὶ τῶν γε θεραπευτήρων ἐπιψαυόντων καὶ ἐπαφωμένων ὑπομένειν
καὶ κούφως φέρειν, καὶ κεχηνέναι καθιέντων
ἐκείνων ⟨τὰς χεῖρας⟩[2] καὶ τοὺς ὀδόντας σφίσι
καθαιρόντων καὶ τὰ ἐσδυόμενα τῶν σαρκίων
ἐξαιρούντων. ἤδη μέντοι καὶ μαντικῆς μετειληχέναι τοὺς προειρημένους[3] κροκοδίλους Αἰγύπτιοί
φασι, καὶ τὸ μαρτύριον ἐκεῖνο προάγονται.
Πτολεμαίου (ὁπόστος δὲ ἦν οὗτος ἐκείνους

[1] κεκόσμητο.　　　　[2] ⟨τὰς χεῖρας⟩ add. H.
[3] προτιμοτέρους.

[a] At Ortygia, in Syracuse.

182

upon them so as to avoid repaying them for past favours.

4 (i). It seems that even Fishes are both tame and _{Tame Fishes} tractable, and when summoned can hear and are ready to accept food that is given them, like the sacred eel in the Fountain of Arethusa.[a] And men tell of the moray belonging to Crassus[b] the Roman, which had been adorned with earrings and small necklaces set with jewels, just like some lovely maiden; and when Crassus called it, it would recognise his voice and come swimming up, and whatever he offered it, it would eagerly and promptly take and eat. Now when this fish died Crassus, so I am told, actually mourned for it and buried it. And on one occasion when Domitius[c] said to him 'You fool, mourning for a dead moray!' Crassus took him up with these words: 'I mourned for a moray, but you never mourned for the three wives you buried.'

(ii). I have heard that the Egyptians assert that _{Tame Crocodiles} the sacred Crocodiles are tame, and if their keepers at any rate touch and handle them they submit and do not object; and they keep their jaws open when the keepers insert their hands and cleanse their teeth and pick out bits of flesh that have got between them. Further, the Egyptians assert that the aforesaid Crocodiles are endowed with prophecy, and adduce the following evidence. Ptolemy (which of

[b] M. Licinius Crassus, defeated Spartacus, 73 B.C.; triumvir with J. Caesar and Pompey, 60 B.C.; defeated by the Parthians at Carrhae, 53 B.C., and later slain.

[c] Cn. Domitius Ahenobarbus, Censor with Crassus, 92 B.C. See Suet. *Nero* 2.

ἔρεσθε) καλοῦντος τὸν πραότατον[1] τῶν κροκοδίλων μὴ ὑπακοῦσαί φασι καὶ τροφὰς ὀρέγοντος μὴ
προσίεσθαι· συνεῖναι[2] δὲ τοὺς ἱερέας ὅτι τὸ τέλος
τῷ Πτολεμαίῳ προσιὸν εἰδὼς ὁ κροκόδιλος εἶτα
μέντοι τὴν ἐξ αὐτοῦ τροφὴν ἠτίμασε λαβεῖν.

5. Οἰωνοῖς μαντευομένους ἀκούω τινὰς καὶ ἐπ'
ὄρνισι καθημένους ἐξετάζειν πτήσεις τε αὐτῶν καὶ
ἕδρας. καὶ ᾄδονταί γε ἐπὶ ταύτῃ τῇ σοφίᾳ
Τειρεσίαι τε καὶ Πολυδάμαντες καὶ Πολύειδοι
καὶ Θεοκλύμενοι καὶ ἄλλοι πολλοί. σπλάγχνων
δὲ ἄρα θέσεις[3] καταγνῶναι δεινοὶ ἦσαν καὶ
Σιλανοὶ καὶ Μεγιστίαι καὶ Εὐκλεῖδαι καὶ ἐπὶ
τούτοις πολὺς κατάλογος. ἀκούω μέντοι τινῶν
λεγόντων ὅτι καὶ ἀλφίτοις μαντεύονταί τινες καὶ
κοσκίνοις καὶ τυρίσκοις. πέπυσμαι δὲ καὶ κώμην
τινὰ Λυκιακὴν μεταξὺ Μύρων καὶ Φελλοῦ,
Σοῦρα[4] ὄνομα, ἐν ᾗ μαντεύονταί τινες ἐπ' ἰχθύσι
καθήμενοι, καὶ ἴσασιν ὅ τι καὶ νοεῖ ἥ τε ἄφιξις
αὐτῶν κληθέντων καὶ ἡ ἀναχώρησις, καὶ ὅταν μὴ
ὑπακούσωσι τί δηλοῦσι, καὶ ὅταν ἔλθωσι πολλοὶ
τί σημαίνουσιν. ἀκούσει δὲ τὰ μαντικὰ τῶν
σοφῶν ταῦτα καὶ πηδήσαντος ἰχθύος καὶ ἀναπλεύ-

---

[1] προτιμότερον.
[2] συνιέντας MSS, Jac retains, marking a lacuna after λαβεῖν.
[3] θέσεις καὶ φύσεις (or φέσεις).
[4] Σύρραν MSS, Σοῦραν Schn.

---

[a] Polydamas, Trojan hero, learned divination from his
father Panthous; see Hom. Il. 12. 210.—Polyeidus; see 5. 2 n.
—Theoclymenus at Hom. Od. 20. 350 foretells the downfall of
the suitors of Penelope.

the line it was, you must ask them) was calling to the tamest of the Crocodiles, but it paid no attention and would not accept the food he offered. And the priests realised that the Crocodile knew that Ptolemy's end was approaching and consequently declined to take food from him.

5. I have heard that some people practise divination by birds and devote themselves to their study and scrutinise their flight and the quarters of the sky where they appear. And seers like Teiresias, Polydamas,[a] Polyeidus, Theoclymenus and many another are celebrated for their knowledge of this art, while men such as Silanus,[b] Megistias, Euclides and the long tale of their successors were skilled in deciding upon the dispositions of entrails. Again, I have heard people assert that some divine by means of barley-corns, of sieves, and of small cheeses. And I have ascertained that there is a village in Lycia between Myra and Phellus called Sura[c] where there are those who devote themselves to divination by means of fish, and they understand what it purports if the fish come at their call or withdraw, and what it signifies if they pay no attention, and what it portends if they come in numbers. And you shall hear these prophetic utterances of the sages when a fish leaps out of the water or comes floating up from the

Divination by Fishes

[b] Silanus of Ambracia, soothsayer to Cyrus II; see Xen. *An.* 1. 7. 18.—Megistias claimed descent from Melampus; died fighting at the battle of Thermopylae of which he had foretold the issue; see Hdt. 7. 221, 228.—Euclides of Phlius divined Xenophon's lack of money and advised him to sacrifice to Zeus the Merciful; see Xen. *An.* 7. 8. 1.

[c] A few miles W of Myra on the sea-coast.

σαντος ἐκ βυθοῦ[1] καὶ τροφὴν προσεμένου καὶ αὖ πάλιν μὴ λαβόντος.

6. Ἦν δὲ ἄρα εὐχείρωτα καὶ αἱρεῖν ῥᾷστα ὄνοι μὲν τοῖς λύκοις, τοῖς μέροψι δὲ αἱ μέλιτται, ταῖς γε μὴν χελιδόσιν οἱ τέττιγες, τοῖς δὲ ἐλάφοις οἱ ὄφεις. ἡ πάρδαλις δὲ αἱρεῖ τῇ ὀσμῇ[2] τὰ πλεῖστα, καὶ ἔτι μᾶλλον τὸν πίθηκον.

7. Μεγασθένους ἀκούω λέγοντος περὶ τὴν τῶν Ἰνδῶν θάλατταν γίνεσθαί τι ἰχθύδιον, καὶ τοῦτο μὲν ὅταν ζῇ ἀθέατον εἶναι, κάτω που νηχόμενον καὶ ἐν βυθῷ, ἀποθανὸν δὲ ἀναπλεῖν. οὗ τὸν ἁψάμενον ἐκθνήσκειν[3] τὰ πρῶτα, εἶτα μέντοι καὶ ἀποθνήσκειν. τὸν δὲ χέλυδρον[4] πατήσας τις καὶ εἰ μὴ δηχθείη, ὡς Ἀπολλόδωρός φησιν ἐν τῷ Θηριακῷ λόγῳ, ἀποθνήσκει[5] πάντως· ἔχειν γάρ τι σηπτικὸν καὶ τὴν μόνην τοῦ ζῴου ἐπίψαυσιν λέγει. καὶ μέντοι καὶ τὸν πειρώμενον θεραπεύειν καὶ ἐπικουρεῖν ἀμωσγέπως τῷ ἀποθνήσκοντι φλυκταίνας ἴσχειν ἐν ταῖς χερσίν, ἐπεὶ μόνον τοῦ πατήσαντος προσέψαυσεν. Ἀριστόξενος δέ πού φησιν ἄνδρα ταῖς χερσὶν ὄφιν τινὰ ἀποκτεῖναι καὶ μὴ δηχθέντα ὅμως[6] ἀποθανεῖν· καὶ τὴν ἐσθῆτα δὲ αὐτοῦ, ἣν ἔτυχε φορῶν ὅτε τὸν ὄφιν ἀνῄρει, καὶ ἐκείνην σαπῆναι οὐ μετὰ μακρόν.

8. Ἀμφισβαίνης δὲ τὴν δορὰν βακτηρίᾳ περικειμένην ἐλαύνειν λέγει Νίκανδρος τοὺς ὄφεις

depths, and when it accepts the food or on the other hand rejects it.

6. It seems that donkeys are easily overcome and seized by wolves, and bees by bee-eaters, cicadas by swallows, and snakes by deer. And the leopard captures most animals, especially the monkey, by its odour.

Hunters and hunted

7. From Megasthenes I learn that a small fish occurs in the Indian Ocean, and that when alive it[1] is invisible, since presumably it swims down in the depths, but that when dead it floats to the surface. Anyone who touches it faints to begin with and later on dies.[2] And if one treads upon the chelydrus even without being bitten, as Apollodorus says in his work *Of Poisonous Animals*, death is inevitable. For he says that mere contact with the creature produces sepsis. And what is more, if anyone tries to administer medical treatment or help of any kind to the dying man he gets blisters on his hands, simply from having touched the man who trod on the snake. And Aristoxenus says somewhere that a man killed a snake with his hands and, though unbitten died[3] notwithstanding. And his very clothes which he happened to be wearing at the time when he slew the snake, turned in a short while to putrefaction.

Animals poisonous to the touch

8. Nicander asserts that the slough of the Amphisbaena if wrapped round a walking-stick drives

The Amphisbaena

---

[1] ἐκ βυθοῦ] *Schn* : νεκροῦ.
[2] λειποθυμεῖν καὶ ἐκθν-.
[3] *Ges* : ἀποθνήσκειν.
[4] *Jac* : θεωμένη.
[5] *OSchn* : χέρσυδρον.
[6] ὅμως θιγόντα.

πάντας καὶ τὰ ἄλλα ζῷα, ὅσα μὴ δακόντα μὲν
παίσαντα δὲ ἀναιρεῖ.

9. Κύων ὑπὸ πλήθους ὀχλούμενος οἶδε πόαν ἐν
ταῖς αἱμασιαῖς φυομένην, ἧσπερ οὖν γευσάμενος
ἐμεῖ πᾶν τὸ λυποῦν μετὰ φλέγματος καὶ χολῆς,
ὑποχωρεῖ δὲ αὐτῷ καὶ τῶν σκυβάλων πάμπολλα
καὶ πορίζει σωτηρίαν ἑαυτῷ, δεηθεὶς ἰατρῶν
συμμάχων οὐδὲ ἕν. καὶ μελαίνης μέντοι χολῆς
ἐκκρίνει πλῆθος, ἧπερ οὖν μείνασα λύτταν ἐργάζεται κυσὶ νόσημα ἀργαλέον. ἑλμίνθων δὲ πεπλη-
ρωμένοι τοῦ σίτου τοὺς ἀθέρας ἐσθίουσιν, ὡς
Ἀριστοτέλης λέγει. τρωθέντες δὲ ἔχουσι τὴν
γλῶτταν φάρμακον, ἧπερ οὖν περιλιχμώμενοι τὸ
τρωθὲν μέρος ἐς ὑγίειαν ἐπανάγουσιν, ἐπίδεσμα
καὶ σπληνία καὶ κράσεις φαρμάκων μακρὰ [1]
χαίρειν εἰπόντες.[2] κύνα δὲ καὶ ἐκεῖνο οὐ διαλέληθεν, ὅτι ἄρα τῆς † μελίας † [3] ὁ καρπὸς τοὺς μὲν ῦς
πιαίνει, αὐτῷ δὲ ἄλγημα ἰσχίου προξενεῖ· καὶ
ὁρῶν ἐμφορουμένην τοῦ προειρημένου τὴν ῦν,
ἀφίσταται αὐτῇ πάνυ ἐγκρατῶς καὶ τοῦ δοκοῦντος
ἡδέος. ἄνθρωποι δὲ τῶν πειθόντων ἄκοντας
ἐσθίειν ἡττῶνται πολλάκις πάνυ ἀκρατῶς.

10. Οὐκ ἄν ποτε ῥᾳδίως τοὺς ἐλέφαντας ἐνέδρα
λάθοι. ὅταν γοῦν [4] γένωνται τῆς τάφρου πλησίον,
ἣν εἰώθασιν ὑπορύττειν οἱ θηρῶντες αὐτούς, εἴτε

---

[1] Cobet : μακράν MSS, H. [2] ἀπολιπόντες.
[3] μελίας corrupt. [4] οὖν.

away all snakes and other creatures which kill not
by biting but by striking.[a]

9. A Dog burdened with a full stomach knows of a The Dog and
herb that grows on dry stone walls, and if he eats it its medicines
he vomits all that is paining him, mixed with phlegm
and bile, and a great deal of excrement also passes
off; so he restores his health without any need of
medical assistance. Further, he voids a quantity of
black bile which if retained causes madness, a troublesome disease in Dogs. And when infected by worms
Dogs eat the awns of corn, according to Aristotle
[*HA* 612 a 31]. When wounded they have their
tongue as a medicine, and with their tongue they lick
the wounded place and restore it to a healthy condition; bandages, compresses, and the compounding
of medicines they scorn. And another thing which
Dogs have not failed to observe is that the fruit of
the . . . fattens swine indeed but causes Dogs a
pain in their haunches. And though a Dog may
see a sow gorging itself with the aforesaid fruit,
with great self-control it leaves it to the sow for all
its seeming sweetness. Men however yield to those
who prevail upon them to eat against their will, often
to an altogether immoderate degree.

10. Elephants would not easily fail to notice an An Elephant
ambush. For instance, when they come near to the hunt
pit which elephant-hunters are in the habit of

---

[a] Nicander (*Th.* 373–83) says no more than that it is good
for chilblains. The discrepancy is explained by Wellmann
(*Hermes* 26. 335), who considers that Ael. was copying some
work based upon Apollodorus in which Nic. was mentioned,
and that he mistakenly ascribed to N. a statement made by A.

ἐννοίᾳ τινὶ φυσικῇ εἴτε μαντικῇ ναὶ μὰ Δία ἀπορρήτῳ τοῦ μὲν περαιτέρω χωρεῖν ἀναστέλλονται, ἑαυτοὺς δὲ ἐπιστρέψαντες εἶτα μέντοι ὡς ἐν πολέμῳ ἀνθίστανται μάλα καρτερῶς, καὶ ἀνατρέψαι πειρῶνται τοὺς θηρατὰς καὶ δι' αὐτῶν ὠσάμενοι φυγῇ πορίσασθαι τὴν σωτηρίαν, κρείττους γενόμενοι τῶν ἀντιπάλων. γίνεται τοίνυν ἐνταῦθα τοῦ καιροῦ μάχη καρτερὰ καὶ φόνος καὶ τῶν καὶ τῶν. ὁ μέντοι τρόπος τῆσδε τῆς μάχης τοιοῦτός ἐστιν. οἱ μὲν ἄνθρωποι δόρατα ἰσχυρὰ [1] ἀφιᾶσι στοχαζόμενοι αὐτῶν, οἱ δὲ ἐλέφαντες τὸν παραπεσόντα ἁρπάζουσι, καὶ τῇ γῇ προσαράξαντες πατοῦντές τε καὶ τοῖς κέρασι τιτρώσκοντες οἰκτίστῳ περιβάλλουσι τέλει [2] καὶ ἀλγεινοτάτῳ. ἐπίασι δὲ οἱ θῆρες ὑπὸ τοῦ θυμοῦ τὰ ὦτα ἐκπετανύντες ὡς ἱστία δίκην τῶν στρουθῶν τῶν μεγάλων, αἵπερ οὖν τὰς πτέρυγας ἁπλώσασαι ἢ φεύγουσιν ἢ ἐπίασιν· ἐπισιμώσαντες δὲ καὶ τὴν προβοσκίδα οἱ ἐλέφαντες καὶ ὑπὸ τοῖς κέρασι πτύξαντες ὥσπερ οὖν νεὼς ἔμβολον σὺν πολλῷ τῷ ῥοθίῳ φερομένης ἐμπεσόντες ῥύμῃ σφοδροτάτῃ πολλοὺς ἀνατρέπουσι βοῶντες διάτορόν τε καὶ ὀξὺ δίκην σάλπιγγος. πατουμένων δὲ τῶν ἁλισκομένων καὶ ἀλοωμένων τοῖς γόνασιν ἄραβος πολὺς ὀστῶν [3] συντριβομένων ἀκούεται καὶ πόρρωθεν, τὰ πρόσωπα δὲ ἐκθλιβομένων τῶν ὀφθαλμῶν καὶ τῆς ῥινὸς συνθλωμένης καὶ ῥηγνυμένου τοῦ μετώπου τὸ ἐναργὲς τοῦ εἴδους ἀπόλλυσι, καὶ ἀγνῶτες γίνονται πολλάκις καὶ τοῖς ἐγγυτάτω προσήκουσι. σώζονται δὲ παραδόξως ἄλλοι τὸν τρόπον τοῦτον. συνείληπται μὲν ὁ θηρατής, ὑφ' ὁρμῆς δὲ τὸ θηρίον ὑπερῆλθεν αὐτόν, καὶ τὰ γόνατα ἐς τὴν γῆν

secretly digging, whether by some natural instinct or by some altogether mysterious faculty of divination they restrain themselves from going any further, and turn back and put up a most strenuous resistance as in war and try to overthrow their hunters and, thrusting their way through them, to seek safety in flight after overcoming their adversaries. So then there ensues a fierce battle and there is a slaughter of hunters and hunted. And this is how the battle is fought. The men take aim and hurl stout spears at them, while the Elephants seize upon any man that has fallen in their way, dash him to earth, trample upon him, and wounding him with their tusks inflict upon him a most pitiful and agonising death. And the animals attack, their ears in passion spread wide like sails, after the manner of ostriches which open their wings to flee or to attack. And the Elephants bending their trunk inwards and folding it beneath their tusks, like the ram of a ship driving along with a great surge, fall upon the men in a tremendous charge, overturning many and bellowing with a piercing, shrill note like a trumpet. And as those who are caught are trampled or smashed by the beasts' knees, a great sound of bones being crushed can be heard even at a distance, and men's faces, with eyes knocked out, nose battered, and forehead split, lose their distinctive features, and frequently become unrecognisable even by their nearest relatives. Others however escape contrary to expectation, in the following manner. A hunter has been caught, but the Elephant in its forward rush has overpassed him and has planted its knees upon the earth and

---

¹ ἰσχυρὰ λόγχας.   ² τῷ τέλει.   ³ τῶν ὀστέων.

AELIAN

ἀπήρεισε,[1] καὶ προσκατέπηξε τὰ κέρατα ἐς
θάμνον ἢ ἐς ῥίζαν ἢ ἄλλο τι τοιοῦτο, καὶ ἔχεται,
καὶ μόγις ἀνασπᾷ καὶ ἐξαιρεῖ· ἐν δὲ τῷ τέως
διεκδὺς ὁ κυνηγέτης ἀπαλλάττεται. οὐκοῦν ἐν
τῇ τοιαύτῃ μάχῃ πολλάκις μὲν κρατοῦσιν οἱ
ἐλέφαντες, πολλάκις δὲ καὶ ἡττῶνται δείματα ἐξ
ἐπιβουλῆς καὶ δέα ποικίλα ἐπαγόντων. καὶ γὰρ
σάλπιγγες ᾄδουσι, καὶ δοῦπόν τε καὶ κτύπον
ἐργάζονται πρὸς [2] τὰς ἀσπίδας ἀράττοντες τὰ
δόρατα, καὶ πῦρ τὸ μέν τι ἐπὶ τῆς γῆς ἐξάπτουσι,[3]
τὸ δὲ μετέωρον [4] αἴρουσι, καὶ ἄλλο σφενδονῶσι [5]
δαλοὺς διαπύρους ἀκοντίζοντες καὶ δᾷδας μακρὰς
πυρὸς ἐνακμάζοντος τοῖς θηρίοις κατὰ προσώπου
βιαίως ἐπισείοντες. ἅπερ οὖν τὰ θηρία δεδιότα
καὶ δυσωπούμενα ὠθεῖται, ⟨καὶ⟩[6] ἔστιν ὅτε καὶ
ἐκνικᾶται ἐμπεσεῖν [7] ἐς τὴν τάφρον, ἣν τέως
ἐφυλάττετο.

11. Ἡγήμων ἐν τοῖς Δαρδανικοῖς μέτροις περὶ
Ἀλεύα τοῦ Θετταλοῦ φησι καὶ ἄλλα μέν, ἐν δὲ
τοῖς καὶ ὅτι ἠράσθη δράκων αὐτοῦ. καὶ ὅτι μὲν
εἶχε κόμην χρυσῆν ὅδε ὁ Ἀλεύας, λέγων τερατεύεται,[8] ἐμοὶ δὲ ἔστω ξανθή. καὶ βουκολεῖν μὲν
αὐτὸν ἐν τῇ Ὄσσῃ φησὶν ὡς ἐν τῇ Ἴδῃ τὸν
Ἀγχίσην, παρὰ δὲ τῇ κρήνῃ νέμειν τὰς βοῦς τῇ
καλουμένῃ Αἱμονίᾳ. Θετταλὴ δ' ἂν καὶ ἡ κρήνη
εἴη. δράκοντα οὖν μεγέθει μέγιστον ἐρασθῆναι
τοῦ Ἀλεύα, καὶ ἀνέρπειν ἐς αὐτόν, καὶ τὴν κόμην
οἱ καταφιλεῖν καὶ τῇ γλώττῃ περιλιχμώμενον

[1] ἐπήρεισεν.　　　[2] Reiske : καὶ πρός.
[3] ἐξάπτοντες.　　　[4] ὑψοῦ μετέωρον.

has besides fixed its tusks in a thicket or in a tree-root or some similar object, and is held fast and can only with difficulty withdraw and pull them out. Meanwhile the hunter slips out and escapes. In such a battle therefore it often happens that the Elephants are victorious, often however that they are defeated through the men designedly applying various means of scaring them. For instance, trumpets are sounded; the hunters make a din and a clash by beating their spears on their shields; now they light a fire on the ground, now they lift it up in the air; or again they launch burning firebrands like javelins and violently brandish great torches in full blaze before the faces of the animals. And as the animals dread and are dazzled by these things they are pushed back and sometimes forced to fall into the pit which till then they have kept clear of.

11. Hegemon in his poem, the *Dardanica*, among other things touching Aleuas the Thessalian, says that a snake was enamoured of him. And when he says that this Alcuas had ' golden ' hair he is romancing; let me call it ' flaxen.' And he says that he was a neatherd on mount Ossa, as Anchises was on Ida, and that he pastured his cattle near the spring called Haemonia. (The spring also would be in Thessaly.) Now a snake of enormous size fell in love with Aleuas and crept up to him and kissed his hair and with its tongue licked and washed the face of its

Love of beauty in animals

---

⁵ αἴροντες . . . σφενδονῶντες.
⁶ ⟨καί⟩ add. *Reiske.*
⁷ *Ges* : ἐκπεσεῖν.
⁸ τερατεύεται ὁ Ἡγήμων δηλονότι.

τὸ πρόσωπον τοῦ ἐρωμένου καθαίρειν, καὶ δωροφο-
ρεῖν αὐτῷ θηρῶντα πάμπολλα. εἰ δὲ Γλαύκης
τῆς κιθαρῳδοῦ κριὸς ἥττητο [1] καὶ ἐν Ἰασῷ
δελφὶς ἐφήβου,[2] τί κωλύει καὶ δράκοντα ἐρασθῆναι
νομέως ὡραίου, τὸν ὀξυωπέστατον κάλλους διαπρε-
ποῦς ἀγαθὸν κριτὴν γεγενημένον; ἦν δὲ ἄρα
ἴδιον ζῴων καὶ ἐρασθῆναι μὴ μόνον τοῦ συννόμου
τε ἅμα καὶ συμφυοῦς, ἀλλὰ καὶ τοῦ προσήκοντος
ἥκιστα, ὡραίου μέντοι.

12. Ὁ παρείας ἢ παρούας (οὕτω γὰρ Ἀπολ-
λόδωρος ἐθέλει) πυρρὸς τὴν χρόαν, εὐωπὸς τὸ
ὄμμα, πλατὺς τὸ στόμα, δακεῖν οὐ σφαλερός,
ἀλλὰ πρᾶος. ἔνθεν τοι καὶ τῷ θεῶν φιλανθρωποτάτῳ [3] ἱερὸν ἀνῆκαν [4] αὐτόν, καὶ ἐπεφήμισαν
Ἀσκληπιοῦ θεράποντα εἶναι οἱ πρῶτοι [5] ταῦτα
ἀνιχνεύσαντες.

13. Ἐν Αἰθιοπίᾳ τοὺς καλουμένους Σιβρίτας
σκορπίους (οὕτω δὲ αὐτοὺς ὡς εἰκὸς οἱ ἐπιχώριοι
φιλοῦσιν ὀνομάζειν) ἀκούω σιτεῖσθαι καὶ σαύρους
καὶ ἀσπίδας καὶ σφονδύλας καὶ τίφας καὶ πᾶν
ἑρπετόν, τὸν δὲ ἐπιβάντα αὐτῶν τοῖς περιττώμασιν
ἕλκουσθαι πέπυσμαι. περὶ Κέρκυραν δὲ γίνονται
αἱ καλούμεναι ὕδραι, αἵπερ οὖν τοὺς διώκοντας

____

[1] ἥττητο καὶ Πτολεμαίῳ γε τῷ Φιλαδέλφῳ ἀντήρα.
[2] Jac : ἑτέρου.
[3] τῷ φιλανθρωποτάτῳ θεῶν.
[4] ἀφῆκαν.
[5] πρῶτοί μου.

loved one and brought him as presents many of the
spoils of its hunting.

Now if a ram was overcome by love of Glauce the
harpist, and a dolphin of a youth at Iassus,[a] what is
there to prevent a snake also from falling in love with
a handsome shepherd, or the most keen-sighted of
creatures from being a good judge of conspicuous
beauty? So it seems that it is in fact a characteristic
of animals to fall in love not only with their companions and kin but even with those who bear no
relation to them at all but are yet beautiful.

12. The *Pareas* or *Paruas* [b] (for this is the form The
preferred by Apollodorus) is of a red colour, has 'Pareas'
sharp eyes and a wide mouth; its bite is not injurious
but gentle. That, you see, is the reason why those
who first made these discoveries consecrated it to the
god who is the kindest to man and gave it the name
of 'servant to Asclepius.'

13. I have heard that in Ethiopia the Scorpions The
known as *Sibritae* (that is what the inhabitants [c] 'Sibritae'
commonly call them, as is natural) feed upon lizards, Scorpions
asps, sphondylae,[d] cockroaches, and all creeping
things, but I have ascertained that anyone who
treads upon their excrement develops ulcers.

In Corcyra there occur water-snakes, as they are Various
called, which round upon their pursuers and by snakes

---

[a] See 6. 15.
[b] *Coluber longissimus* (or *Aesculapii* or *flavescens*), a
beneficent snake, kept in the temple of Asclepius at Epidaurus.
[c] The Sibritae were an Ethiopian tribe dwelling between
the upper arms of the Nile and the Red Sea.
[d] Perh. a kind of beetle; one of the *Cerambycidae* or long-
horn beetles (Gossen § 52).

ἐπιστραφεῖσαι καὶ φυσήσασαι πνεῦμα ἄτοπον εἶτα
ἀναστέλλουσι τῆς ὁρμῆς καὶ ἀποστρέφουσι. τὸν
τύφλωπα δέ, ὃν καὶ τυφλίνην καλοῦσι καὶ κωφίαν
προσέτι,[1] κεφαλὴν μὲν παραπλησίαν ἔχειν μυραίνῃ
λέγει τις λόγος, ὀφθαλμοὺς δὲ ἄγαν βραχίστους.
καὶ θάτερον μὲν τοῖν ὀνομάτοιν ἐντεῦθεν εἴληφε,
τόν γε μὴν[2] κωφίαν, ἐπεὶ νωθής ἐστι τὴν ἀκοήν.
δορὰν δὲ ἰσχυρὰν ἔχει καὶ διακοπτομένην βραδύ-
τατα. τὸν δὲ ἀκοντίαν χέρσυδρον εἶναί φασι,
χρόνου δὲ[3] ἐν ξηρῷ ποιεῖσθαι τὴν διατριβὴν
πολλοῦ καὶ ἐλλοχᾶν ζῷον πᾶν. ἡ δὲ σοφία τῆς
ἐπιβουλῆς τῆς ἐξ αὐτοῦ τοιάδε ἐστίν. ἐν ταῖς
λεωφόροις που λαθὼν ὑποκρύπτεται, πολλάκις δὲ
καὶ ἐπί τι δένδρον ἀνερπύσας εἶτα ἑαυτὸν συνειλήσας καὶ τὴν κεφαλὴν ἐν τῇ σπείρᾳ ὑποκρύψας
τοὺς παριόντας ἡσυχῇ[4] ὑποβλέπει· εἶτα ἑαυτὸν
ἀφίησιν ἐς τὸ παριόν, εἴτε ἄλογον εἴη ζῷον εἴτε
ἄνθρωπος. ἔστι δὲ ἁλτικὸν θηρίον καὶ διαπηδῆσαι
καὶ εἴκοσιν εἰ δέοι πήχεις οἷόν τε· ἁλλόμενόν τε
παραχρῆμα ἐνέφυ.

14. Λύκοι βοΐ ἐς τέλμα βαθὺ ἐμπεσόντι ἐάν
πως περιτύχωσι, ταράττουσι μὲν αὐτὸν ἔξωθεν
καὶ φοβοῦσι, διανήξασθαι καὶ ἐπιβῆναι τῆς γῆς
οὐκ ἐπιτρέποντες, ἀναγκάζουσι δὲ τῷ χρόνῳ
στρεβλούμενον καὶ ἰλυσπώμενον ἀποπνιγῆναι. εἶτα
εἰς αὐτῶν ὁ τελεώτατος ἐμπηδήσας τῷ ὕδατι καὶ
προσνεύσας ἐλάβετο τῆς οὐρᾶς τοῦ βοὸς καὶ
ἕλκει ἐς τὸ ἔξω, καὶ ἕτερος τῆς ἐκείνου λαβόμενος
αὐτὸν ἕλκει, καὶ τὸν δεύτερον ὁ τρίτος, καὶ

---

[1] δὲ προσέτι. [2] τὸν μὲν δή.
[3] γάρ. [4] ἡσυχῇ καὶ λανθάνων.

blasts of foul breath make them pause in their
attack and deter them. According to one account
the *Typhlops* (blind-eyes),[a] which people also call
Typhline and *Cophias* as well, has a head nearly
resembling the moray, but very small eyes. And
the second of its two names, that is *Cophias*, it has
derived from the fact that it is dull of hearing. But
its skin is hard and takes a long time to cut through.
And the *Acontias* (javelin-snake), they say, is amphibious and spends much time on dry land, lying in
wait for every kind of living creature. And it shows
skill in its fell designs, thus. It lurks hidden it may
be in thoroughfares; often it crawls up some tree and
coils itself up and concealing its head in its coils, spies
quietly upon the passers-by. Then it launches
itself on whatever is passing, be it brute beast or
man. The creature is good at leaping and is
capable of jumping as much as twenty cubits, if
need be. And where it leaps it instantly fastens
on.

14. If by chance Wolves come upon an Ox that has   Wolves
fallen into a deep pond, they harass and terrify him   and Ox
from the bank, never allowing him to swim across
and get out on to land, and compel him after long
torment and floundering to drown. Then the
strongest Wolf in the pack leaps into the water and
swimming up to the Ox, seizes its tail and begins to
drag it to the bank; and a second wolf seizes the
tail of the first and drags it, then a third drags the

---

[a] 'Probably *Pseudopus pallasi*,' Thompson on Arist. *HA*
567 b 25 (Eng. tr.). It is a limbless lizard and is known as a
'glass-snake.' Other interpretations are *Anguis fragilis*
(Brenning), *Typhlops vermicularis* (Gossen–Steier).

τοῦτον ὁ τέταρτος, καὶ δρᾶται τὸ εἰρημένον μέχρι
τοῦ τελευταίου, ὅσπερ οὖν ἔξω τοῦ ὕδατος
ἕστηκε. καὶ τὸν τρόπον τοῦτον ἐξαγαγόντες τὸν
βοῦν ποιοῦνται δεῖπνον. βοὸς δὲ μόσχον πεπλανη-
μένον ἐλλοχήσαντες εἶτα αὐτῷ προσπηδῶσι, καὶ
τοῦ μυκτῆρος λαβόμενοι ἕλκουσιν· ὁ δὲ ἀντισπᾷ,[1]
καὶ ἅμιλλα ὑπὲρ τούτου πολλή, τῶν μὲν ἐκβιάσα-
σθαι πειρωμένων, τοῦ δὲ μὴ εἶξαι ἀγώνισμα
ποιουμένου. ὅταν δὲ αὐτὸν οὕτως θεάσωνται
σφόδρα ἀντιτείνοντα, μεθῆκαν· καὶ ἐκεῖνος ὑπὸ
τῆς ἐς τοὐπίσω βίας ἀνατέτραπται, καὶ οἱ λύκοι
ἐμπεσόντες ἀνέρρηξαν τὴν νηδὺν καὶ ἐσθίουσιν
αὐτόν.

15. Ὅταν ὑπερβῆναι τάφρον οἱ ἐλέφαντες μὴ
δύνωνται, εἰς ὁ μέγιστος ἑαυτὸν ἐς αὐτὴν ἐμβάλλει,
καὶ πλάγιος ἵσταται, καὶ γεφυροῖ τὸ κενόν, καὶ
κατ᾽ αὐτοῦ βαίνοντες ἐς τὸ ἀντιπέρας ἴασι καὶ
ἀποδιδράσκουσι, πρότερον μέντοι καὶ ἐκεῖνον
ἀνασώσαντες. ὁ δὲ τῆς σωτηρίας τρόπος οὗτός
ἐστιν. ἄνωθέν τις τὸν πόδα προτείνει, καὶ
ἐκείνῳ παρέχει τὴν προβοσκίδα περιπλέξαι· οἱ
δὲ ἄλλοι φρύγανα ἐμβάλλουσι καὶ ξύλα ὤκιστα,
ὧν ἐπιβαίνων, εἰλημμένος ⟨τε⟩[2] τοῦ ποδὸς μάλα
ἐγκρατῶς τε καὶ εὐλαβῶς ἀνασπᾶται ῥᾶστα.

Ἔστι δὲ ἐν τοῖς Ἰνδοῖς ἄρουρα, καὶ κέκληται
Φαλάκρα. τὸ δὲ αἴτιον τοῦ ὀνόματος, ὁ γευσάμενος
τῆς ἐνταῦθα γινομένης[3] πόας καὶ τὰς τρίχας
ἀποβάλλει καὶ τὰ κέρατα. οὐκοῦν οἱ ἐλέφαντες
ἑκόντες εἶναι οὐ προσίασι τῇδε τῇ ἀρούρᾳ, ἀλλ᾽

---

[1] ἀντισπᾷ ἑαυτόν.   [2] ⟨τε⟩ add. Reiske.

second, and a fourth the third, and this is repeated
up to the last Wolf, which is standing out of the
water. And having hauled out the Ox in this way,
they enjoy a feast. They lie in wait for a strayed
Calf and leap upon it, and seizing it by the nose drag
it along. But the Calf pulls against them and there
is a fierce struggle for it, the Wolves trying to overcome it by force, the Calf fighting hard not to yield.
And when they see it resisting with all its might in
this way, they let go; whereupon the Calf by
straining in the opposite direction is upset, and the
Wolves leap upon it, tear open its belly, and devour
it.

15. When Elephants are unable to cross a ditch Elephants
the largest one in the herd throws himself into it cross a ditch
and standing transversely bridges the gap, while the
rest tread on his back, cross to the far side, and make
off, but not until they have rescued him. And the
way in which they rescue him is as follows. One of
them on the bank puts his foot forward and allows
the large Elephant to wrap his trunk round it.
Meantime the others throw undergrowth and timber
into the trench as fast as they can. And he mounts
on these and clinging firmly with all his might to the
other's foot is drawn up without difficulty.

There is in India a tract of land called *Phalacra*
(bald). And the reason for the name is that any
creature which eats the grass growing there loses its
hair and its horns. Accordingly Elephants do not
willingly go near this tract, but if they have drawn

---

[3] γενομένης.

ἀποστρέφονται πλησίον γενόμενοι, πᾶν τὸ βλάπτον
φεύγοντες ὡς ἄνθρωποι φρόνιμοι οἱ ἐλέφαντες.

16. Τὴν σπογγιὰν ἰθύνει βραχὺ ζῷον, οὐ
καρκίνῳ τὴν ἰδέαν παραπλήσιον, ἀλλὰ ἀράχνῃ
μᾶλλον. οὐ γὰρ ἄψυχον οὐδὲ αἵματος ἄμοιρον ἡ
σπογγιὰ κύημά ἐστι θαλάττης,[1] ἀλλὰ[2] ταῖς
πέτραις προσφύεται, ὥσπερ οὖν καὶ ἕτερα, ἔχει
δέ τινα κίνησιν ἰδίαν, δεῖται δὲ ὡς ἂν εἴποις τοῦ
ὑπομνήσοντος αὐτὴν ὅτι ἔμψυχός ἐστιν. ἀτρε-
μοῦσα γὰρ ὑπό[3] τινος συμφυοῦς μανότητος καὶ
ἡσυχάζουσα τοῖς τρήμασιν αὐτῆς ὅταν προσπέσῃ
τι, ἐνταῦθα ὑπὸ τοῦ ἀραχνώδους ζῴου νύττεται,
καὶ συλλαμβάνει τὸ ἐμπεσόν, καὶ τροφὴν ἴσχει.
ὅταν δὲ ἄνθρωπος προσίῃ ἐπ' ἐκτομῇ αὐτῆς,
κεντουμένη[4] ὑπὸ τοῦ ζῴου τοῦ συντρόφου φρίττει
καὶ ἑαυτὴν συστρέφει, καὶ αἰτία πόνου τε καὶ
καμάτου γίνεται τῷ θηρατῇ ναὶ μὰ Δία πολλοῦ.

17. Εἴρηται μὲν οὖν ἡμῖν περὶ ἐλεφάντων ἰδίᾳ,
τὰ δὲ καὶ εἰρήσεται. † οὑτωσὶ κρατοῦσι μὲν
βίου †[5] σωφροσύνης δὲ ὅπως μετειλήχασιν, εἰπεῖν
πρεπωδέστατον. οὐ γὰρ ὡς ὑβρίζοντες οὐδὲ ὡς
λάγνοι ἐπὶ τὴν ὁμιλίαν τὴν πρὸς τὴν θήλειαν
ἔρχονται, ἀλλ' ὥσπερ οὖν οἱ γένους διαδοχῆς
δεόμενοι καὶ παιδοσποροῦντες, ἵνα μὴ αὐτοὺς
ἐπιλίπῃ ἡ ἐπιγονὴ ἡ ἐξ ἀλλήλων, ἐάσωσι δὲ
σπέρμα. ἅπαξ γοῦν ἐν τῷ βίῳ τῷ σφετέρῳ
μνημονεύουσιν ἀφροδίτης, ὅταν ἡ θήλεια ὑπομένῃ
καὶ αὐτή· εἶτα ἐμπλήσας ἕκαστος τὴν σύννομον

___

[1] θαλάττης καὶ πέφυκεν εἶναι ζῷον.   [2] καί.

near to it they move away, since Elephants, like
prudent men, avoid anything that is harmful.

16. The Sponge is directed by a small animal The Sponge
resembling a spider rather than a crab. For the
Sponge is no lifeless or bloodless object engendered
by the sea, but clings to the rocks like other creatures
and has a certain power of movement in itself,
though it needs, as you might say, someone to remind
it that it is a living creature, for owing to some
natural porosity it remains motionless and at rest,
until something encounters its pores; then the
spider-like creature pricks it, and it seizes what
has fallen in and makes a meal. But when a man
approaches to cut it off, the Sponge is pricked by
the animal that lives in it, shudders, and contracts,
and the trouble and labour that this causes to the
fisherman is considerable, and no mistake.

17. I have indeed spoken of Elephants in a separate The
chapter, but I shall add the following . . . it is Elephant, its
continence
most fitting to state that they have been gifted with
temperance. For they seek intercourse with the
female not as though minded to commit an outrage
or from lust, but like men desiring a succession to
their family and to beget children, in order that
their common offspring may not fail but that they
may leave their seed after them. At any rate once
only in a life-time do their thoughts turn to love,
when the female herself submits. Then when each
one has impregnated its mate, thereafter it knows

---

[3] ὡς ὑπό. [4] Schn: ἐκκεντουμένη.
[5] οὑτωσὶ . . . βίου corrupt.

τὸ ἐντεῦθεν[1] οὐκ οἶδεν αὐτήν. συμπλέκονται δὲ
οὐκ ἀνέδην οὐδὲ ἐν τῇ τῶν ἄλλων ὄψει ἀλλ᾽
ἀναχωρήσαντες· καὶ ἑαυτῶν προβάλλονται ἢ
δένδρα δασέα ἢ ὕλην τινὰ συμφυῆ ἢ χῶρον κοῖλον
καὶ βαθὺν τοῦ λαθεῖν αὐτοῖς παρέχοντα ἀφθονίαν.
ὡς μὲν οὖν εἰσι δίκαιοι ἄνω εἶπον, καὶ τὸ ἀνδρεῖον
αὐτῶν καὶ τοῦτο ἤδη λέλεκται· τὸ[2] σῶφρον δὲ
ἀποδέδεικται[3] τὰ νῦν ταῦτα. ἀλλὰ καὶ τὸ
μισοπόνηρον ὅτῳ σχολὴ μανθάνειν, οὗτος ὑπέχων
τὰ ὦτα ἀκουέτω. ἐλέφαντι ἡμέρῳ πωλευτὴς ἦν,
καὶ εἶχε γυναῖκα ἀφηλικεστέραν μέν, πλουσίαν
δέ. οὐκοῦν ἑτέρας ἐρῶν καὶ τὰ τῆς συνοικούσης
σπεύδων ἐκείνης γενέσθαι[4] ταύτην μὲν ἀποπνίγει
καὶ τῆς τοῦ ἐλέφαντος φάτνης κατορύττει πλησίον
ὁ θερμόβουλος ἄνθρωπος, ἄγεται δὲ τὴν ἄλλην.
ἐνταῦθα οὖν ὁ ἐλέφας τῇ προβοσκίδι λαβόμενος
τὴν νεωστὶ ἀφιγμένην ἄγει τῆς νεκρᾶς[5] πλησίον, καὶ τοῖς κέρασιν ἀνορύξας καὶ γυμνώσας τὸ
σῶμα, ἃ εἰπεῖν οὐκ ἠδύνατο, ταῦτα ἐπεδείκνυε δι᾽
αὐτῶν τῶν ἔργων, τὴν γυναῖκα τὸν τρόπον τοῦ
γήμαντος αὐτὴν ἐκδιδάσκων ὁ μισοπόνηρος ἐλέφας.

18. Ἐγγραύλεις, οἱ δὲ ἐγκρασιχόλους καλοῦσιν
αὐτάς, προσακήκοά γε μὴν καὶ τρίτον ὄνομα
αὐτῶν, εἰσὶ γὰρ οἱ καὶ λυκοστόμους αὐτὰς
ὀνομάζουσιν. ἔστι δὲ μικρὰ ἰχθύδια, καὶ πολύγονα
φύσει, λευκότατα ἰδεῖν. ἐσθίουσί γε μὴν μάλιστα
οἱ ἀγελαῖοι τῶν ἰχθύων αὐτά. δείσαντα οὖν[6]
συνθεῖ πρὸς[7] ἄλληλα, καὶ ἐχόμενον τοῦ πλησίον

---

[1] Schn : τὸ ἐντεῦθεν ἐπὶ τούτοις.
[2] καὶ τό.
[3] ἀπολέλεκται.

her no more. And they do not couple without reserve or in the sight of others but withdraw and screen themselves in thick trees or in some close-growing forest or in some deep hollow, which affords them ample means of hiding.

Now I said above that they were just, and I have already spoken of their valour. Their continence has been displayed in the present instance. Further, anyone who has leisure to learn of their detestation of evil should lend an ear and listen to this. The trainer of a tame Elephant had a somewhat elderly but rich wife. Now he was in love with another woman, and desiring that his wife's property should become hers, he strangled his wife and buried her, rash man that he was, close by the Elephant's manger, and married the other woman. So then the Elephant seizing hold of the new arrival with its ^{reveals} trunk led her up to the dead body, dug it up, and ^{murder} laid it bare with its tusks, showing by its mere action what it could not express in words, and enlightening the woman as to the conduct of him who had wedded her; such was the Elephant's hatred of evil.

18. Anchovies (*engrauleis*, which some call *en-* The *crasicholi*, and I have even heard a third name ^{Anchovy} applied to them, for some call them ' wolf-mouths ') are a tiny fish, prolific by nature, and pure white in appearance. They are principally eaten by fish which swim in shoals, and so when scared they rush to one another, and as each clings to its neighbour,

---

⁴ γίνεσθαι.   ⁵ νεκροῦ.
⁶ Schn : μήν.   ⁷ εἰς.

AELIAN

ἕκαστον τῇ σφίγξει τὸ ῥᾳδίως ἐπιβουλεύεσθαι
διαπέφευγε. τοσαύτη δὲ ἄρα αὐτῶν ἡ ἕνωσις
γίνεται συνδραμόντων, ὡς καὶ πορθμίδας ἐπιθεού-
σας μὴ διασχίζειν αὐτά· καὶ μέντοι καὶ κώπην ἢ
κοντὸν εἴ τις αὐτῶν διεῖναι θελήσειε, τὰ δὲ οὐ
διαξαίνεται, ἀλλὰ ἔχεται ἀλλήλων ὡς συνυφασμένα. καθεὶς δὲ τὴν χεῖρα ὡς ἐκ σωροῦ πυρῶν
ἢ κυάμων λάβοις [1] ἂν βιαίως ἀποσπάσας, ὡς καὶ
διασπᾶσθαι πολλάκις, καὶ τὰ μὲν ἡμίτομα τῶν
ἰχθυδίων λαμβάνεσθαι, τὰ δὲ ὑπολείπεσθαι.[2] καὶ
γὰρ [3] τὸ μὲν οὐραῖον καθέξεις, μένει δὲ σὺν [4] τοῖς
ἄλλοις ἡ κεφαλή· ἢ κεφαλὴν κομιεῖς οἴκαδε,[5]
μένει δὲ ἐν τῇ θαλάττῃ τὸ λοιπόν. καλεῖται δὲ
αὐτῶν ἡ πυκνή τε καὶ συνεχὴς νῆξις βόλος, καὶ
πεντήκοντα ἁλιάδας πολλάκις ἐπλήρωσεν εἷς
βόλος, ὥς φασιν οἱ θαλαττουργοί.[6]

19. Ἡ ὗς γνωρίζει τοῦ συβώτου τὴν φωνήν,
καὶ ὑπακούει καλοῦντος, κἂν ᾖ πλανηθεῖσα·
πλησίον δὲ τούτου τὸ μαρτύριον. τῇ γῇ τῇ
Τυρρηνίδι κακοῦργοι ναῦν λῄστειραν προσέσχον,
καὶ προελθόντες [7] αὐλίῳ περιτυγχάνουσι, καὶ ἦν
συβωτῶν τὸ αὔλιον, καὶ εἶχε πολλὰς ὗς. ταύτας
οὖν συλλαβόντες ἐς τὴν ναῦν ἐνέβαλον, καὶ
ἀπολύσαντες τὰ πείσματα εἴχοντο τοῦ πλοῦ. οἱ
τοίνυν συβῶται παρόντων μὲν τῶν λῃστῶν ἡσύχα-
ζον, ἐπεὶ δὲ ἔτυχον τῆς γῆς ἀποσαλεύσαντες,
ὅσσον τε γέγωνε βοήσας, ἐνταῦθά τοι τὰς σῦς

---

[1] λάβοι.
[2] ἀπολείπεσθαι.
[3] καὶ γὰρ ἐν τῷ πλήθει.
[4] ἐν.
[5] οἴκαδε σὺν τοῖς ἄλλοις.
[6] φησιν ὁ θαλαττουργός.

204

by their close cohesion they avoid falling an easy prey to plots upon their life. And so united is their mass when they have rushed together that even ships which run into them do not cleave it. Moreover should someone wish to drive an oar or a pole through them, they are not torn apart, but cling to each other as though woven together. But if you put your hand down and pull hard as if you were drawing grains of wheat or beans from a heap, you may catch some, with the result that they are often torn to pieces and that fragments of fish are caught, while the rest is left behind. For though you may get possession of the tail, yet the head remains with the other fish; or you may take home a head, but the rest of the fish remains in the sea. Their swimming in a dense, compact mass is called a 'draught,' and a single draught often fills fifty fishing-boats, as toilers of the sea inform us.

19. The Sow recognises the voice of the swine- Pigs and herd, and attends to his call even though it has pirates wandered away. Evidence for this statement is to hand. Some miscreants beached their pirate vessel on the shore of Etruria, and proceeding inland came upon a fold belonging to some swineherds and containing a large number of Sows. These they seized, put them on board, loosed their cables, and continued on their voyage. Now so long as the pirates were on the spot the swineherds kept quiet, but when they were off shore in the roadstead 'and as far as a cry might carry,'[a] then the swineherds with their

[a] Hom. *Od.* 5. 400.

---

[7] *Jac* : προσελθόντες MSS, *perh.* προσσχόντες *H*.

τῇ συνήθει βοῇ ὀπίσω [1] παρὰ σφᾶς ἀπεκάλουν οἱ
συβῶται. αἱ δὲ ὡς ἤκουσαν, ἐπὶ θάτερα τοῦ
πλοίου ἑαυτὰς συνώσασαι ἀνέτρεψαν αὐτό. καὶ
οἱ μὲν κακοῦργοι παραχρῆμα διεφθάρησαν, αἱ δὲ
ὗς παρὰ τοὺς ἑαυτῶν δεσπότας ἀπενήξαντο.

20. Ζηλότυπον δὲ εἶναι καὶ τὸν πελαργόν
φασιν. ἐν γοῦν Κραννῶνι τῆς Θετταλίας Ἀλκι-
νόην ὄνομα γυναῖκα ὡραίαν ὁ γήμας ἀπολιπὼν
οἴκοι ἔς τινα ἐστείλατο ἀποδημίαν. ἡ τοίνυν
Ἀλκινόη ὡμίλει τῶν θεραπόντων τινί. τοῦτο
συνιδὼν ὁ πελαργὸς ὁ οἰκέτης οὐχ ὑπέμεινεν,
ἀλλὰ ἐτιμώρησε τῷ δεσπότῃ. προσπηδῶν γοῦν
ἐπήρωσε τῆς ἀνθρώπου τὴν ὄψιν. ἀνωτέρω [2]
μὲν ἐμνήσθην πορφυρίωνος ζηλοτυπίας εἶτα κυνὸς
τοιούτου, νῦν γε μὴν πελαργοῦ τὰ ἴσα ἐκείνοις [3]
ἐς νοσοῦντα γάμον.

21. Μεταβάλλει δὲ τὰς χρόας τὰ πρόβατα ἐκ
τῆς περὶ τὸ πῶμα ἀλλαγῆς κατὰ τὴν τῶν ποταμῶν
ἰδιότητα· ἡ δὲ ὥρα τοῦ ἔτους, καθ' ἣν ἀπαντᾷ
τοῦτο αὐτοῖς, ὁ τῆς μίξεως καιρός ἐστιν. γίνεται
οὖν καὶ ἐκ λευκῶν μέλανα, καὶ ἔμπαλιν τρέπει
τὴν χρόαν. φιλεῖ δέ πως ταῦτα γίνεσθαι περί τε
τὸν ἐν Ἀντανδρίᾳ ποταμὸν [4] καὶ τὸν ἐν Θρᾴκῃ,
οὗ τὸ ὄνομα ἐροῦσιν οἱ πάροικοι Θρᾷκες. ὁ δὲ
ἐν Τροίᾳ Σκάμανδρος ἐπεὶ ξανθὰς ἀποφαίνει

[1] ὀπίσω καί.   [2] Jac: ἀνωτάτω.
[3] Perh. some word like θυμωθέντος (H 1858) has been lost after
ἐκείνοις.

accustomed cry called the Swine back to them. And when the Swine heard it they pressed together to one side of the vessel and capsized it. And the miscreants were drowned forthwith, but the Swine swam away to their masters.

20. They say that the Stork also is subject to jealousy.[a] At any rate at Crannon in Thessaly a man who had married a beautiful wife of the name of Alcinoe left her at home and went away on his travels. So Alcinoe had intercourse with one of the servants. The Stork that was about the house got to know of this and would not tolerate it, but avenged its master. At any rate it sprang upon the woman and blinded her eyes.

A Stork punishes adulteress

I have earlier on spoken of jealousy on the part of a Purple Coot, then of a Dog in like case, and now of a Stork equally affected over a marriage that went wrong.

21. Sheep change their colour as their drink varies with the character of the rivers. The season of the year in which this occurs is the season of mating. So from being white they become black, and the contrary change of colour occurs. This commonly takes place near the river of Antandria [b] and the river in Thrace whose name the neighbouring Thracians will tell you. And since the Scamander in the Troad turns the sheep that drink of it yellow,

Waters that change the colour of Sheep

---

[a] See INDEX II, s.v. 'Jealousy.'

[b] Antandrus, town at the head of the gulf of Adramyttium in Mysia; the river was the Satniois.

---

[4] τὼ . . . ποταμώ *Gron, comp.* Arist. *HA* 519 a 16.

πινούσας τὰς οἷς, πρὸς τῷ Σκαμάνδρῳ τῷ ἐξ
ἀρχῆς ἄλλο ὄνομα ἢ τῶν προβάτων ἐπίκτητος
χρόα ἔθετο αὐτῷ τὸ Ξάνθου.

22. Χάριτος δὲ ἀπομνησθῆναι τὰ ζῷα καὶ κατὰ
τοῦτο ἀγαθά. ἐν Τάραντι γίνεται γυνὴ τά τε
ἄλλα σπουδῆς ἀξία καὶ οὖν καὶ σώφρων πρὸς τὸν
ἄνδρα· Ἡρακληὶς ὄνομα αὐτῇ. περιεῖπε μὲν οὖν
ζῶντα τὸν γεγαμηκότα εὖ μάλα κηδεμονικῶς·
ἐπεὶ δὲ τὸν βίον οὗτος κατέστρεψε, τὰς ἀστικὰς
ἡ προειρημένη γυνὴ μισεῖ διατριβὰς καὶ τὴν
οἰκίαν, ἐν ᾗ τὸν ἄνδρα νεκρὸν ἐθεάσατο, καὶ ὡς
εἶχε λύπης ἐς τοὺς τάφους μετοικίζεται, καὶ τοῖς
ἠρίοις τοῦ ποτε ἀνδρὸς τλημόνως παρέμεινε,
πιστὴν ἑαυτὴν [1] τῷ κατὰ γῆς ὄντι ἀποφαίνουσα.
καὶ ποτε ἦν ὥρα θέρειος, καὶ πελαργῶν ἔτι
νεοττῶν πρόπειραν τῆς ἑαυτῶν πτήσεως λαμβα-
νόντων εἷς ὁ μάλιστα νεαρὸς ἀκρατὴς ὢν ἔτι τῶν
ταρσῶν κατώλισθε, καὶ τοῖν σκελοῖν συντρίβει τὸ
ἕτερον. ἡ τοίνυν Ἡρακληὶς θεασαμένη τὸ πτῶμα
καὶ τοῦ ποδὸς τὸ πάθος καταμαθοῦσα οἰκτείρει
τὸν νεοττόν, καὶ ἀναλαβοῦσα σὺν πολλῇ τῇ φειδοῖ
κατειλεῖ τὴν πληγήν, καὶ θεραπεύει καταιονήμασι
καὶ ἐπιπλάσμασι, καὶ τροφὴν προσέφερε καὶ
ποτὸν ὤρεγε, χρόνῳ δὲ τῷ εἰκότι ῥωσθέντα καὶ
φύσαντα τὰ ὠκύπτερα ἐλεύθερον εἶναι μεθῆκεν.
ὁ δὲ εἰδὼς ἐννοίᾳ τινὶ φυσικῇ καὶ θαυμαστῇ
ὀφείλων ζωάγρια ᾤχετο ἀπιών. εἶτα ἐνιαυτοῦ
διελθόντος ἡ μὲν ἔτυχεν ἦρος ὑπολάμποντος ἐν
ἡλίῳ θερομένη, ὁ δὲ πελαργὸς ὁ ἰαθεὶς ὑπ' αὐτῆς
ἰδὼν τὴν εὐεργέτιν ὑφῆκε τῆς τῶν πτερῶν ὁρμῆς,
καὶ ἑαυτὸν χθαμαλωτέρᾳ τῇ πτήσει κατάγων

the colour which the flocks acquire has caused the name *Xanthus* (yellow) to be added to its original name of ' Scamander.'

22. In this respect also animals are good, viz at remembering to be grateful. There was a woman in Tarentum, admirable in other ways and particularly as a faithful wife. Her name was Heracleïs. So long as her husband lived she cared for him with the utmost devotion. But when he died the woman took a dislike to life in the city and to the home in which she had seen her husband dead, and such was her grief that she went to dwell among the tombs and was content to remain by her late husband's sepulchre, constant to him who was beneath the soil. And once in summer when some storks, still fledglings, were essaying their first flight, one of them, the youngest, not having sufficient strength of wing, fell and broke one of its legs. So Heracleïs seeing its fall and finding how its leg was injured, took pity on the nestling and picking it up very gently wrapped up the wound, and tended it with fomentations and plasters, brought it food, gave it drink, and, when in due course it was strong and had grown its quill-feathers, set it free. And the stork, knowing by some strange instinct that it owed her the price of its life, departed. Later when a year had passed and spring was just beginning to brighten, the woman chanced to be warming herself in the sun, and the Stork which had been healed by her, seeing its benefactress, checked the speed of its wings and sinking nearer to earth came close, opened its bill,

Woman of Tarentum and Stork

---

¹ ἑαυτὴν καὶ σώφρονα.

πλησίον γίνεται, καὶ χανὼν ἀνεμεῖ λίθον ἐς τὸν
τῆς Ἡρακλῆιδος κόλπον, καὶ ἀναπετασθεὶς ἐπὶ
τοῦ τέγους ἑαυτὸν ἐκάθισεν. ἡ δὲ τὰ πρῶτα ὡς
εἰκὸς ἐθαύμασέ τε καὶ ἐκταραχθεῖσα ἠπόρει, τί
εἴη τὸ πραχθὲν συμβαλεῖν οὐκ ἔχουσα· τὴν δ᾿
οὖν λίθον ἔνδον που[1] κατέθετο, εἶτα νύκτωρ
διυπνισθεῖσα ὁρᾷ αὐγήν τινα καὶ αἴγλην ἀφιεῖσαν,
καὶ κατελάμπετο ὁ οἶκος ὡς ἐσκομισθείσης δᾳδός·
τοσοῦτον ἄρα ἐκ τῆς βώλου τὸ σέλας ἀνῄει τε
καὶ ἐτίκτετο.[2] συλλαβοῦσα δὲ τὸν πελαργὸν καὶ
ἐπαφωμένη κατενόησε τὴν ἐκ τῆς πληγῆς οὐλήν,
καὶ ἐγνώρισε τοῦτον ἐκεῖνον εἶναι τὸν ὑπ᾿ αὐτῆς
οἴκτου τε καὶ θεραπείας τετυχηκότα.

23. Ἀστακὸν εἰ λάβοις καὶ πορρωτάτω κομί-
σειας, σημεῖον καταλιπὼν ἔνθεν αὐτὸν τεθήρακας,
εὑρήσεις τὸν αὐτὸν ἐνταῦθα, ὅθεν καὶ συνείληπται.
λέγω δέ, εἰ παρὰ τὴν θάλατταν κομίσας εἶτα
καταθεῖο αὐτόν που πλησίον, ὡς ἑρπύσαι δυνηθῆναι
ἐς τὴν θάλατταν.

24. Ἀγρεὺς τὸ ὄνομα, τὴν φύσιν πτηνός, τὸ
γένος κοσσύφων φράτωρ,[3] μέλας τὴν χρόαν,
μουσικὸς τὴν γλῶτταν. κέκληται δὲ ἀγρεύς, καὶ
δικαίως· τῷ γάρ τοι μέλει τῶν ἄλλων ὀρνέων
αἱρεῖ τὰ ἁπαλὰ προσπετόμενα τῇ τῆς εὐμουσίας
θέλξει. εἰδὼς οὖν τὸ συμφυὲς αὐτῷ πλεονέκτημα,
ἔοικε χρῆσθαι τῷ παρὰ τῆς φύσεως δώρῳ ἐς
ἡδονὴν ἅμα καὶ τροφήν· ἀκούων μὲν γὰρ ἑαυτοῦ

[1] ποι.
[2] ἐτίκτετο, καὶ ἦν μέγα τίμιος.
[3] φράτωρ καὶ συγγενής.

and disgorged a stone into the lap of Heracleïs, and then flew off and settled on the roof. At first, naturally enough, she was amazed and startled out of her wits, and was at a loss to conjecture what this action could mean. And so she put the stone away somewhere indoors; later being woken in the night she saw that it diffused a brightness and a gleam, and the house was lit up as though a torch had been brought in, so strong a radiance came from, and was engendered by, the lump of stone. And when she had taken hold of the Stork and handled it she recognised the scar left by the wound, and knew that it was the very bird which had been the object of her pity and her ministrations.

23. If you catch a Smooth Lobster and remove it to a great distance, leaving a mark at the place where you caught it, you will find the self-same Lobster at the spot where it was captured: I mean, if you take it along the seashore and put it down somewhere near enough for it to be able to crawl into the sea.

The Smooth Lobster

24. 'Hunter'[a] is its name; Nature has given it wings; it is allied to the tribe of thrushes; its colour is black; it has a musical voice. And it is called 'the Hunter,' and rightly so; for with its song it captivates the small birds that fly to it beneath the spell of its sweet music. Knowing therefore the natural advantage that it possesses, it appears to employ this gift of Nature to please itself and also to feed itself, for it delights to listen to its

The Indian Mynah

[a] The *Mynah* of India.

AELIAN

εὐφραίνεται, θηρῶν δὲ τὰ προσιόντα ἐμπίπλαται.
τοῦτον εἴ τίς ποτε ἐθήρασε καὶ ἐν οἰκίσκῳ
καθεῖρξεν, οὐδὲν αὐτῷ πλέον τὸ τῆς σπουδῆς·
ἔχει γὰρ ἄφωνον ὄρνιν, ὥσπερ οὖν τὸν θηράσαντα
ὑπὲρ τῆς δουλείας ἀμυνόμενον τῇ σιωπῇ.

25. Ἀνωτέρω εἶπον ἣν οἱ τροχίλοι κατατίθενται
ἐς τοὺς κροκοδίλους εὐεργεσίαν, ἧσπερ[1] ἐν τοῖς
Αἰγυπτίοις μέμνηται καὶ Ἡρόδοτος λόγοις· ὁ
δὲ οὐκ εἶπον εἰδώς, ⟨τοῦτο⟩[2] εἰρήσεται νῦν, ἵνα
καὶ ἄλλος μάθῃ. ὁ μὲν τροχίλος ὄρνις ἐστὶ τῶν
ἑλείων εἷς, καὶ παρὰ τὰς ὄχθας τῶν ποταμῶν
ἀλᾶται καὶ ὅ τι ἂν τύχῃ παρεκλέγων βόσκεται,
τρέφει δὲ αὐτὸν καὶ ὁ κροκόδιλος οἷς εἶπον. καὶ
ἐκεῖνος αὐτὸν ἀμείβεται καθεύδοντος προμηθῶς
ἔχων καὶ ὑπεραγρυπνῶν αὐτοῦ· κειμένῳ μὲν γὰρ
καὶ ὑπνώττοντι[3] ἐπιβουλεύει ὁ ἰχνεύμων, καὶ
ἐμφὺς τῇ δέρῃ πολλάκις ἀπέπνιξεν αὐτόν· ἀλλ᾽
ὅ γε τροχίλος βοᾷ, καὶ παίει κατὰ τῆς ῥινὸς
αὐτόν, καὶ ἀνίστησι καὶ πρὸς τὸν ἐχθρὸν ὑποθήγει.
εἰ μὲν οὖν χρὴ τὸν ὄρνιν ἐπαινεῖν οὕτως ἔχοντα
φροντιστικῶς ζῴου παμβόρου καὶ ἀδηφάγου,
εἰσόμεθα·[4] τὸ δ᾽ οὖν ἴδιον τῶνδε τῶν ζῴων εἶπον.

26. Ἡ τρυγὼν (οὔ φημι νῦν τὴν ὑπαέριον, ἀλλὰ
τὴν ἐν τῇ θαλάττῃ) ὅτε βούλεται, νήχεται, καὶ αὖ
πάλιν ἀρθεῖσα πέτεται. ἔχει δὲ κέντρον, οὗ καὶ
ἀνωτέρω μνήμην ἐποιησάμην, θανατηφόρον. τὸ

---

[1] εὐεργεσίαν τὴν ἐκ τῶν βδελλῶν ὥσπερ.
[2] ⟨τοῦτο⟩ add. H.
[3] Jac: ὑπερυπνώττοντι.
[4] Perh. ⟨ἄλλοτε⟩ εἰσόμεθα, or ἐῶ H, ⟨ἀλλαχοῦ⟩ Grasberger.

212

own voice, and pursues the birds that approach it and takes its fill of them. Anyone who hunts this bird and confines it in a cage, gets nothing for his pains, for he possesses a bird that refuses to sing, seeming by its silence to punish its captor for enslaving it.

25. I have spoken above [a] of the benefit which the Egyptian Plovers confer upon Crocodiles, and Herodotus mentions it in his Account of Egypt [2. 68]. But what I did not mention, though I knew it, I will mention now, in order that others also may learn the facts. *The Egyptian Plover*

The Egyptian Plover is one of the marsh-fowls, and ranges along the banks of rivers, feeding upon whatever it chances to pick up here and there, while the Crocodile provides it with the food that I spoke of. And the bird repays it by taking care of it and keeping watch on its behalf while it sleeps. For as it lies asleep the Ichneumon has designs upon it, and fastening on its throat has often throttled it. But the Egyptian Plover utters its cry, beats the Crocodile on the nose, rouses it, and eggs it on against its enemy. Now whether we should applaud the bird for its solicitude on behalf of an omnivorous and gluttonous animal, we shall know later. It is the special characteristics of these creatures that I have mentioned.

26. The *Trygon* (I am not speaking of the one that lives in the air [*i.e.* the Turtle-dove] but of the one in the sea [*i.e.* the Sting-ray]) swims when it wants to, or again raises itself and flies. Its sting, of which I *The Sting-ray*

[a] See 3. 11.

μὲν οὖν κεντεῖν καὶ ζῷα ἄλογα καὶ ἀνθρώπους καὶ
παραχρῆμα ἀπολλύναι, οὔπω παράδοξόν ἐστιν·
ὃ δὲ ἄξιον ἐκπεπλῆχθαι, τοῦτο εἰρήσεται. δένδρῳ
τῷ μεγίστῳ καὶ πάνυ εὐθαλεῖ καὶ εὐερνεῖ καὶ
λίαν τεθηλότι τὴν χλόην εἰ προσαγάγοις τὸ
κέντρον καὶ νύξειας[1] τὸ δένδρον, οὐ μετὰ μακρὸν
ἐκβάλλει τὰ φύλλα· καὶ ἐκείνων καταρρεόντων ἐς
τὴν γῆν τὸ πᾶν πρέμνον αὐαίνεται καὶ ἔοικεν
ἡλιοβλήτῳ.[2]

27. Τίκτεται ἐλέφας κατὰ τὴν κεφαλὴν ἐκπη-
δῶν, τὸ δὲ μέγεθός ἐστι τοῦ τικτομένου κατὰ
δέλφακα τὴν μεγίστην. μιᾷ δὲ μητρὶ πλείω
ἐλεφαντίσκια ἔπεται, φασίν. εἰ δὲ βούλοιο τῶν
βρεφῶν νεογόνων ὄντων προσάψασθαι, αἱ μητέρες
οὐδὲν ἀγανακτοῦσιν ἀλλὰ ἐῶσι· συνιᾶσι γὰρ ὅτι
μήτε ἐπὶ λύμῃ τις ἐπιψαύει[3] αὐτῶν μήτε ἐπὶ
κολάσει, ἀλλὰ φιλοφρονούμενοι πάντες καὶ κολα-
κεύοντες. ἐπεὶ τίς ἂν τὸ τηλικοῦτον βλάψειεν;
ὅταν δὲ θηρώμενοι ἐμπέσωσιν ἐς τὴν τάφρον,
καὶ ἴδωσιν ὅτι λοιπὸν ἄφυκτα αὐτοῖς ἐστι, τοῦ
μὲν τέως θυμοῦ τοῦ σὺν τῇ ἐλευθερίᾳ λήθην
λαμβάνουσι, καὶ ὀρεγόντων σιτία ἑτοίμως προσ-
ίενται, καὶ ὕδωρ προτεινόντων πίνουσι, καὶ
οἶνον ἐγχεόντων ἐς τὰς προβοσκίδας οἱ δὲ τὴν
φιλοτησίαν οὐκ ἀναίνονται.

28. Τὸν ἰχθὺν τὸν ἔλλοπα ἱερὸν ἰχθὺν ὑπὸ τοῦ
ποιητοῦ κληθῆναι νομίζουσι. λέγει δέ τις λόγος[4]

---

[1] νύξεις MSS, νύξαις Schn.
[2] ἡλιοβλήτῳ ὑπ' αὐχμοῦ βιαίου ξηρῷ γεγενημένῳ.

have spoken above, is deadly.[a] Yet that it should
sting brute beasts and men and kill them on the spot
is no matter for wonder. But what is startling is
this which I am about to mention. If you apply the
sting to the largest tree when in a thriving state,
flourishing, and in full foliage, and stab the tree, in a
short while it sheds its leaves, and as they float down
to earth the entire stem withers and seems as though
scorched by the sun.

27. An Elephant emerges head first at birth, and The young
the size of it when born is that of the largest sucking- Elephant
pig. Several small Elephants follow a single mother,
so they say. And if you want to touch the little ones
when new-born, the mothers do not resent it but
permit it. For they know that no one will lay hands
on them to do them harm or punish them, but that
everyone has kindly intentions and would pet them.
For who would hurt such a little creature? But
when they are hunted and fall into the pit and see
that there is no escape for them, they forget the
spirit that possessed them when they were free and
readily go for any food that is held out to them and
drink the water that is offered, and if wine is poured
into their trunks they do not refuse that loving-cup.

28. Our great poet is supposed to call the Stur- The
geon (?) a 'sacred fish' [b] [Il. 16. 407]. According to Sturgeon (?)
one account it is rare, but is caught in the sea off

[a] See 1. 56; 2. 36, 50.
[b] See Leaf's note ad loc. The word ἔλλοψ does not occur
in our texts of Homer.

[3] Perh. -ψαύσει H.   [4] λόγος τις.

AELIAN

σπάνιον μὲν αὐτὸν εἶναι, ἐν δὲ τῷ κατὰ Παμφυλίαν
πελάγει θηρᾶσθαι, γλίσχρως δὲ καὶ ἐκεῖθι. ἐὰν
δὲ ἁλῷ, στεφάνοις μὲν αὐτοὶ σφᾶς αὐτοὺς ὑπὲρ
τῆς εὐερμίας ἀγλαΐζουσι, στεφανοῦσι δὲ καὶ τὰς
ἁλιάδας, καταίρουσί τε κρότῳ καὶ αὐλοῖς τὸ
θήραμα μαρτυρόμενοι. οἱ δὲ οὐ τοῦτον ἀλλὰ τὸν
ἀνθίαν νομίζουσιν ἱερόν. τὸ δὲ αἴτιον, ἔνθα ἂν
ὅδε φανῇ τῆς θαλάττης, ἀνάγκη δήπου τὸν χῶρον
ἄθηρον εἶναι σπονδάς τε [1] ἰχθύσι πρὸς [2] πᾶν
ὅσον ὑδροθηρικόν, καὶ αὐτοὶ δὲ οἱ ἰχθύες θαρροῦντες ἀποτίκτουσι. φύσεως δὲ ἀπόρρητα ἐλέγχειν οὐκ ἐμόν, καὶ εἰκότως, ἐπεὶ καὶ ἀλεκτρυόνα
δέδοικε λέων καὶ τὸν αὐτὸν βασιλίσκος καὶ
μέντοι καὶ ὗν ἐλέφας.[3] τὰς δὲ αἰτίας ὅσοι σχολὴν
ἄγουσι πολλὴν ζητοῦντες τοῦ μὲν χρόνου κατα-
φρονήσουσιν, οὐ μὴν ἐς τέλος ἀφίξονται τῆς
σπουδῆς.

[1] Schn : δέ.   [2] Reiske : εἰς.
[3] ὁ ἐλέφας.

Pamphylia, though even there hardly at all. But if it is caught, the fishermen deck themselves with garlands to celebrate their good luck; they garland the fishing-boats as well, and put into port, as with cymbals and flutes they summon people to bear witness to their catch.

Others however consider that the Anthias, and The Anthias not this fish, is sacred. And the reason is that in whatever part of the sea it appears, that spot is presumably bound to be free from savage creatures and there is peace between fish and everything that seeks its prey in the waters, while the fish themselves bring forth their young without fear.

But it is no business of mine to explore the mysteries of Nature, and rightly so, since the lion goes in fear of the cock, and so does the basilisk, moreover the elephant dreads a pig. But those who have much leisure to spend in seeking the reasons for these things will take no account of time, and for all that, will never come to the end of their researches.

BOOK IX

Θ

1. Ὁ λέων ἤδη προήκων τὴν ἡλικίαν καὶ γήρᾳ βα-
ρὺς γεγενημένος θηρᾶν μὲν ἥκιστός ἐστιν, ἀσμένως
δὲ ἀναπαύεται ἐν ταῖς ὑπάντροις ἢ λοχμώδεσι
καταδρομαῖς, καὶ τῶν θηρίων οὐδὲ τοῖς ἀσθενεστά-
τοις ἐπιθαρρεῖ, τόν τε αὑτοῦ χρόνον ὑφορώμενος
καὶ τὸ τοῦ σώματος ἐννοῶν ἀσθενές. οἱ δὲ ἐξ
αὑτοῦ γεγενημένοι θαρροῦντες τῇ τῆς ἡλικίας
ἀκμῇ καὶ τῇ ῥώμῃ τῇ συμφυεῖ προΐασι μὲν ἐπὶ
θήραν, ἐπάγονται δὲ καὶ τὸν ἤδη γέροντα, ὠθοῦντες
αὐτόν· εἶτα ἐπὶ μέσης τῆς ὁδοῦ ἧς ἐλθεῖν δεῖ
καταλιπόντες, ἔχονται τῆς ἄγρας αὐτοί, καὶ
τυχόντες τοσούτων ὅσα ἀποχρήσει καὶ αὑτοῖς καὶ
τῷ γεγεννηκότι [1] σφᾶς, βρυχησάμενοι γενναῖόν
τε καὶ διάτορον καλοῦσιν [2] ὡς δαιτυμόνα ἑστιά-
τορες ἐπὶ θοίνην οἱ νέοι τὸν γεγηρακότα, τὸν πατέρα οἱ παῖδες. ὁ δὲ ἡσυχῇ καὶ βάδην καὶ οἷον
ἕρπων ἔρχεται, καὶ περιβαλὼν τοὺς παῖδας, καὶ
τῇ γλώττῃ μικρὰ ὑποσήνας, ὥσπερ οὖν ἐπαινῶν
τῆς εὐθηρίας, ἔχεται τοῦ δείπνου, καὶ σὺν τοῖς
υἱέσιν ἑστιᾶται. καὶ Σόλων μὲν τοῖς λέουσιν οὐ
κελεύει ταῦτα, [3] διδάσκει δὲ ἡ φύσις, ᾗ νόμων
ἀνθρωπικῶν οὐδὲν μέλει· γίνεται δὲ ἄτρεπτος
αὐτὴ νόμος.

[1] Schn : γεγενηκότι.
[2] καλοῦσιν τὸν πατέρα.
[3] ταῦτα νομοθετῶν τρέφειν τοὺς πατέρας ἐπάναγκες.

BOOK IX

1. When the Lion is advanced in years and heavy with age he is quite incapable of hunting and is glad to take his ease in caves or lairs in the jungle; nor has he the spirit to attack even the weakest of animals, for he mistrusts his age and is conscious of his bodily infirmity. Whereas his offspring confident in the vigour of their youth and their natural strength go out to hunt and bring the old one with them by pushing him along. Then, when they have come half the necessary distance, they leave him behind and give themselves to the chase. And when they have obtained enough for themselves and for their sire, with a magnificent and thrilling roar, even as banqueters summon a guest, so do these young children summon their aged father to the feast. And he comes softly, step by step, and almost crawling, and embraces his children, fawning upon them a little with his tongue as though he applauded their success, and attacks the meal and feasts with his sons. This is no order of Solon's to the Lions: it is Nature that teaches them—Nature that ' recks nought of laws ' [Eur. *fr.* 920 N] made by man. But she is a law that does not change.

2. Τὸν ἀετὸν τὸν τῶν ὀρνίθων βασιλέα οὐ μόνον περιόντα[1] καὶ ζῶντα δέδοικε τὰ ὄρνεα καὶ καταπτήσσει φανέντος, ἀλλὰ καὶ τὰ πτερὰ ἐκείνου ἐάν τις τοῖς τῶν ἄλλων συναναμίξῃ, τὰ μὲν τοῦ ἀετοῦ μένει ὁλόκληρα καὶ ἀνεπιβούλευτα, τὰ δὲ ἕτερα κατασήπεται τὴν πρὸς ἐκεῖνα κοινωνίαν οὐ φέροντα.

3. Οἱ μύες εἰσὶ μὲν καὶ ἄλλως πολύγονον ζῷον, καὶ ἀθρόᾳ τῇ ὠδῖνι πολλὰ τίκτουσιν· εἰ δέ πως καὶ ἁλὸς γευσάμενοι τύχοιεν, ἐνταῦθα δήπου καὶ πάμπολλα ἀποκυΐσκουσι καὶ πλείω τῆς συνηθείας πολλῷ. οἱ δὲ κροκόδιλοι, ὅταν τέκωσι, τὸ γνήσιον καὶ τὸ νόθον τόνδε τὸν τρόπον ἐλέγχουσιν. ἐάν τι παραχρῆμα ἐκγλυφεὶς ἁρπάσῃ, τελεῖ τὸ λοιπὸν ἐς τὸ γένος, καὶ φιλεῖται τοῖς γειναμένοις, καὶ πεπίστευται κροκοδίλων εἷς εἶναι καὶ ἠρίθμηται· ἐὰν δὲ ἐλινύσῃ καὶ βλακεύσῃ καὶ μὴ λάβῃ ποθὲν ἢ μυῖαν ἢ σέρφον ἢ ἔντερον γῆς ἢ σαῦρον τῶν νεαρῶν, διέσπασεν ὁ πατὴρ αὐτὸν ὡς ἀδόκιμόν τε καὶ κίβδηλον καὶ προσήκοντά οἱ οὐδὲ ἕν. καὶ δοκοῦσιν, ὡς οἶδε οἱ θῆρες, καὶ οἱ ἀετοὶ βασανίζοντες καὶ ἐκεῖνοι τὰ γνήσια τῇ ἀκτῖνι τοῦ ἡλίου κρίσει φιλεῖν τὰ ἔκγονα καὶ οὐ πάθει.

4. Ἀκούω δὲ τοὺς ὀδόντας τῆς ἀσπίδος, οὓς ἂν ἰοφόρους τις εἴποι καλῶν ὀρθῶς, ἔχειν οἱονεὶ χιτῶνας περικειμένους ἄγαν λεπτοὺς καὶ ὑμέσι παραπλησίους, ὑφ' ὧν περιαμπέχονται. ὅταν οὖν ἐμφύσῃ τινὶ τὸ στόμα ἡ ἀσπίς, διαστέλλεσθαι[2] μέν φασι τὰ ὑμένια, ἐκχεῖσθαι δὲ τὸν ἰόν, καὶ

---

[1] παρόντα MSS, H would read ζ. καὶ ἔτι περιόντα, cp. 11. 39.

2. Not only when he is alive and active do birds The Eagle's
feathers dread the Eagle, the king of birds, and cower down when he appears, but if one mixes his feathers with those of other birds, the Eagle's remain entire and untainted, while the others, unable to endure the association, rot away.

3. Mice, besides being prolific creatures, bring The Mouse forth many offspring at a single birth; and if by some means they happen to eat salt, then they bring forth a great number and far more than is customary. And when Crocodiles give birth they test the legiti- The Croco-
dile and its
young mate and the bastard offspring in this manner. If on being hatched a young Crocodile immediately seizes something, it is henceforward reckoned among the family and is loved by its parents, is believed to be, and is counted as, one of the Crocodiles. If however it remains inactive and is lazy and fails to seize some fly or gnat or earthworm or young lizard, the sire tears it to pieces as a poor creature, spurious, and no kin of his. And as these creatures act, even so do Eagles appear to test their legitimate offspring by the rays of the sun [a] and to love them as the result of judgment and not of any feeling.

4. I have heard that the Asp's fangs, which one The Asp,
its fangs would be correct in styling ' poison-carriers,' have an exceedingly thin coating, so to say, round them, like membrane, covering them all over. So when the Asp fastens its mouth on a man, they say that these membranes part and the poison is ejected,

[a] See 2. 26.

---

[2] στρέφεσθαι.

πάλιν συντρέχειν ἐκεῖνα καὶ ἑνοῦσθαι. τοῦ γε μὴν
σκορπίου τὸ κέντρον ἔχειν τινὰ κολπώδη διπλόην
ὑπὸ τῆς ἄγαν λεπτότητος οὐ πάνυ τι [1] σύνοπτον.
καὶ εἶναι μὲν τὸ φάρμακον καὶ τίκτεσθαι λέγουσιν
ἐνταῦθα, ἅμα δὲ τῇ κρούσει προϊέναι διὰ τοῦ
κέντρου καὶ ἐκρεῖν. ὀπὴν δὲ εἶναι δι' ἧς ἔξεισιν
οὐδὲ ταύτην ὄψει θεωρητήν. ἀνθρώπου δὲ σιάλῳ
καταπτύοντος ἀμβλύνεσθαι τὸ κέντρον καὶ μαλ-
κίειν καὶ ἐς τὴν πληγὴν ἀδύνατον γίνεσθαι.

5. Ἡ κύων εἰ καὶ πολλὰ τίκτει σκυλάκια, ἀλλὰ
γοῦν τὸ πρῶτον τῆς μήτρας [2] προελθὸν καὶ τῆς [3]
ὠδῖνος πρεσβύτατον ὂν κατηγορεῖ τὸν πατέρα.
ἐκείνῳ γοῦν ὁμοιότατον τίκτεται πάντως, τὰ δὲ
ἄλλα ὡς ἂν τύχῃ. ἔοικε δὲ φιλοσοφεῖν ἐν τῷδε ἡ
φύσις, προτιμῶσα τοῦ ὑποδεχομένου τὸ σπεῖρον.

6. Τῶν ὀστρακονώτων τε καὶ ὀστρακοδέρμων
καὶ τοῦτο ἴδιον. κενώτερά πως ταῦτα καὶ κουφό-
τερα ὑποληγούσης τῆς σελήνης φιλεῖ γίνεσθαι. καὶ
τῶν μὲν ὀστρακονώτων ἐλέγχουσιν ὃ λέγω πορφύραι
καὶ κήρυκες καὶ σφόνδυλοι καὶ τὰ τούτοις ὁμοφυῆ·
τῶν δὲ ἑτέρων πάγουροί τε καὶ κάραβοι καὶ
ἀστακοὶ καὶ καρκίνοι καὶ εἴ τι τούτων συγγενές.
λέγεται δὲ καὶ τῶν ὑποζυγίων τὰ τικτόμενα ληγού-
σης τῆς σελήνης ἀδυνατώτερα τῶν ἄλλων εἶναι
καὶ ἀσθενέστερα, καὶ μέντοι καὶ συμβουλεύουσιν

---

[1] πάντη.   [2] Reiske: μητρός MSS, H.
[3] καὶ ἐκείνης τῆς.

---

[a] See Thompson, Gk. fishes, s.v. σπόνδυλος, O. Keller, Ant.
Tierwelt 2. 561.

and then again they close and unite. Again, the
sting of the Scorpion has a kind of hollow core, so very   The
fine as to be hardly visible. That is where they say   Scorpion,
the poison resides and is engendered, and directly the   its sting
Scorpion strikes, the poison shoots forward along the
sting and flows out. And this opening also, through
which it passes, is so fine as to be invisible to the eye.
But if a man spits upon it the sting is blunted and
numbed and becomes incapable of wounding.

5. Even if a Bitch produces a number of puppies,   Puppies
it is nevertheless the one that issues first from the
womb and the eldest of the litter that declares the
sire. At any rate it bears the closest resemblance
to him in every respect, while the rest are born as
chance may dictate. In this matter Nature appears
to pursue reason in setting the male which sows
above the female which receives.

6. Here is another characteristic of Testaceans   The Moon,
and Crustaceans. As the moon wanes they are in   its influence
the habit of somehow becoming both emptier and   on Shellfish
lighter. Among Testaceans the purple shellfish,   and Animals
whelks, red thorny oysters,[a] and those of the same
species prove my statement; among Crustaceans,
edible crabs, crayfish,[b] lobsters, crabs in general,[c]
and all their kin. It is also asserted that the young
of beasts of burden born when the moon is on the
wane are less capable and feebler than others, and
what is more, those who have knowledge of these

---

[b] At 11. 37 κάραβοι are included among *Testacea*.
[c] Καρκίνος is the generic term for crabs of all kinds, πάγουρος
the common or edible crab.

οἱ τούτων ἐπιστήμονες τὰ ἐν τούτῳ τῷ μέρει τοῦ
μηνὸς γεννώμενα μὴ τρέφειν· μὴ γὰρ εἶναι
σπουδαῖα αὐτά. κατὰ τὴν νουμηνίαν δὲ τὰ ζῷα,
ὡς πυνθάνομαι, ἢ φθέγγεταί τι τῇ συντρόφῳ
φωνῇ ἢ πίπτει· λέων δὲ ἄρα μόνος, ὡς Ἀριστοτέλης φησίν, οὐδέτερον[1] αὐτοῖν δρᾷ.

7. Ἀκοὴν[2] ὀξύτατον τὸν λάβρακα Ἀριστοτέλης εἶναί φησι καὶ μέντοι καὶ τὴν χρόμιν καὶ
τὴν σάλπην καὶ τὸν κεστρέα. πυνθάνομαι δὲ[3] τὸν
λάβρακα σαφῶς εἰδέναι ὅτι ἄρα ἐν τῇ κεφαλῇ
αὐτοῦ λιθίδιόν ἐστι. καὶ χειμῶνος τοῦτο ψυχρό-
τατον γίνεται, καὶ λυπεῖ αὐτὸν ἰσχυρῶς. ταύτῃ
τοι καὶ κατ' ἐκείνην τὴν ὥραν τοῦ ἔτους ἀλεαίνειν
αὐτόν, καὶ ἐπινοεῖν τῇ ψύξει τῇ ἐκ τοῦ λίθου
φάρμακον τοῦτο καὶ μάλα γε ἀντίπαλον. καὶ
χρόμιν δὲ τὸ αὐτὸ ποιεῖν καὶ φάγρον καὶ σκίαιναν
πέπυσμαι· ἔχειν γάρ τοι[4] ὅμοιον λίθον καὶ ταῦτα.

Παράσιτοι δὲ ἄρα καὶ ἐν ἰχθύων γένει ἦσαν.
ὁ γοῦν φθεὶρ οὕτω λεγόμενος παρατρώγει τῶν τοῦ
δελφῖνος θηραμάτων· ὁ δὲ ἥδεται αὐτῷ καὶ ἑκὼν
μεταδίδωσιν. ἔνθεν τοι καὶ πιότατός ἐστιν, ὥσπερ
οὖν ἐκ πλουσίας καὶ ἀμφιλαφοῦς ἑστιάσεως
ἐμπιπλάμενος. καὶ ὁ μὲν τοῦ Μενάνδρου Θήρων
μέγα φρονεῖ, ὅτι ῥινῶν ἀνθρώπους φάτνην αὐτοὺς
ἐκείνους εἶχε· Κλείσοφος δὲ[5] καὶ τὸν ὀφθαλμὸν
τὸν ἕτερον δεσμῷ κατελάμβανε, Φιλίππῳ χαριζό-

---

[1] οὐθέτερον.
[2] ἀκοὴν ἀγαθὸν καί ὅ.
[3] τε.
[4] τόν.
[5] εἶχεν· καὶ ὁ Στρουθίας τοιοῦτος. Κλείδημος δέ ὁ Φιλίππου.

matters recommend that animals born in this part of the month should not be reared on the ground that they are not of good quality. Whereas animals born at the new moon, as I learn, either utter their natural sound or drop. The Lion alone, as Aristotle says,[a] does neither.

7 (i). Aristotle asserts [*HA* 534 a 9] that the Basse is extremely quick of hearing, and so too are the Chromis,[b] the Saupe, and the Mullet. I have ascertained also that the Basse knows full well that there is in fact a small stone [c] in its head, and this in winter becomes intensely cold and causes it severe pain. This is why at that season of the year it warms itself [d] and devises this highly effective remedy against the cold due to the stone. And the Chromis, the Sea-bream, and the Maigre, I learn, do the same, for these fish also have a similar stone.

The Basse and its otolith

(ii). It seems that among fishes also there exist parasites.[e] At any rate the Sucking-fish, as it is called, nibbles what the dolphin catches, and the dolphin is glad that he should, and willingly allows him a share. That is why the fish is exceedingly plump, like one gorged with a rich and abundant feast. And Theron in Menander's play [*frr.* 895, 937 K] boasts that he has led men by the nose and used them as his manger. And Cleisophus [f] covered one of his eyes with a bandage out of compliment to

Fishes and their 'parasites'

---

[a] Not in any extant work; *fr.* 236 (Rose, p. 254).
[b] Perhaps identical with σκίαινα, *Maigre*; Thompson, *Gk. fishes*, s.v. σκίαινα.
[c] The otolith.
[d] See 9. 57.
[e] In the Greek sense of 'hangers-on.'
[f] See Ath. 6. 248 D, and Ael. *frr.* 107, 108.

μενος ἐν τῇ τῆς Μεθώνης πολιορκίᾳ τὸν ἕτερον
ἐκκοπέντι. φιλία δὲ ἐμοὶ δοκεῖν καὶ συντροφία
τῷ φθειρὶ πρὸς τὸν δελφῖνά ἐστι· κολακεύειν μὲν
γὰρ ὡς καὶ ἄλλα κακὰ ἄνθρωπος οἶδε, τὰ δὲ
ἄλογα οὐκ οἶδεν.

8. Ἐλέφαντος δὲ ἄρα ἐς τὰ τέκνα καὶ ἐκεῖνο
φίλτρον ἰσχυρόν. οἱ τούτων θηραταὶ τάφρους
ὀρύττουσιν, ἐς ἃς ἐμπίπτει [1] τὰ ζῷα ταῦτα, καὶ
τὰ μὲν ἁλίσκεται, τὰ δὲ ἀναιρεῖται. καὶ τίς ὁ
τρόπος τῆς τοιᾶσδε ταφρεύσεως καὶ τὸ σχῆμα
ὁποῖον καὶ ὁπόσον [2] τὸ βάθος [3] καὶ ἔσοδοι ποταπαί,
ἀλλαχόθεν εἴσεσθε· ἐκκαλύψων δὲ ἔγωγε καὶ
ἐλέγξων τὴν στοργὴν ἔρχομαι. ἡ μήτηρ θεασαμένη τὸ ἑαυτῆς βρέφος ἐς μίαν τῶν τάφρων
ἐμπεσόν, οὔτε ἐμέλλησεν οὔτε βλακεύουσα διέ-
τριψεν, ἀλλὰ ὡς εἶχεν ὁρμῆς ἐκθύμως καὶ περιπα-
θῶς ἐπιδραμοῦσα, κατὰ τοῦ παιδὸς αὑτὴν ἔωσεν
ἐς κεφαλήν, καὶ ἄμφω κατὰ ταὐτὸν τὸ τέλος
εἰχέτην· ὁ μὲν γὰρ ἐκ τοῦ μητρῴου βάρους
πιεσθεὶς [4] συνετρίβη, ἡ δὲ ⟨κατὰ⟩ κεφαλὴν
ἄξασα. . . [5] γελοῖοι τοίνυν εἰσὶν οἱ διαποροῦντες
εἰ φυσικὴ πρὸς τὰ ἔκγονα στοργή ἐστιν.

9. Αἱ δὲ φῶκαι τίκτουσι μὲν ἐπὶ τῆς γῆς, κατὰ [6]
μικρὰ δὲ ὑπάγουσιν ἑαυτῶν τὰ σκυλάκια ἐς τὸ
νοτερόν, καὶ ἀπογεύουσι τῆς θαλάττης, εἶτα
ἐπανάγουσιν ἐς τὸν τῆς ὠδῖνος τόπον τὸν ἐξ
ἀρχῆς, καὶ αὖ πάλιν κατάγουσιν ἐς τὴν θάλατταν,

---

[1] ἐμπίπτουσι.   [2] Schn : ὁποῖον.
[3] Gron : πάθος.   [4] Reiske : ἐκπιεσθείς.

Philip who had lost an eye at the siege of Methone.[a]
Sucking-fish and dolphin are in my opinion friends
and messmates, for whereas man understands
flattery like other vices, brute beasts do not.

8. Here again is an example of the Elephant's The
strong affection for its young. Elephant-hunters Elephant and its
dig trenches and these animals fall into them, and young
while some are captured, others are killed. You
will learn from other sources how they dig these
trenches, how they are shaped, how deep, and what
the entrances to them are like. I however propose
to reveal and demonstrate the Elephant's affection.
When the mother sees her young one has fallen into
one of the trenches, she does not hesitate, does not
waste time, but rushing up at full speed, all courage
and passion, hurls herself upon the head of her child,
and the pair meet one and the same end, for the
young one is crushed by the mother's weight; she
falls on her head . . . So those who doubt whether
Elephants have a natural affection for their offspring
are absurd.

9. Seals give birth on land, but by degrees lead The Seal
their cubs down to the water and give them a taste
of the sea. Then they lead them back to the
original place of their birth, and again bring them
down to the sea, and quickly lead them out, and by

[a] On the NW coast of the Thermaic gulf; taken by
Philip II after a prolonged siege, 352 B.C.

---

[5] ⟨κατὰ⟩ κεφαλὴν ἄξασα . . .] a main verb wanting. Gow, τὴν
κ. ἀΐξασα MSS, ⟨κατὰ⟩ τὴν κ. ᾄ. H.
[6] Reiske : καὶ κατά.

καὶ ταχέως ἐξάγουσι· καὶ ὅταν πολλάκις τοῦτο
δράσωσι, τελευτῶσαι νηκτικώτατα ἀπέφηναν αὐτά.
ῥᾳδίως δὲ ἐς τὸν θαλάττιον βίον ὑπολισθάνει, τῆς [1]
διδασκαλίας μὲν προαγούσης αὐτά, βιαζομένης δὲ
τῆς φύσεως τῶν μητρῴων καὶ ἠθῶν καὶ ἐθῶν ἐρᾶν.

10. Ἀετὸς ζῷον πλεονεκτικόν, καὶ δι᾽ ἁρπαγῶν
ποιοῦνται ⟨τὰς⟩ [2] τροφάς, καὶ σαρκῶν ἐσθίουσι·
καὶ γὰρ λαγὼς ἁρπάζουσι καὶ νεβρὸν καὶ χῆνα ἐξ
αὐλῆς καὶ ἄλλα. μόνος δὲ ἄρα ἐν αὐτοῖς ὅσπερ
οὖν καὶ Διὸς κέκληται κρεῶν οὐχ ἅπτεται, ἀλλὰ
ἀπόχρη οἱ πόα· καὶ Πυθαγόρου τοῦ Σαμίου
διακούσας οὐδὲ ἕν, ὅμως ἐμψύχων ἀπέχεται.

11. Εἰ τοῦ φαλαγγίου καὶ μόνον ἐφάψαιτό τις,
ἀπέκτεινεν αὐτὸν μηδὲ ὀδυνηθέντα φασὶ [3] ἰσχυρῶς.
ἀλλὰ καὶ τὸ τῆς ἀσπίδος δῆγμα πραότατον εἶναι
ἤλεγξε Κλεοπάτρα, ὅτε τοῦ Σεβαστοῦ προσιόντος [4]
ἀνώδυνον [5] θάνατον ἐν τοῖς συμποσίοις ἐβασάνιζε,
καὶ τὸν μὲν διὰ τοῦ ξίφους εὕρισκεν ἀλγεινόν, τῶν
τιτρωσκομένων τοῦτο ὁμολογούντων, τὸν δὲ διὰ
τῶν φαρμάκων λυπηρόν· σπασμὸν γάρ τινα
ἐμποιεῖν καὶ καρδιώττειν ἀναγκάζειν· τὸν δὲ ἐκ
τοῦ δήγματος τῆς ἀσπίδος πρᾶον εἶναι καὶ ἵνα
Ὁμηρείως [6] εἴπω ἀβληχρόν. ἔστι δὲ ἃ [7] καὶ
μόνον ἀψαμένους ἀπέκτεινε καὶ προσερυγόντα δέ,
ὥσπερ οὖν ὁ κεντρίνης [8] καὶ ἡ φρύνη.

12. Σὺ μέν μοι λέξεις [9] πανοῦργον εἶναι ζῷον
ἀλώπεκα, ταύτην δὴ τὴν ἐκ τῆς γῆς τρεφομένην·

---

[1] Reiske : καὶ τῆς.
[2] ⟨τάς⟩ add. H.
[3] Ges : φησίν.
[4] Schn : προϊόντος.

doing this many times they end by making them excellent swimmers. And they easily slide into life in the sea: their instruction affords an inducement, while Nature forces them to love the haunts and the habits of their mothers.

10. The Eagle is a predatory bird: it feeds upon what it can rob, and eats flesh. For it seizes hares, fawns, and geese from the courtyard, and other creatures. Only the Eagle which is called ' Zeus's bird ' does not touch meat: for it, grass is sufficient. And though it has never heard of Pythagoras of Samos, for all that it abstains from animal food. The Eagle
'Zeus's
Eagle'

11. If one merely touches a Malmignatte, it kills, they say, without any violent pain. Moreover Cleopatra established that the bite of an Asp is exceedingly gentle, when as Augustus was approaching she made enquiries at her banquets for a form of death that should be painless: death by the sword, she was told, entailed suffering, as was confessed by those who were wounded; death by drinking poison caused distress, for it produced convulsions and pains in the stomach; whereas death from the bite of an Asp was gentle (πρᾶος), or to use Homer's word [Od. 11. 135] ἀβληχρός (faint, mild). And there are some creatures that kill by a belch those that only touch them, as for instance the dipsas and the toad. The
Malmignatte
and the Asp,
their bites

12. You will tell me that the Fox is a creature full of guile; this is the fox that lives on the land. But The Fox-
shark

---

[5] αἱρεθεῖσα ἀνώδυνον.   [6] Ὁμήρῳ ἰδίως or ὁμοίως.
[7] Jac: ὅτε.   [8] κεντρίτης.   [9] λέγεις.

ἄκουε δὲ καὶ ⟨τὰς⟩ [1] τῆς θαλαττίας μηχανάς, καὶ
ὁποῖα δρᾷ καὶ ἐκείνη πυνθάνου. ἢ γὰρ οὐ πρόσεισι
τῷ ἀγκίστρῳ τὴν ἀρχήν, ἢ καταπιοῦσα παραχρῆμα ἑαυτῆς τὸ ἐντὸς μετεκδῦσα ἔστρεψεν [2] ἔξω,
ὥσπερ οὖν χιτῶνα τὸ σῶμα ἀνελίξασα, καὶ τοῦτον
δήπου τὸν τρόπον ἐξεώσατο τὸ ἄγκιστρον.

13. Ἴυγγας ἐρωτικὰς ἄνθρωποί φασιν εἶναί
τινας, μίξεως δὲ ἀφροδισίου σύνθημα ὁ βάτραχος
ἀφίησι πρὸς τὴν θήλειαν βοήν τινα, ὡς ἐραστὴς
ᾠδήν τινα κωμαστικήν, καὶ κέκληται ἥδε ἡ βοὴ
ὀλολυγών, ὥς φασιν. ὅταν δὲ τὴν θήλειαν προσ-
αγάγηται, μένουσιν ἄμφω τὴν νύκτα· ἐν μὲν γὰρ
τῷ ὕδατι συνελθεῖν οὐ δύνανται, μεθ’ ἡμέραν δὲ
ἐπὶ γῆς συμπλακῆναι ὀρρωδοῦσι. νυκτὸς δὲ ἐπιστάσης κατὰ πολλὴν τὴν ἄδειαν προελθόντες [3]
ἀλλήλων ἀπολαύουσιν.

Ὅταν δὲ βάτραχοι γεγωνότερον φθέγγωνται καὶ
τῆς συνηθείας λαμπρότερον, ἐπιδημίαν δηλοῦσιν
ὑετοῦ.

14. Εἴ τις προσάψαιτο τῆς νάρκης ὅτι τὸ ἐκ
τοῦ ὀνόματος πάθος τὴν χεῖρα αὐτοῦ καταλαμβάνει,
τοῦτο καὶ παιδάριον ὢν ἤκουσα τῆς μητρὸς
λεγούσης πολλάκις. σοφῶν δὲ ἀνδρῶν ἐπυθόμην
ὅτι καὶ τοῦ δικτύου ἐν ᾧ τεθήραται [4] εἴ τις προσάψαιτο, ναρκᾷ πάντως. εἰ δέ τις ἐς σκεῦος
αὐτὴν ἐμβάλοι ζῶσαν, καὶ ἐπιχέοι θαλαττίου
ὕδατος, ἐὰν ἐγκύμων ᾖ καὶ ὁ καιρὸς τῆς ὠδῖνος
ἀφίκηται, τίκτει. καὶ τὸ ἐν τῷ σκεύει ὕδωρ εἴ τις

---

[1] ⟨τὰς⟩ add. H.   [2] ἔστρεψεν οὕτως.

listen also to the wiles of the Fox-shark and learn the kind of things it does. Either it will not come near the hook at all, or else it swallows it and immediately turns itself inside out, reversing its body just like a garment, and in this way no doubt it gets rid of the hook.

13. Men say that there are certain spells to cause love; the Frog as a signal for sexual intercourse emits a certain cry to the female, like a lover singing a serenade, and this cry is called its croak, so they say. And when it attracts the female to itself they wait for the night. They cannot copulate under water, and they shun mutual embraces on land in the daytime. But when night descends they emerge with complete fearlessness and take their pleasure of one another. *Frogs and their mating*

Whenever Frogs utter their cry more loudly and more clearly than is their wont, it signifies that rain is coming.

14. I have often heard my mother say, when I was a child, that if a man touches a Torpedo, his hand is seized with the affliction corresponding to its name (torpor). And I have learnt from persons of experience that if a man touches even the net in which it has been captured his entire body is numbed. And if one throws it alive into a vessel and pours salt water upon it, and if the fish happens to be pregnant and the time of its delivery is at hand, then it gives birth. And if one pours the water in *The Torpedo*

---

³ *Schn* : προσελθόντες.　　⁴ θηρᾶται.

καταχέοι [1] χειρὸς ἀνθρώπου ἢ ποδός,[2] ναρκᾶν τὴν χεῖρα ἢ τὸν πόδα ἀνάγκη.

15. Τὰ ζῷα οὔτε ἐν ταῖς πληγαῖς οὔτε ἐν τοῖς δήγμασιν ἀεὶ τὴν αὐτὴν δύναμιν ἴσχει, ἀλλ' ἐπιτείνεται πολλάκις ἔκ τινος αἰτίας. ὁ γοῦν σφὴξ γευσάμενος ἔχεως χαλεπώτερός ἐστι τὴν πληγήν, καὶ ἡ μυῖα τοιούτῳ τινὶ προσελθοῦσα πικροτέρα δακεῖν ἐστι καὶ ὀδύνας ἔδωκε, καὶ μέντοι καὶ τῆς ἀσπίδος τὸ δῆγμα γίνεται παντελῶς ἀνήκεστον, ἐὰν βατράχου φάγῃ. ὁ δὲ κύων ὑγιαίνων μὲν ἐὰν δάκῃ, τραῦμα εἰργάσατο καὶ ἀλγηδόνα ἐξῆψεν· ἐὰν δὲ λυττῶν, διέφθειρεν.[3] ἀκέστρια δὲ ἀκουμένη χιτώνιον ῥαγὲν ὑπὸ λυττῶντος κυνός, δακοῦσά πως τῷ στόματι τὸ χιτώνιον, ἵνα ἀποτείνῃ αὐτό, ἐλύττησε καὶ ἀπέθανεν. ἀνθρώπου δὲ ἀσίτου δῆγμα χαλεπὸν καὶ δυσίατον. λέγονται δὲ οἱ Σκύθαι πρὸς τῷ τοξικῷ, ᾧ τοὺς ὀιστοὺς ἐπιχρίουσι, καὶ ἀνθρώπειον ἰχῶρα ἀναμιγνύναι φαρμάττοντες, ἐπιπολάζοντά πως αἵματι, † ὅνπερ ἴσασιν ἀπόκριμα αὐτοῖς †.[4] τεκμηριῶσαι τοῦτο καὶ Θεόφραστος ἱκανός.

16. Ὅταν ἀποδύσηται τὸ γῆρας ὁ ὄφις (ὑπαρχομένου δὲ τοῦ ἦρος δρᾷ τοῦτο), ἐνταῦθά τοι καὶ τῶν ὀφθαλμῶν τὴν ἀχλὺν καὶ τὸ ἀμβλὺ τῆς ὄψεως ῥύπτεται καὶ ἐκεῖνο ὡς γῆρας ὀφθαλμῶν, τῷ δὲ

---

[1] Lobeck : καταχέει.
[2] χειρί . . . ποδί.
[3] διέφθειρεν· ὕδωρ τε δεδιέναι κατηνάγκασε πρῶτον, καὶ ὁ μετριάσαι δοκῶν πάλιν ἐξάπτεται εἰς τὴν ὀδύνην καὶ ὑλακτήσας ἀπέθανεν.

the vessel over a man's hand or foot, the hand or foot is inevitably numbed.

15. Neither in the stings nor in the bites which they inflict do animals always retain the same force, but it is often augmented from some cause. For instance, if a Wasp has tasted a viper's flesh its sting is fiercer; and if a Fly has been near something of the same kind its bite is sharper and causes pain; the bite of an Asp too is rendered quite incurable if it eats of a frog. If a healthy Dog bites a man, it causes a wound and a burning pain, but if the Dog is mad, the bite is deadly. A sempstress was mending a shirt that had been torn by a mad Dog, when she somehow bit it with her mouth in order to stretch the shirt: she went mad and died. The bite of a human being when fasting is dangerous and hard to cure. And the Scythians are even said to mix serum from the human body with the poison that they smear upon their arrows to drug them. This serum somehow floats on the surface of the blood ⟨and they know a means of separating it?⟩.[a] Theophrastus [b] is a sufficient witness to the fact.

The stings and bites of various creatures

16. When a Snake sloughs its old skin (it does so at the beginning of spring), then is the time when it purges away the mist over its eyes and the dullness of its sight and what I may call the ' old age ' of its

The Snake and its eye-sight

[a] The text is corrupt and the translation conjectural; cp. [Arist.] *Mirab.* 845 a 5. Post's conjecture might be rendered ' which is a secretion that comes when they agitate the blood.'
[b] Not in any extant work.

4 ὅνπερ . . . αὐτοῖς *corrupt*: ὃν περι⟨σεί⟩σασιν ἀ. αὐ. *conj.* Post.

AELIAN

μαράθῳ ὑποθήγων ¹ τε καὶ παραψήχων τὸ ὄμμα
ἑκάτερον, εἶτα ἐξάντης τοῦδε τοῦ πάθους γίνεται.
ἀμβλυώττει δὲ ἄρα διὰ τοῦ χειμῶνος φωλεύσας ἐν
μυχῷ καὶ σκότῳ. οὐκοῦν μαλκίουσαν ἐκ τῶν
κρυμῶν τοῦ ζῴου ² τὴν ὄψιν ὑποθερμαῖνον τὸ
μάραθον καθαίρει, καὶ ὀξυωπέστερον ἀποφαίνει.

17. Ἡ ἀλκυὼν ὅταν αἴσθηται ἑαυτῆς κυούσης,
τηνικαῦτά τοι ³ ἐς τὴν τῶν νεοττῶν ὑποδοχὴν
καλιὰν ἐργάζεται, οὔτε πηλοῦ καὶ ὀρόφου ὡς ἡ
χελιδὼν δεομένη καὶ οἴκων, καὶ ἄκλητος ἐσιοῦσα
ξένη, καὶ λυποῦσα τὰ ἑωθινὰ τῷ ⁴ λάλῳ καὶ
μέντοι καὶ διακόπτουσα τῶν ὕπνων τὸν ἥδιστον,
οὔτε πάλιν τῷ σώματι . . . ⁵ μόνῳ ἐν ἐλευθέροις
χωρίοις ἔχεται τοῦ προειρημένου, συμπλέκουσα δὲ
καὶ ἀθροίζουσα τὰς τῆς βελόνης ἀκάνθας, δεσμῷ
τινι ἀπορρήτῳ τῆς εὐθημοσύνης περιλαμβάνει τὸ
ποίημα. τὰς μὲν γὰρ ἐς τὸ εὐθὺ ⁶ κατέδησεν
αὐτῶν, τὰς δὲ ἐπικαρσίας (ὑφαντικῆς ἐπιστήμονα
γυναῖκα εἴποις ἂν ⁷ τῷ στήμονι τὴν κρόκην ἐπι-
πλέκειν), στρογγύλον δὲ ἡσυχῇ τὸ ἔργον ἀποφαίνει
καὶ κολπῶδες,⁸ οἱονεὶ πλέγμα κύρτου δημιουρ-
γοῦσα. καὶ ὅταν ἐξυφήνῃ τὸ εἰρημένον, κομίζει
πρὸς τὴν θάλατταν, ἔνθα τοῦ κύματος ἐπιπολάζον-
τος ἡσυχῇ ⁹ τὸ κλύσμα ἐπιὸν ἐλέγχει τῇ ἀλκυόνι
τὸ ἔργον· τὸ γάρ τοι μὴ στεγανὸν μέρος τὸ ὕδωρ

¹ προσυποθήγων.
² τῶν ζῴων.
³ μέντοι.
⁴ ἐν τῷ.
⁵ Lacuna : ⟨χρωμένη, ἀλλὰ τῷ στόματι⟩ conj. Schn.
⁶ ἰθύ.

236

eyes; and as it sharpens either eye by rubbing fennel along the edges it rids itself of this affliction. You see, after hibernating through the winter in some dark hole, it is short-sighted. And so the gentle warmth of the fennel cleanses the creature's vision which the frosts have numbed, and makes its sight keener.

17. When the Halcyon realises that it is pregnant it builds itself a nest[a] to receive its brood; but it has no need of mud and a roof and houses, like the swallow which entering as an uninvited guest saddens the dawn with its twitter and even disturbs our slumbers at their sweetest; nor yet ⟨does it use⟩ its body ⟨but its beak⟩ alone as it applies itself to the aforesaid task in places away from man, weaving together and collecting the spines of the gar-fish, and by some mysterious means it binds together and encloses the fabric of its careful contriving. For some of the bones it fixes upright, others cross-wise (one would say that it was some woman skilled in weaving that was interlacing the woof with the warp), and makes the nest approximately round and bellying in shape, as though it were plaiting a weel. And when it has woven the aforesaid nest it takes it down to the sea, and there, as the waves flow gently in, the advancing surf puts the Halcyon's labour to a test. For the water encountering any part that is

The Halcyon and its nest

---

[a] Cp. Ar. *HA* 616 a 19–32 and Thompson's notes.

---

[7] ἂν αὐτήν.
[8] καί τι καὶ κολπῶδες ὑπόμηκες.
[9] *Reiske*: εἶτα ἡσυχῆ.

τὸ ἐμπῖπτον . . . [1] ἀκεῖται αὖθις. τὰ δὲ ἡρμοσμένα [2] εἴγε παίοις λίθῳ, οὐκ ἂν διατρήσειας αὐτά.
εἰ δὲ καὶ διακόψαι σιδήρῳ ἐθέλοις, τὰ δὲ οὐκ ἂν
εἴξειε, καλῶς τε καὶ εὖ διυφασμένα, τοῦ θώρακος
τοῦ λινοῦ οὐ μεῖον, ὅνπερ οὖν ἀναθεῖναι τῇ Ἀθηνᾷ
τῇ Λινδίᾳ Ἄμασιν ᾄδουσι. τὸ στόμα δὲ τοῦ
κύρτου τοῦδε ἄλλῳ μὲν οὔτε ἐσβατὸν οὔτε πάνυ
τι [3] σύνοπτον, δέχεται δὲ ἐκείνην μόνην. οὐκ ἂν
δὲ ἐσρεύσειε δι' αὐτοῦ οὐδὲ τῆς θαλάττης ἔσω
οὐδὲ ἕν· οὕτω τοι στεγανόν ἐστιν. ἐνταῦθά τοι
⟨καὶ⟩ [4] τοὺς νεοττοὺς τρέφει κατὰ τῶν κυμάτων ἡ
ἀλκυὼν φερομένη, ὥς φασιν.

18. Τοῦ Νείλου πλησίον πόα γίνεται, καὶ
καλεῖται λυκοκτόνος, καὶ οὐκ ἔστι ψευδώνυμος,
[καὶ εἰκότως]· [5] ὅταν γὰρ αὐτῆς ἐπιβαίνῃ [6]
λύκος, σπώμενος ἀποθνήσκει. ἔνθεν τοι καὶ οἱ
σέβοντες Αἰγυπτίων τοῦτο τὸ ζῷον ἐς τὴν ἑαυτῶν
χώραν κωλύουσι ταύτην τὴν πόαν κομίζεσθαι.

19. Τῶν κατὰ τὴν οἰκίαν ὄρνις ἐὰν ἐς οἶνον
ἐμπέσῃ καὶ ἀποπνιγῇ, οὐδὲν λυμαίνεται οὔτε τοῦ
οἴνου φασὶν οὔτε τῶν ἔνδον· ἐὰν δὲ ἐς ὕδωρ
κατενεχθῇ, δυσῶδες ἀπέφηνε τὸ ὕδωρ, καὶ κακοσ-
μίαν περὶ τὸν ἀέρα ἐργάζεται. γαλεώτης δὲ ἐὰν
ἐς οἶνον κατολισθὼν [7] εἶτα ἀποπνιγῇ, [8] λυπεῖ οὐδὲ
ἕν· ἐὰν δὲ ἐς ἔλαιον ἐμπέσῃ καὶ ἀποθάνῃ, δυσῶδες

---

[1] Lacuna.
[2] Jac : ἡρμοσμένα ἐᾷ καλῶς συνυφασμένα καί.
[3] πάντη.
[4] ⟨καὶ⟩ add. H.
[5] [καὶ εἰκότως] condemned by H.
[6] Jac : ἐπιβαίῃ.

not watertight ⟨penetrates the nest, and the Halcyon
seeing this ?⟩,[a] repairs it. But if you strike with a
stone the parts which have been closely fitted, you
will not pierce them. And if you try to cut them
with steel, so well and truly have they been inter-
woven that they will not yield, any more than that
linen corslet which they say Amasis [b] gave as an
offering to Athena of Lindus.[c] And the mouth of
this weel no other creature can enter or indeed
detect at all: it admits the Halcyon alone. But not
even a drop of sea water could trickle in, so watertight
is the nest. And there, they say, rocked on the
waves the Halcyon rears its young.

18. By the Nile there grows a herb, and it goes by
the name of ' Wolf's-bane,' [d] and it is truly named.
For when a wolf treads upon it he dies in convulsions.
That, you see, is why those Egyptians who worship
this animal prevent this herb from being introduced
into their country.

The herb
Wolf's-bane

19. If a bird of the household falls into a vessel of
wine and is drowned, they say that neither the wine
nor any of the inmates of the house suffers any harm;
whereas if it sinks in water, it causes the water to
smell, and diffuses a foul odour in the surrounding
air. But if a Gecko falls into wine and is drowned,
it does no harm. If however it falls into oil and dies,

Dead bodies
in wine
and oil

---

[a] Lacuna; the translation is conjectural.
[b] King of Egypt, 6th cent. B.C. See Hdt. 2. 182.
[c] Town on the E coast of Rhodes.
[d] Aconite.

---

[7] κατολισθήσας.   [8] ἀποπνιγῇ ἢ εἰς ὕδωρ.

τὸ ἔλαιον ἀποφαίνει, καὶ ὁ γευσάμενος αὐτοῦ
φθειρσὶν ἐξέζεσεν.

20. Τὸ τοῦ ἐλάφου κέρας θυμιώμενον ὅτι τοὺς
ὄφεις διώκει δῆλόν ἐστιν. λέγει δὲ Ἀριστοτέλης
ὅτι καὶ λίθος ὁ γινόμενος ἐν τῷ Πόντῳ ποταμῷ
(ἔστι δὲ οὗτος ἐν τῇ χώρᾳ τῇ Σιντικῇ [1] τε καὶ
Μαιδικῇ [2]) ἐπιθυμιώμενος διώκει τοὺς αὐτούς,
καὶ μέντοι καὶ φύσιν τοῦ λίθου περιηγεῖται τοιάνδε.
ὕδατος μὲν εἴ τις αὐτοῦ [3] καταχέοι, ἐξάπτεται·
καόμενον δὲ ὑπερεξάψαι ῥιπίδι εἰ θελήσειας,[4] ὁ δὲ
κατασβέννυται. θυμιώμενον δὲ αὐτὸν ὀσμὴν ἀφιέναι ἀσφάλτου βαρυτέραν φασί. τούτοις ὁμολογεῖ
καὶ Νίκανδρος.

21. Ἡ Φάρος ἡ νῆσος πάλαι (λέγουσι δὲ Αἰγύπτιοι οἷα μέλλω λέγειν) ἐπεπλήρωτο [5] ὄφεων
πολλῶν τε καὶ διαφόρων. ἐπεὶ δὲ Θῶνις ὁ τῶν
Αἰγυπτίων βασιλεὺς λαβὼν παρακαταθήκην τὴν
Διὸς Ἑλένην (ἔδωκε δὲ αὐτὴν ἄρα καὶ περὶ τὴν
ἄνω [6] Αἴγυπτον καὶ περὶ τὴν Αἰθιοπίαν πλανώμε-
νος ὁ Μενέλεως) εἶτα ἠράσθη αὐτῆς ὁ Θῶνις,
βίαν [7] αὐτοῦ προσφέροντος τῇ Ἑλένῃ ἐς ὁμιλίαν
ἀφροδίσιόν φησιν ὁ λόγος [8] τὴν τοῦ [9] Διὸς αὐτὰ [10]
εἰπεῖν ἕκαστα πρὸς τὴν τοῦ Θώνιδος γαμετὴν (Πο-
λύδαμνα ἐκαλεῖτο), τὴν δὲ δείσασαν μή ποτε ἄρα
ὑπερβάληται ἡ ξένη τῷ κάλλει αὐτήν,[11] ὑπεκθέσθαι
τὴν Ἑλένην ἐς Φάρον, πόαν δὲ τῶν ὄφεων τῶν

---

[1] Gron : Ἰνδικῇ.   [2] Schn : Παιονικῇ.
[3] αὐτῷ.   [4] θελήσεις.
[5] πεπλήρωτο.
[6] Reiske : ἄνω καὶ περὶ τὴν Αἴ.

it makes the oil smell nasty, and on anyone who tastes it lice at once break out.

20. It is clear that the burning of a Stag's horn expels snakes. And Aristotle asserts [*Mir.* 481 a 27] that the stone [a] which occurs in the river Pontus (it is in the territory of the Sinti and Maedi) [b] if burnt also chases away snakes. Moreover he describes the nature of the stone as follows. If you pour some water upon it, it lights; and if when burning you hope to kindle it into a bigger blaze by fanning it, it goes out. They say that as it burns it gives off a smell more oppressive than bitumen. And Nicander [*Ther.* 45] agrees with this.

The 'Thracian Stone'

21. The island of Pharos (what I am about to tell you is reported by the Egyptians) was once infested with a great variety of snakes. But when Thonis the Egyptian King took under his charge Helen the daughter of Zeus (because Menelaus entrusted her to him while he was wandering through Upper Egypt and Ethiopia), he fell in love with her, and when he attempted to force her to lie with him, the story goes that the daughter of Zeus repeated the whole tale to the wife of Thonis (Polydamna was her name), and she on her side, anxious lest this alien should prove more beautiful than she, removed Helen to the safety of Pharos and gave her a herb disliked

Helen of Troy and Snakes in Pharos

[a] The 'Thracian stone,' Θράκιος λίθος, is perhaps quicklime.
[b] It is the river Strymon which flows through that part of Paeonia inhabited by the S. and M.

---

[7] καὶ βίαν.   [8] λόγος δείσασαν.
[9] τοῦ del. H.   [10] ταῦτα.
[11] αὐτὴν ἅμα τε καὶ οἰκτείρασαν.

AELIAN

ἐκεῖθι ἐχθρὰν δοῦναι, ᾗσπερ οὖν αἴσθησιν λαβόντας
τοὺς ὄφεις εἶτα καταδῦναι. τὴν δὲ αὐτὴν κατα-
φυτεῦσαι, καὶ χρόνῳ ἀναθῆλαι καὶ ἀφεῖναι [1]
σπέρμα ἐχθρὸν ὄφεσι, καὶ μέντοι καὶ ἐν τῇ Φάρῳ
θηρίον τοιόνδε οὐκέτι [2] γενέσθαι. κληθῆναι δὲ τὴν
πόαν ἐλένιον λέγουσιν οἱ ταῦτα εἰδέναι δεινοί.

22. Θαλάττιον ζῷον οἱ ἀστέρες, καὶ εἰσὶ καὶ
οὗτοι μαλακόστρακοι, ἐχθροὶ δὲ τοῖς ὀστρέοις·
δειπνοῦσι γὰρ αὐτά. καὶ ὁ τρόπος τῆς ἐπιβουλῆς
τῆς κατ᾽ αὐτῶν ἐκεῖνός ἐστι. τὰ μὲν κέχηνε πολλάκις ψύχους δεόμενα καὶ ἄλλως εἴ τί σφισιν
ἐμπέσοι τούτῳ τραφησόμενα· οἱ τοίνυν ἀστέρες
μέσον τῶν ὀστράκων διείρουσιν ἓν κῶλον τῶν
σφετέρων ἕκαστος [3], καὶ ἐμπίπλανται τῶν σαρκῶν,
διειργομένων συνελθεῖν τῶν ὀστράκων αὖθις.
ἴδιον μὲν δὴ καὶ ἀστέρων θαλαττίων εἰρήσθω
ἡμῖν τοῦτο.

23. Τὴν μὲν ὕδραν τὴν Λερναίαν τὸν ἆθλον τὸν
Ἡράκλειον ᾀδέτωσαν ποιηταὶ καὶ μύθων ἀρχαίων
συνθέται, ὧνπερ οὖν καὶ Ἑκαταῖος ὁ λογοποιός
ἐστιν· ᾀδέτω δὲ καὶ Ὅμηρος Χιμαίρας φύσιν
κεφαλὰς ἐχούσης τρεῖς, τέρας τοῦτο Λύκιον
Ἀμισωδάρου τοῦ Λυκίων βασιλέως, ἐπὶ λύμῃ [4]
πολλῶν θρέμμα ποικίλον τε καὶ ἀπρόσμαχον, ναὶ
μὰ Δία. καὶ ταῦτα μὲν ἔοικεν ἐς τοὺς μύθους
ἀποκεκρίσθαι· ἡ δὲ ἀμφίσβαινα ὄφις δικέφαλός
ἐστι, καὶ τὰ ἄνω καὶ ὅσα ἐς τὸ οὐραῖον· προϊοῦσα
δέ, ὅπως ἂν ἐς τὴν ὁρμὴν ἐπαγάγῃ τῆς προόδου

[1] ἀφιέναι.   [2] οὐ.

by the snakes there; so as soon as they were aware
of this, the snakes went underground. But Helen
planted the herb and in time it flourished and
produced seed disagreeable to the snakes, and in
Pharos such creatures have never recurred. Experts in these matters say that this herb is called
Helenion.[a]

22. Starfishes are marine creatures, and they too The Starfish
and Oysters
have a soft shell, but are the enemies of oysters, for
they feed on them. And their method of assailing
the oysters is as follows. The latter frequently
open for coolness' sake and anyhow in order to feed
themselves on whatever comes their way. Accordingly the Starfishes insert one of their limbs between
the shells and take their fill of the flesh, the oysters
being precluded from closing again. So much then
for this characteristic of Starfishes.

23. Poets and the compilers of ancient legends, The Amphis
baena
among whom is Hecataeus the chronicler, may sing
of the Hydra of Lerna, one of the Labours of
Heracles; and Homer may sing of the Chimaera
with its three heads [*Il.* 6. 181 ; 16. 328], the monster
of Lycia kept by Amisodarus the Lycian king for
the destruction of many, of varied nature, and
absolutely invincible. Now these seem to have been
relegated to the region of myths. The Amphisbaena however is a snake with two heads, one at the
top and one in the direction of the tail. When it
advances, as need for a forward movement impels

---

[a] Elecampane, *Inula helenium*; cp. Diosc. 1. 29.

---

[3] εἰς ἕκαστον.   [4] λύπη.

ἡ χρεία αὐτήν, τὴν μὲν ἀπέλιπεν οὐρὰν εἶναι, τὴν
δὲ ἀπέφηνε κεφαλήν. καὶ μέντοι καὶ πάλιν εἰ
δεηθείη τὴν ὀπίσω ἰέναι, κέχρηται ταῖς κεφαλαῖς
ἐς τὸ ἐναντίον ἢ τὸ πρόσθεν ἐχρήσατο.

24. Ἦν δὲ ἄρα τι βατράχου γένος, καὶ καλεῖται
τοῦτο ἁλιεύς, καὶ καλεῖται [1] τὸ ὄνομα ἐξ ὧν δρᾷ.
δελέατα ἐκεῖνος [2] ὑπεράνω τῶν ὀφθαλμῶν ἔχει
προμήκεις [3] ὡς ἂν εἴποις τινὰς βλεφαρίδας, εἶτα
ἑκάστῃ βραχὺ [4] σφαιρίον προσπέφυκε. σύνοιδεν
οὖν ἑαυτῷ τούτοις ἐφολκοῖς [5] ὑπὸ τῆς φύσεως ἐς
τοὺς ἄλλους ἰχθῦς παρεσκευασμένῳ τε καὶ τεθηγ-
μένῳ προσέτι. οὐκοῦν ὑποκρύψας ἑαυτὸν ἐν τοῖς
θολερωτέροις τε καὶ ἰλύος μᾶλλον πεπληρωμένοις
ἡσυχάζει, προτείνων τὰς τρίχας τὰς προειρημένας.
τὰ τοίνυν βράχιστα τῶν ἰχθύων προσνεῖ ταῖσδε
ταῖς βλεφαρίσι, τὰς ἐπ᾽ ἄκρου σφαιροειδεῖς
περιφορὰς οἰόμενα δέλεαρ εἶναι, ὁ δὲ ἐλλοχῶν ἀτρεμεῖ, πλησίον δὲ ἐκείνων γεγενημένων, ὑπάγει τὰς
τρίχας ἐς ἑαυτόν (αἱ δὲ ἐσάγονται κρυπταῖς τισιν [6]
ὁδοῖς καὶ ἀφανέσι), γειτνιάσαντά τε ὑπὸ λαιμαργίας
τὰ ἰχθύδια δεῖπνόν ἐστι ⟨τῷ⟩ [7] βατράχῳ τῷ
προειρημένῳ.

25. Κάραβος πολύποδι ἐχθρός. τὸ δὲ αἴτιον,
ὅταν αὐτῷ τὰς πλεκτάνας περιβάλῃ, τῶν μὲν ἐπὶ
τοῦ νώτου ἐκπεφυκότων [8] αὐτῷ κέντρων ποιεῖται
οὐδεμίαν ὥραν, ἑαυτὸν δὲ περιχέας αὐτῷ ἐς πνῖγμα
ἄγχει. ταῦτα ὁ κάραβος σαφῶς οἶδε, καὶ ἀποδι-

---

[1] κέκτηται Reiske.
[2] Ges : ἐκεῖνα.
[3] προμήκεις τρίχας.
[4] Jac : τραχύ.
[5] ἐφοδίοις τὴν τροφήν.
[6] Schn : τισι ταῖς.

it, it leaves one end behind to serve as tail, while the
other it uses as a head. Then again if it wants to
move backwards, it uses the two heads in exactly
the opposite manner from what it did before.[a]

24. There is, it seems, a species of frog which
bears the name of ' Angler,' and is so called from
what it does. It possesses baits above its eyes: one
might describe them as elongated eyelashes, and at
the end of each one is attached a small sphere. The
fish is aware that nature has equipped it and even
stimulated it to attract other fish by these means.
Accordingly it hides itself in spots where the mud is
thicker and the slime deeper, and extends the aforesaid hairs without moving. Now the tiniest fishes
swim up to these eyelashes, imagining that the
round, swinging objects at the end are edible; meanwhile the Angler lies in wait, never stirring, and
when the little fishes are near to him, he withdraws
the hairs towards himself (they are drawn in by some
secret and invisible means), and the little fishes,
whose gluttony has brought them close up, provide
a meal for the aforesaid frog.

The Fishing-frog

25. The Crayfish is the enemy of the Octopus.
The reason is this: when the Octopus throws its
tentacles round it, it cares nothing for the spines
that spring from the back of the Crayfish, but wraps
itself round and throttles it till it suffocates. This

Crayfish and Octopus

[a] See Gow-Scholfield on Nic. *Th.* 372.

7 ⟨τῷ⟩ add. H.
8 Ges : εἰσπεφυκότων MSS.

AELIAN

δράσκει αὐτόν. καράβου δὲ ἡ φύσις ἐκείνη ἐστίν.[1]
ὅταν ἀδεὴς ᾖ, πορεύεται ὅδε ὁ ἰχθὺς πρόσω,
πλαγιάσας δεῦρο καὶ ἐκεῖσε τὰ κέρατα, ἵνα μὴ
πρὸς ἐναντίαν τὴν νῆξιν τὸ ὕδωρ ἰὸν εἶτα ἀνα-
στέλλῃ οἱ τὰ κέρατα καὶ ἐμποδίζῃ[2] πρόσω χωρεῖν·
εἰ δὲ φεύγοι, τὴν ὀπίσω ἰὼν παρῆκεν αὐτὰ τελέως.
τὸ δὲ αἴτιον, ὡς κώπαις ἐρέττων καὶ ὑποκινῶν
δίκην πορθμίδος πολὺ ἀποσπᾷ. εἰ δὲ γένοιτο
μάχη καράβων πρὸς ἀλλήλους, τὰ κέρατα ἐγεί-
ροντες εἶτα ὡς κριοὶ ἐμπίπτοντες προσαράττουσι
τὰ μέτωπα. ἀγῶνα δὲ μυραίνης καὶ καράβου
ἀνωτέρω εἶπον.

26. Ἐλαύνει δὲ ἰσχυρῶς[3] τοὺς ὄφεις ἡ ἔνδροσός
τε καὶ νοτερὰ καλαμίνθη φασὶ καὶ ὁ ἄγνος. τοῦ-
τόν τοι καὶ ἐν Θεσμοφορίοις ἐν ταῖς στιβάσι τὰ
γύναια τὰ Ἀττικὰ ὑποστόρνυται. καὶ δοκεῖ μὲν
καὶ ἐχθρὸς εἶναι τοῖς δακετοῖς ὁ ἄγνος, ἤδη δὲ καὶ
ὁρμῆς ἀφροδισίου κώλυμά ἐστι, καὶ ἔοικε τό γε
ὄνομα λαβεῖν ἐντεῦθεν. δέδοικε δὲ ἄρα τε αὐτὰ
δακετὰ καὶ τὴν καλουμένην λιβανωτίδα πόαν.[4]

27. Θεοφράστου δὲ ἔγωγε ἀκούω καὶ ἐκεῖνα.
πόαν τινὰ ἱστορεῖ ὅδε ὁ ἀνήρ, καὶ ὄνομα θηλυφόνον
καλεῖ αὐτήν, ἥνπερ οὖν εἴ τις σκορπίῳ κατὰ νώτου
θεὶς ἐάσειεν,[5] ὁ δὲ παραχρῆμα αὖός ἐστιν. ὁ δὲ
αὐτὸς λέγει τὸν αὐτὸν ἀναστήσεσθαι, εἴπερ οὖν
λευκοῦ ἑλλεβόρου καταπάσειας αὐτοῦ. ἐγὼ δὲ

---

[1] ἡ φύσις· νῆξει.
[2] ἀναστέλληται . . . ἐμποδίζηται.
[3] ἰσχυρῶς τῇ φυγῇ.
[4] πόαν λιβανωτίδα.   [5] ἐάσαι or ἐάσει.

246

the Crayfish knows full well, and makes its escape. The nature of the Crayfish is as follows. When it has nothing to fear, this fish moves in a forward direction, turning its feelers[a] to either side, in order that the water encountering it as it swims may not thrust them back and hinder its advance. But if it is trying to escape, it goes backwards, relaxing its feelers completely, in order that, like one rowing with oars and moving lightly like a boat, it may withdraw to a great distance. If Crayfish fight with one another they raise their feelers, fall upon each other like rams, and butt their foreheads together. But a struggle between a moray and a Crayfish I have described earlier on.[b]

26. They say that the dewy Water-mint and the Agnus-castus are a potent means of expelling snakes. The latter, you know, is strewn by the women of Attica on their pallets at the Thesmophoria. And it appears that the Agnus-castus is offensive to noxious creatures, and at the same time represses sexual appetite; from this fact it appears to derive its name. And the same noxious creatures have a dread of the herb known as rosemary frankincense.

Snakes and certain herbs

27. From Theophrastus [*HP* 9. 18. 2] I learn the following. This great man mentions a certain herb and calls it by the name of ' Female-killer ';[c] and if one puts it on a scorpion's back and lets it lie, the creature immediately shrivels. But the same writer says that it revives if you sprinkle some white

The Aconite

---

[a] Lit. ' horns.'
[b] See 1. 32.
[c] One of several names for aconite; see Nic. *Al.* 36 ff.

ἐπαινῶ μὲν τὸ θηλυφόνον, τὸν δὲ λευκὸν ἑλλέβορον
ἥκιστα. τὸ δὲ αἴτιον, μισῶ μὲν σκορπίους, φιλῶ
δὲ ἀνθρώπους. Καλλίμαχος δὲ ἄρα ἐν τῇ γῇ τῇ
Τραχινίᾳ ᾄδει δένδρον τι φύεσθαι καὶ καλεῖσθαι
σμῖλον, ᾧ τὰ ἑρπετὰ γειτνιάσαντα καὶ παραψαύ-
σαντα ἀρχὴν εἶτα ἀποθνήσκει.

28. Τὴν ὗν κρέα ἔχειν τῶν ἄλλων κρεῶν ἡδίω
ἐκ πολλοῦ πεπίστευται. καὶ ἐκεῖνο δὲ ἡ πεῖρα
διδάσκει¹ καὶ μάλα γε ἐναργῶς.² ὅταν ποτὲ
σαλαμάνδραν φάγῃ, αὐτὴ μέν ἐστιν ἀπαθής, τούς
γε μὴν αὐτῆς γευσαμένους ἀποκτείνει.

29. Εὐφράτης, ὅσπερ οὖν Πάρθων καὶ Σύρων
ῥεῖ μέσος,³ ὅ τι μὲν⁴ καὶ ἕτερον τῶν ἄλλων ἔχει
ποταμῶν περιττὸν ἐρῶ ἄλλοτε, ὃ δὲ αὐτῷ συνίσασι
Πάρθοι τε καὶ Σύροι καὶ ἔστι συμμελὲς τοῖς
λόγοις τοῖσδε, τοῦτο εἰρήσεται. πρὸς ταῖς πρώ-
ταις ἀνατολαῖς τοῦδε τοῦ ποταμοῦ φύονταί τινες
ὄφεις, καὶ μάλα γε ἀνθρώπων ἐχθροί, ἀλλ' οὐ
τῶν ἐπιχωρίων καὶ συντρόφων, τῶν δὲ ξένων καὶ
προσηκόντων οὐδὲ ἕν. καὶ τιμῶνταί γε τὴν
ἐπιδημίαν θανάτου αὐτοῖς.

30. Λέων ὅταν βαδίζῃ, οὐκ εὐθύωρον πρόεισιν,
οὐδὲ ἐᾷ τῶν ἰχνῶν ἑαυτοῦ ἁπλᾶ εἶναι τὰ ἰνδάλματα,
ἀλλὰ πῇ μὲν πρόεισι, πῇ δὲ ἐπάνεισι, καὶ αὖ
πάλιν τοῦ πρόσω ἔχεται, καὶ μέντοι καὶ ἵεται
⟨ἐς⟩⁵ τοὐμπαλιν. εἶτα προφορεῖται τὴν ὁδόν, καὶ

---

¹ Schn : διδάξει.
² ἐναργής.
³ μέσος ποταμός.
⁴ μέν τοι.

hellebore upon it. Now I am in favour of Female-killer, but not at all of white hellebore. The reason is that I detest scorpions but love mankind. Callimachus [*fr.* 100 f. 48 P] relates how a tree that goes by the name of yew grows in Trachis, and if creeping things go near and touch it at all they die.

28. It is generally believed that the flesh of the Pig is sweeter than all others. And the fact is quite clearly proved by experiment. Whenever it eats a salamander, the Pig itself is unaffected, but kills those who taste its flesh. Flesh of the Pig

29. In what respect the Euphrates, which flows between Parthia and Syria, is superior to other rivers I will explain some other time; but what the Parthians and Syrians know about it, and what is relevant to the present discourse, that I will now tell. Near to the spot where the river first rises certain Snakes breed which are deadly enemies to men, not however to the natives who have been brought up in their midst, but to strangers who have no connexion whatever with them. And they even punish visitors with death. Snakes at the source of the Euphrates

30. The Lion when walking does not move straight forward, nor does he allow his footprints to appear plain and simple, but at one point he moves forward, at another he goes back, then he holds on his course, and then again starts in the opposite direction. Next he goes to and fro, effacing his tracks so as to The Lion's tracks

---

[5] ⟨ἐς⟩ *add. H.*

ἀφανίζει [1] τοῖς θηραταῖς ἰέναι κατὰ στίβον τὸν
ἑαυτοῦ καὶ ῥᾳδίως τὴν κοίτην ἔνθα ἀναπαύεται καὶ
οἰκεῖ σὺν τοῖς σκύμνοις εὑρίσκειν. καὶ ταῦτα μὲν
λεόντων ἐστὶν ἴδια δῶρα φύσεως.[2]

31. Ποιμένα μοι νόει νομευτικὴν [3] ἀγαθόν.
οὐκοῦν ὁ νομεὺς φιλεῖ μὲν τὰς οἶς, φιλεῖ δὲ καὶ τὰς
αἶγας, μισεῖ δὲ λύγγα. νόσημα ⟨δὲ⟩ [4] τοῦτο
ἀνθρώπῳ πολλάκις ἐμπῖπτον, ἄγει δὲ ἡ πλησμονὴ
καὶ ταῖσδε [5] τὴν λύγγα. οὐκοῦν πόαν τινὰ ἐχθρὰν
τῷ πάθει τῷδε τοῖς τῶν προειρημένων σηκοῖς οἱ
νομεῖς παραφυτεύουσι, καὶ ἥδε ἡ πόα ἀνείργει τὸ
κακὸν αὐταῖς.[6] λέγουσι δὲ οἱ πεπειραμένοι ὅτι
ἄρα καὶ ἀνθρώποις ἐς τὸ αὐτὸ πάθος ἐστὶν ἀγαθὸν
ἡ πόα αὕτη.

32. Ὑοσκύαμον καὶ ὀπὸν ὅσοις ἔργον τρυγᾶν,
οὗτοι περισκάπτουσι μὲν γύρους καὶ ὑποκινοῦσι
τὰς ῥίζας, οὐ μὴν διὰ χειρῶν τῶν σφετέρων
ἀνασπῶσιν,[7] ἀλλὰ τῶν ζῴων πτηνὸν ὅ τι οὖν
θηράσαντες ἢ πριάμενοι τοῖν ποδοῖν τὸν ἕτερον
προσέδησαν τῇ πόᾳ. τὸ δὲ ἰλυσπώμενον εἶτα
μέντοι ἀνασπᾷ αὐτήν. καὶ ἔστι λυσιτελῆ ἑκάτερα
ἐς ἃ δέονται ἄνθρωποι. εἰ δὲ μὴ ταύτῃ τις
ἀνασπάσειεν,[8] ἔχει ἄλλως ὅπερ οὖν οἴεται καλῶς
καὶ ἐς δέον [9] θησαύρισμα εἰληφέναι.

---

[1] ἀφανίζει corrupt.
[2] φύσεως ἄνωθεν αὐτοῖς δοθέντα; cp. 12. 32 fin.
[3] Schn : νομευτικόν.
[4] ⟨δέ⟩ add. H.
[5] τοῖσδε.
[6] αὐτοῖς.
[7] ἔτι ἀνασπῶσιν.

prevent [a] hunters from following his path and easily discovering the lair where he takes his rest and lives with his cubs. These habits of the Lion are Nature's special gifts.

31. Consider what makes a good shepherd. Now the herdsman loves both his sheep and his goats, but he abhors the hiccups. This affliction often befalls man, and a surfeit induces hiccups in sheep and goats also. Accordingly herdsmen plant round the pens of the aforesaid animals a certain herb which counters this complaint, and the herb protects them against it. And those who have had experience maintain that this herb is beneficial to man also in the same affliction.[b]

32. Those whose business it is to gather Henbane and the juice of Silphium [c] dig trenches round the plants and stir the roots a little; they do not however pull them up with their hands, but capture or buy some bird and fasten one leg to the herb. And as the bird flutters it pulls up the herb. Both are serviceable to man's needs. But if a man has not these means to pull them up, then the treasure which he fancies he has found so happily and in answer to his needs is of no service.

[a] The sense is clear, but the text is faulty.
[b] The herb (whose name A. does not disclose) is *Alyssum* or *madwort*; cp. Plut. *Mor.* 2. 648A.
[c] Ὀπός, the common term for *juice of silphium*, cannot be right here, unless Ael. attaches some other meaning to the word.

---

8 ἀνασπάσει.   9 *Kühn* : εἰλέον.

33. Τὸ ἀβρότονον ὅσα ἀγαθὰ δρᾷ καὶ ὅπως ὁδοὺς πνεύματι δίδωσι καὶ μέντοι καὶ πνεύμονός ἐστι καθαρτήριον οὐ νῦν λέγειν καιρός· ζῴῳ ⟨δ'⟩ [1] οὖν πονηρῷ πολέμιόν ἐστι, καὶ ἀναιρεῖ τὴν ἕλμινθα, ἥπερ οὖν ἐπὶ πλέον ἰοῦσα [2] θηρίον γίνεται σπλάγχνοις μὲν ἐντικτόμενον, ἀνθρωπείαις δὲ νόσοις ἐναριθμούμενον, καὶ ταῦτα ταῖς ἄγαν ἀνιάτοις τε καὶ ὑπὸ χειρὸς θνητῆς [3] ἐς ἄκεσιν ἥκειν ἀδυνάτοις. τεκμηριῶσαι τοῦτο καὶ Ἵππυς ἱκανός. ὃ δὲ λέγει ὁ συγγραφεὺς ὁ Ῥηγῖνος, τοιοῦτόν ἐστι. γυνὴ εἶχεν ἕλμινθα, καὶ ἰάσασθαι αὐτὴν ἀπεῖπον οἱ τῶν ἰατρῶν δεινοί. οὐκοῦν ἐς Ἐπίδαυρον ἦλθε, καὶ ἐδεῖτο τοῦ θεοῦ [4] ἐξάντης γενέσθαι τοῦ συνοίκου πάθους. οὐ παρῆν ὁ θεός· οἱ μέντοι ζάκοροι κατακλίνουσι τὴν ἄνθρωπον ἔνθα ἰᾶσθαι ὁ θεὸς εἰώθει τοὺς δεομένους. καὶ ἡ μὲν ἄνθρωπος ἡσύχαζε προσταχθεῖσα, οἵ γε μὴν ὑποδρῶντες τῷ θεῷ τὰ ἐς τὴν ἴασιν αὐτῆς [5] ἐποίουν, καὶ τὴν κεφαλὴν μὲν ἀπὸ τῆς δέρης ἀφαιροῦσι, καθίησι δὲ τὴν χεῖρα ὁ ἕτερος, καὶ ἐξαιρεῖ τὴν ἕλμινθα, θηρίου μέγα τι χρῆμα. συναρμόσαι δὲ καὶ ἀποδοῦναι τὴν κεφαλὴν ἐς τὴν ἀρχαίαν ἁρμονίαν οὐκ ἐδύναντο οὐκέτι. ὁ τοίνυν θεὸς ἀφικνεῖται, καὶ τοῖς μὲν ἐχαλέπηνεν ὅτι ἄρα ἐπέθεντο ἔργῳ δυνατωτέρῳ τῆς ἑαυτῶν σοφίας· αὐτὸς δὲ ἀμάχῳ τινὶ καὶ θείᾳ δυνάμει ἀπέδωκε τῷ σκήνει τὴν κεφαλήν, καὶ τὴν ξένην ἀνέστησε. καὶ οὔ τι που, [6] ὦ βασιλεῦ καὶ θεῶν φιλανθρωπότατε Ἀσκληπιέ, ἀβρότονον ἔγωγε ἀντικρίνω

---

[1] ⟨δ'⟩ add. H.
[2] ἰοῦσα καὶ αὐξανομένη.
[3] Ges : θνητῆς οὐ δυναμέναις.

33. This is not the occasion for mentioning all the Intestinal
Worm benefits that accrue from Wormwood, how it eases the windpipe and even cleanses the lungs. But to a troublesome creature it is certainly an enemy: it destroys intestinal worm. This creature grows and grows and becomes a monster bred in the intestines, and is reckoned among the diseases of mankind, and what is more, among those which are hardest to cure and which will not yield to any mortal treatment. Hippys is sufficient witness to this. The account given by the historian of Rhegium is as follows. A woman suffered from an intestinal worm, and the cleverest doctors despaired of curing her. Accordingly she went to Epidaurus and prayed the god *a* that she might be rid of the complaint that was lodged in her. The god was not at hand. The attendants of the temple however made her lie down in the place where the god was in the habit of healing his petitioners. And the woman lay quiet as she was bid; and the ministers of the god addressed themselves to her cure: they severed her head from the neck, and one of them inserted his hand and drew out the worm, which was a monstrous creature. But to adjust the head and to restore it to its former setting, this they always failed to do. Well, the god arrived and was enraged with the ministers for undertaking a task beyond their skill, and himself with the irresistible power of a god restored the head to the body and raised the stranger up again. For my part, O King Asclepius, of all gods the kindliest

a There was a famous temple of Asclepius 5 mi. W of Epidaurus in Argolis.

---

[4] τῶν θεῶν.   [5] αὐτῇ.   [6] πω.

τῇ σοφίᾳ τῇ σῇ· μὴ μανείην ἐς τοσοῦτον· ἀλλὰ
ἐπελθὼν[1] ἐμνήσθην εὐεργεσίας τε σῆς καὶ ἰάσεως
ἐκπληκτικῆς. ὡς δὲ καὶ ἥδε ἡ πόα σὸν δῶρόν
ἐστιν οὐδὲ ἀμφιβάλλειν χρή.

34. Ὁ δὲ ναυτίλος πολύπους[2] ἐστὶ καὶ αὐτός,
καὶ κόγχην μίαν ἔχει. ἀναπλεῖ μὲν οὖν[3] τὴν
κόγχην στρέψας περὶ τὰ κάτω, ἵνα μὴ τῆς ἅλμης
ἀρύσηται καὶ ὠθήσῃ αὖθις αὐτόν· γενόμενος δὲ
ἐπὶ τοῖς κύμασιν, ὅταν μὲν ᾖ γαλήνη καὶ εἰρήνη
πνευμάτων, στρέφει τὴν κόγχην ὑπτίαν (ἡ δὲ
ἐπιπλεῖ δίκην πορθμίδος) καὶ παρεὶς δύο πλεκτάνας ἐντεῦθέν τε καὶ ἐκεῖθεν καὶ ὑποκινῶν ἡσυχῇ
ἐρέττει τε καὶ προωθεῖ τὴν συμφυῆ ναῦν. εἰ δὲ
εἴη πνεῦμα, τοὺς ἐρετμοὺς μὲν τοὺς τέως προτείνας
μακροτέρους οἴακας ἐργάζεται, ἄλλας δὲ ἀνατείνας
πλεκτάνας, ὧν μέσος χιτών ἐστι λεπτότατος,
τοῦτον διαστήσας ἱστίον αὐτὸν ἀποφαίνει. πλεῖ
μὲν δὴ τὸν τρόπον τοῦτον ἀδεὴς ὤν· ἐὰν μέντοι
φοβηθῇ τι τῶν ἁδροτέρων, βυθίσας τὴν κόγχην
ἐπλήρωσε, καὶ κατώλισθεν ἐκ τοῦ βάρους, καὶ
ἑαυτὸν ἀφανίσας τὸν ἐχθρὸν ἀπέδρα. εἶτα ἐν
εἰρήνῃ γενόμενος ἀνέθορέ τε καὶ πλεῖ πάλιν. καὶ
ἐκ τούτων ἔχει τὸ ὄνομα.

35. Ἐς τριακοσίας ὀργυιάς φασιν ἀνθρώποις
κάτοπτα εἶναι τὰ ἐν τῇ θαλάττῃ, περαιτέρω γε μὴν
οὐκέτι. εἴτε δὲ ὑπονέουσιν ἰχθύες ἔτι εἴτε καὶ
θηρία, ἢ εἰ καὶ τούτοις μὲν ἄβατά ἐστι, θεοὶ δὲ

[1] ἐπελθών V, ἐπελθόν other MSS, H.
[2] πολύπους μέν.   [3] οὖν ἐκ τῆς θαλάττης.

a Poseidon.

to man, I do not set Wormwood against your skill (heaven forbid I should be so insensate!), but in considering Wormwood I was reminded of your beneficent action and of your astounding powers of healing. And there is no need to doubt that this herb also is a gift from you.

34. The Argonaut also is one of the polyps and The has one shell. Now it rises to the surface by turning ^{Argonaut} its shell upside down to prevent it from taking in salt water and being thrust down again. And when it is on top of the waves, if the weather is calm and the winds are at rest, it turns its shell (which floats like a boat) on its back, and letting down two tentacles, one on either side, with a gentle motion rows and propels its natural vessel. And if there is a wind it extends still further what up till now were oars, using them as rudders, and raises other tentacles between which there is a web of most delicate texture, and this it spreads and turns into a sail. And in this way it navigates so long as it has nothing to fear. If however it is afraid of some of the larger and stronger fish, it submerges and fills its shell and sinks with the weight of water, and by disappearing escapes from its enemy. Then when it has peace again it rises and resumes its sailing. It is from these activities that it derives its name.

35. They say that men have explored the sea to The depth a depth of 300 fathoms, but not as yet beyond that. ^{of the sea} Whether there are fishes and animals swimming at an even greater depth, or whether even to them these regions are inaccessible, although the gods of the sea and also the overlord of the moist world [a]

θαλάττιοι καὶ ἐνάλιοι δαίμονες εἰλήχασι τὸν χῶρον
καὶ μέντοι καὶ ὁ τῆς ὑγρᾶς οὐσίας δεσπότης, οὔτε
ἐγὼ πολυπραγμονῶ οὔτε ἄλλος λέγει.

36. Ἦν δὲ ἄρα πέτραις ἠθὰς καὶ ἐν ταύταις
νεμόμενος γένος κεστρέως ἰχθύς, καὶ ἰδεῖν ξανθός
ἐστι. διαρρεῖ δὲ ἄρα ὑπὲρ αὐτοῦ διπλοῦν ὄνομα [1].
οἱ μὲν γὰρ ἄδωνιν καλοῦσιν, οἱ δὲ ἐξώκοιτον.[2]
ὅταν γάρ τοι τὸ κῦμα ἐν τοῖς ὑπευδίοις καὶ
γαληνοῖς πραϋνθῇ, τηνικαῦτα ἑαυτὸν ἐξοκέλλει,
τοῦ κύματος ἐποχούμενος τῇ ὁρμῇ, καὶ κατὰ τῶν
πετρῶν ἁπλοῖ, καὶ καθεύδει βαθὺν καὶ εἰρηναῖον
εὖ μάλα τὸν ὕπνον. καὶ ἐκ μὲν τῶν ἄλλων
ἁπάντων ἔνσπονδα ὥς ἐστίν οἱ καλῶς οἶδε,
πέφρικε δὲ τοὺς ὄρνιθας ὅσοι θαλάττης ἔντροφοι
καί εἰσι καὶ νομίζονται. ἐὰν οὖν ἐκείνων ἐπιφανῇ
τις, ὁ δὲ ἀναπάλλεται καὶ πηδᾷ χορείᾳ τινὶ
φυσικῇ καὶ ὀρχήσει [3] ὥς ἂν εἴποις μάλα ἀπορρήτῳ,[4] ἔστ' ἂν ἀπὸ τῆς πέτρας ἐξαλλόμενος εἶτα
ἐμπεσὼν τοῖς κύμασι σωθῇ. Ἄδωνιν δ' ἐθέλουσι
λέγειν αὐτόν, ἐπεὶ καὶ γῆν καὶ θάλατταν ἔχει
φίλην, τῶν πρώτων ἐμοὶ δοκεῖν θεμένων τὸ ὄνομα
αἰνιξαμένων ἐς τὸν τοῦ Κινύρου παιδὸς βίον τὸν
διῃρημένον δύο δαίμοσι, τῆς μὲν ὑπὸ γῆς, τῆς δὲ
ἄνω γῆς ἐρώσης αὐτοῦ.[5]

37. Φυτοῦ ἑτέρου κλάδος ἐπιφύεται πρέμνῳ,
προσήκων οἱ μηδὲ ἓν πολλάκις. τὸ δὲ αἴτιον

---

[1] Jac : τὸ ὄνομα MSS.   [2] ἐξώκοιτον αὐτόν.
[3] ὀρχηστική.
[4] Reiske : ἀπορρήτως.
[5] αὐτοῦ ἑκατέρας.

have their allotted dwelling there—these are matters into which I shall not enquire too closely, and no one else informs us.

36. There is, it seems, a fish of the species mullet which is accustomed to live and to feed among rocks, and is yellow in appearance. There are two names for it in common use, for some call it 'Adonis,' others 'Exocoetus'.[a] For, you see, when the waves are lulled in places where the water is calm and smooth, it runs aground, borne forward by the force of the wave, and spreading itself upon the rocks, sleeps a deep and tranquil sleep. And it is well aware that there is peace between it and all other creatures, though it dreads all birds that are or are reputed to be nurslings of the sea. And so if one appears, the fish leaps up and dances as nature has taught it with movements that, one might say, baffle description, until it jumps off the rock, falls into the sea, and is safe. People like to call it 'Adonis' because it loves both land and sea, and those who first gave it this name were hinting (so I think) at the son of Cinyras [b] whose life was divided between two goddesses; one who loved him was beneath the earth, the other above.

The 'Adonis' fish

37. A twig of one tree will grow on the stock of another to which it often bears no relation. And

Grafting of trees

[a] That is, 'sleeping out of the water.' The fish has not been identified.

[b] Adonis was the son of Cinyras by his daughter Myrrha. Aphrodite concealed the baby in a chest which she entrusted to Persephone. On Persephone's declining to give the child back Zeus ordained that he should spend one half of each year with either goddess.

Θεόφραστος λέγει, φυσικώτατα ἀνιχνεύσας ὅτι τὰ
ὀρνύφια τὴν ἄνθην τῶν δένδρων σιτούμενα εἶτα
ἐπὶ τοῖς φυτοῖς καθήμενα τὰ περιττὰ ἀποκρίνει.
οὐκοῦν τὸ σπέρμα ταῖς κοιλάσι [1] καὶ ταῖς ὀπαῖς
αὐτῶν καὶ τοῖς σηραγγώδεσιν ἐμπῖπτον καὶ
ἐπαρδόμενον τοῖς ὄμβροις τοῖς ἐξ οὐρανοῦ, εἶτα
ἀναφύει ἐκεῖνα [2] ἐξ ὧν ἐβλάστησεν.[3] οὕτω τοι
καὶ ἐν ἐλαίᾳ συκῆν κατανοήσεις, καὶ ἐν ἄλλῳ ἄλλο.

38. Φωλεύει δὲ [4] ἐν τοῖς μυχοῖς τῆς θαλάττης
τὸ [5] πρόβατον, καὶ οἱ καλούμενοι ἥπατοι, καὶ
οὕσπερ οὖν φιλοῦσιν ἁλιεῖς ὀνομάζειν πρέποντας.
καὶ μέγιστοι μέν εἰσιν ἰδεῖν τὴν φύσιν, νωθεῖς δὲ
τὴν νῆξιν, καὶ εἰλοῦνται περὶ τοῖς φωλεοῖς, ἔνθεν
τοι οὐδὲ ἀπολείπουσι τὴν σφετέραν ὑποδρομήν.
λοχῶσι δὲ τῶν ἰχθύων τῶν ἀσθενεστέρων τοὺς
παρανέοντας. ἀριθμοῖτο δ' ἂν ἐν τούτοις καὶ ὁ
ὄνος· δέδοικε δὲ μάλιστα ἰχθύων τὴν τοῦ Σειρίου
ἐπιτολὴν οὗτος ὁ ὄνος.

39. Τίκτεται δὲ ἄρα ἐν τοῖς τῶν πυρῶν ληΐοις
καὶ ταῖς αἰγείροις καὶ ταῖς συκαῖς [6] προσέτι τὸ
τῶν κανθαρίδων φῦλον, ὥσπερ οὖν Ἀριστοτέλης
λέγει, ἔν γε μὴν [7] τοῖς ἐρεβίνθοις τὸ τῶν καμπῶν,
ἐν δὲ ⟨τῷ⟩ [8] ὀρόβῳ φαλάγγια ἄττα, ἐν δὲ τοῖς
πράσοις ἡ καλουμένη πρασοκουρίς. τίκτεται δὲ
καὶ ἐν τῇ κράμβῃ σκωλήκων γένος, καὶ ὄνομα

[1] ἐν ταῖς κοιλάσι.
[2] ἐκεῖνο.
[3] ἐβλάστησεν ἀναπείθει MSS, κἀμὲ πείθει Jac.
[4] δὲ καί.
[5] διαιτᾶται τό. [6] Schn : τοῖς σύκοις.
[7] γε μήν] μέν. [8] δὲ ὀρόβῳ γεννᾶται.

Theophrastus, who has traced the cause of this in a thoroughly scientific way, explains the cause [*CP* 2. 17. 5 & 8]: small birds eat the blossoms of trees and then as they sit upon the trees void their excrement. And so the seed dropping into hollows and cracks and cavities, and being watered by the rains of heaven, produces the same wood as that from which it sprang. Thus you will see a fig-tree on an olive-tree, and the same with other trees.

38. The Sea-sheep and the *Hepatus* [a] as it is named, and what fishermen are accustomed to call the *Prepon* [b] have their lairs in the recesses of the sea. They are of enormous size to look at but sluggish swimmers, and range to and fro around their lairs, and so it comes about that they never abandon their hiding-places. But they lie in wait for fish of weaker species that swim past. The Hake too may be reckoned as belonging to this class. More than any other fish does it dread the rising of the Dog-star. *The Sea-sheep, and others*

39. It seems that the family of Blister-beetles is produced in fields of wheat and on poplar-trees and on fig-trees also, as Aristotle says [*HA* 552 b 1]; and Caterpillars are produced among peas, and certain Spiders among bitter vetch, and the Leek-cutter,[d] as it is called, among leeks. And in the cabbage is born a kind of worm which derives its *Insects, etc., born in plants*

---

[a] Unidentified; not the same as the *Hepatus* of 15. 11.

[b] Unidentified.

[c] See D. W. Thompson's note on Arist. *l.c.* (Eng. tr.).

[d] ? 'leaf-maggot' (Hort on Thphr. *HP* 7. 5. 4); 'Prob. milliped' (L-S⁹). The *Hylemyia antiqua* (order *Anthomyidae*) may attack the *bulb* of leeks.

αὐτῷ ἐκ τῶν ἠθῶν, ἐν οἷς διαιτᾶται. καλεῖται
γοῦν κραμβίς. τίκτει ⟨δέ⟩ [1] τι καὶ ἡ μηλέα·
καὶ διαφθείρει μὲν τοῦτο πολλάκις τὸν καρπὸν τοῦ
φυτοῦ τοῦδε, ταῖς δὲ ἔτι τοῦ τίκτειν ἐχούσαις
ὥραν γένοιτο ἂν καὶ ἐς κύησιν ἀγαθόν. καὶ τὸν
τρόπον ἐρεῖ ἄλλος.

40. Οἶδε δὲ ἄρα τῶν ζῴων ἕκαστον ἐν ᾧ μέρει
κέκτηται τὴν ἀλκήν, καὶ τούτῳ θαρρεῖ, καὶ ἐπι-
βουλεῦον μὲν χρῆται ὡς ὅπλῳ, κινδυνεῦον δὲ ὡς
ἀμυντηρίῳ. ὁ γοῦν ξιφίας ἀμύνεται [2] τῷ ῥύγχει
ὡς ξίφει, ἔνθεν τοι καὶ κέκληται· ἡ δὲ τρυγὼν τῷ
κέντρῳ, ἡ δὲ μύραινα τοῖς ὀδοῦσι, καὶ μάλα γε
εἰκότως· ἔχει [3] γὰρ αὐτῶν διστοιχίαν.

41. Οἱ μὲν [4] μῦς οἱ κατὰ τὴν οἰκίαν δειλὸν καὶ
ἀσθενὲς ζῷόν εἰσι,[5] καὶ φοβοῦνται κτύπον, καὶ
τὴν γαλῆν πεφρίκασι κρίξασαν· δειλοὶ δὲ καὶ οἱ
ἀρουραῖοι. τῶν γε μὴν οἰκετῶν θρασύτεροι οἱ
θαλάττιοι. μικρὸν μὲν αὐτῶν τὸ σῶμα, τόλμα
δὲ ἄμαχος· καὶ θαρροῦσι δύο ὅπλοις, δορᾷ τε
εὐτόνῳ καὶ ὀδόντων κράτει· μάχονται δὲ καὶ τοῖς
ἰχθύσι τοῖς ἁδροτέροις καὶ τῶν ἁλιέων τοῖς
μάλιστα θηρατικοῖς.

[1] ⟨δέ⟩ add. H.   [2] ἀμύνει.   [3] Ges : ἔχουσι.
[4] μὲν οὖν.   [5] ἐστι.

---

[a] The larvae or caterpillar of the large white butterfly, *Pieris rapae*, injure cabbages, turnips, radishes, etc.

name from its habitat. At any rate it is called the Cabbage-caterpillar.[a] The apple-tree also produces a creature[b] which frequently destroys the fruit of this tree, although it may help women who are still of an age to bear children to conceive. How this happens another shall tell.

40. It seems that every creature knows in which part of its body its strength resides, and this gives it confidence, for when attacking it employs it as a weapon, when in danger as a means of defence. For instance, the Swordfish defends itself with its snout as with a sword; hence its name; and the Sting-ray with its sting, and the Moray with its teeth, and well it may, because it has a double row of them.

Animals know where their strength lies

41. The domestic Mouse is a timorous and feeble creature and is scared by noise and trembles at the squeak of a marten.[c] Field-mice also are timorous, whereas the Sea-mice[d] are bolder than the domestic animal. Though their body is small their courage is irresistible, and this they derive from two weapons, their tough skin and their powerful teeth. And they fight even with fish of greater bulk and with the most skilled fishermen.

The Mouse

The 'Sea-mouse'

[b] The caterpillar of the Codling moth, *Carpocapsa pomonella* L.

[c] With us it would be 'the mew of a cat.'

[d] Oppian (*Hal.* 1. 174) speaks of μυῶν χαλεπὸν γένος as 'confident in their tough hide and close-set teeth,' and as 'contending with men, though not so very large.' This is probably the *Turtle*, whose sharp but toothless jaws can inflict a savage bite. See Thompson, *Gk. fishes*, s.v. μῦς, II, p. 167.

42. Τῆς τῶν ὡρῶν μεταβολῆς ἔχουσιν αἰσθητικῶς οἱ θύννοι καὶ ἴσασι τροπὰς ἡλίου ὀξύτατα,
καὶ δέονται τῶν τὰ οὐράνια εἰδέναι δεινῶν [1] οὐδὲ
ἕν. ὅπου [2] γὰρ ἂν αὐτοὺς χειμῶνος ἀρχὴ κατα-
λάβῃ, ἐνταῦθα ἡσυχάζουσί τε καὶ ἀτρεμοῦσιν
ἀγαπητῶς, καὶ καταμένουσιν ἐς τὴν ἐπιδημίαν
τῆς ἰσημερίας.[3] καὶ τεκμηριοῖ Ἀριστοτέλης τοῦτο·
ὅτι δὲ τῷ ἑτέρῳ τῶν ὀφθαλμῶν ὁρῶσι, τῷ δὲ
ἄλλῳ οὐκέτι, καὶ Αἰσχύλος ὁμολογεῖ λέγων

τὸ σκαιὸν ὄμμα παραβαλὼν θύννου δίκην.

παρίασί τε ἐς τὸν Πόντον, καὶ κατὰ τὴν δεξιὰν
ἑαυτῶν πλευρὰν τὴν γῆν λαμβάνουσι, καθ᾽ ἣν καὶ
βλέπουσιν· ἐξιόντες τε αὖ κατὰ τὴν ἀντιπέρας [4]
νέουσι τῆς γῆς ἐχόμενοι, τὴν φρουρὰν τὴν τοῦ
σώματος κατὰ τὸν ὁρῶντα τῶν ὀφθαλμῶν λαμ-
βάνοντες προμηθέστατα.

43. Τοῖς παγούροις τὸ πρῶτον ἔλυτρον ῥήγνυται,
καὶ ὥσπερ οἱ ὄφεις τὸ γῆρας, οὕτω δήπου καὶ
οὗτοι τὸ ὄστρακον ἀποδύονται. ὅταν δὲ αἴσθωνται
ἀφιστάμενον τῆς σαρκὸς αὐτό, πανταχοῦ φοιτῶσιν
οἰστρούμενοι καὶ μαστεύοντες τροφὴν πλείονα,
ἵνα ὄγκου προσγενομένου αὐτοῖς ὑποπρησθέντες
ἀπορρήξωσιν ἑαυτῶν τὸ ἔλυτρον. ὅταν δὲ διολί-
σθωσιν ἐξ αὐτοῦ καὶ ἐλεύθεροι γένωνται, κεῖνται
παρειμένοι κατὰ τῆς ψάμμου, νεκροῖς εἰκασμένοι·
δεδοίκασι δὲ ὑπὲρ τῆς φυομένης αὐτοῖς δορᾶς
ὑγροτέρας τε οὔσης καὶ ἔτι ἁπαλῆς. κατὰ μικρὰ

---

[1] ποιουμένων or προσπ-. [2] Jac : ὅποι.
[3] τὴν τῆς ἐπιδημίας ἰσημερίαν.
[4] ἀντίπερα.

42. The Tunny is aware of the changes of the The Tunny seasons and knows precisely when the solstices occur and has no need whatsoever of persons skilled in celestial matters. For in whatever place the beginning of winter overtakes these fish, there they are glad to remain at rest without stirring, and there they stay until the coming of the equinox. Aristotle bears witness to this [*HA* 599 b 9]. And that they see with one eye and not with the other is admitted by Aeschylus when he says [*fr*. 308 N]

' Casting his left eye askance like a tunny.'

And they pass into the Euxine, keeping the land on their right, on which side in fact they look out. Contrariwise when issuing from the Euxine they swim along the opposite shore and hug the land, taking the utmost precaution to safeguard their life by means of the eye which sees.

43. The first shell of the common Crab splits and, The common just as snakes slough their ' old age,' so do these Crab creatures put off their shell. And directly they perceive that it is coming away from their flesh they move frantically in every direction in their search for more food, in order that they may become inflated by the additional bulk and so break off their shell. And when they have contrived to slip out of it and are free, they lie on the sand exhausted like dead bodies. But their growing shell causes them anxiety while it is still rather pliable and tender. Gradually however they gather themselves together and come to life, as it were, and begin by eating sand.[a] But as long as their outer covering consists

---

[a] πρώτης . . . ψάμμου ' *verba corrupta*,' H.; but cp. Opp. *Hal.* 1. 96, ψάμμον ἐρεπτόμενοι καὶ ὅσ᾽ ἐν ψαμάθοισι φύονται.

AELIAN

δὲ ἑαυτοὺς ἀθροίσαντες καὶ ἀναβιωσκόμενοι τρόπον
τινά, πρώτης μὲν ἀπογεύονται τῆς ψάμμου. ἐς
τοσοῦτον δὲ ἄτολμοί εἰσι καὶ ἥκιστα θαρραλέοι, ἐς
ὅσον αὐτοῖς ὑμὴν περίκειται στέγασμα [1] ἔξωθεν·
ὅταν δὲ ἄρξηται πήγνυσθαι καὶ ἐς ὀστράκου
φύσιν μεταχωρεῖν, ἀπέρριψαν ἐνταῦθα τὴν δειλίαν,
ὡς ὅπλῳ θαρροῦντες τῇ τῆς περιβολῆς σκέπῃ τε
ἅμα καὶ [2] ὡς ἂν εἴποις πανοπλίᾳ.

44. Τρωγλοδύται γένος ἀνθρώπων ὑμνεῖται, καὶ
τό γε ὄνομα εἴληφεν ἐκ τῆς διαίτης.[3] φοβοῦνται
δὲ αὐτοὺς οἱ ὄφεις. τὸ δὲ αἴτιον, ἐσθίουσιν αὐτοὺς
οἱ ἄνθρωποι. μιγνύμενοι δὲ ἀλλήλοις οἱ ὄφεις
βαρυτάτην ὀσμὴν ἀφιᾶσιν.

45. Ἀγροῦ γειτνιῶντος θαλάττῃ καὶ φυτῶν
παρεστώτων ἐγκάρπων γεωργοὶ πολλάκις κατα-
λαμβάνουσιν ἐν ὥρᾳ θερείῳ πολύποδάς τε καὶ
ὀσμύλους ἐκ τῶν κυμάτων προελθόντας καὶ διὰ
τῶν πρέμνων ἀνερπύσαντας καὶ τοῖς κλάδοις
περιπεσόντας καὶ ὀπωρίζοντας, καὶ δίκην [4] ἐπέθε-
σαν τοῖς φωρσὶ συλλαβόντες αὐτούς. ἀνθ' ὧν δὲ
ἐτρύγησαν οἱ προειρημένοι, ὑπὲρ τούτων ἀντεφεσ-
τιῶσι [5] δι' ἑαυτῶν τοὺς δεσπότας τῶν σεσυλημένων
καρπῶν.

46. Ῥυάδες ὄνομα θαλαττίου ζῴου σοφοῦ διαγ-
νῶναι τὴν τῶν ὡρῶν διάβασιν. ὑπαρχομένου γοῦν
τοῦ χειμῶνος ἡσυχάζουσι τοὺς κρυμοὺς ἀποδιδρά-
σκουσαι, καὶ ἀσμένως ἑαυτὰς τῇ καταμονῇ

[1] τὸ στέγασμα.   [2] καὶ ἐς τὸν βίον.
[3] διαίτης καὶ τοῦ βίου δηλονότι.

264

of membrane, for so long are they timid and utterly lacking in courage. When however the membrane begins to harden and to assume the nature of a shell, then they cast aside their fears, and the protection of their covering and their full suit of armour, as you might call it, gives them the same confidence as a shield would.

44. The race of men known as Troglodytes is famous, and derives its name from its manner of living. Snakes are afraid of them, the reason being that the men eat them.

Snakes when engaged in coupling emit a most offensive odour.

The Troglo-
dytes and
Snakes

45. If a field, or if trees with fruit upon them are close by the sea, farmers often find that in summer Octopuses and Osmyluses[a] have emerged from the waves, have crept up the trunks, have enveloped the branches, and are plucking the fruit. So when they have caught them they punish them. And as quittance for what the aforesaid fish have reaped they provide the owners of the pillaged fruit with a feast.

The Octopus
and fruit-
trees

46. 'Migrants' is the name for marine creatures that are clever at knowing the transition of the Seasons. At any rate at the beginning of winter they escape from the frosts and remain at rest and are glad by so remaining to keep warm, sharing

The migra-
tion of fishes

[a] See 5. 44.

---

[4] τὴν δίκην. [5] ἀνθεστιῶσι V, H.

AELIAN

θάλπουσαι ἰσομοιρίᾳ ἀδελφικῇ· εἶτα ἦρος[1] νεῖν
ὑπάρχονται ἐκεῖναί γε καὶ πορείας τῆς μακροτέρας,
καὶ νέμονται τροφὴν οὐ μόνον τὴν προσπεσοῦσαν,
ἀλλὰ καὶ ἣν ἂν μαστεύσασαι εἶτα ἀνιχνεύσωσιν.

47. Τοὺς ἐχίνους ἔτι ζῶντας καὶ ἐν τοῖς
ὀστράκοις ὄντας καὶ προβεβλημένους τὰ κέντρα εἴ
τις συντρίψας καὶ διαρρίψας ἐς τὴν θάλατταν ἄλλο
ἄλλῃ τρύφος καταλίποι, τὰ δὲ ἄρα συνέρχεται
αὖθις καὶ ἑνοῦται καὶ τὸ συγγενὲς θρύμμα ἀνέγνω,
καὶ προσπλακέντα[2] συνέφυ. καὶ ὁλόκληροι γίνονται φύσει τινὶ θαυμαστῇ καὶ ἰδίᾳ αὖθις.

48. Ὑπὲρ τοῦ πλείονα τὴν ἐπιγονὴν τῶν ζῴων
σφίσι γίνεσθαι οἱ τούτων μελεδωνοὶ[3] τὰ ἄρθρα
τῶν θηλειῶν καὶ οἰῶν καὶ αἰγῶν καὶ ἵππων
ἀνατρίβουσι κατὰ τὸν τῆς ὀχείας καιρὸν ἁλῶν καὶ
λίτρου[4] τὰς χεῖρας ἀναπλήσαντες. ἐκ τούτων
ὄρεξις αὐτοῖς γίνεται περὶ τὴν ἀφροδίτην μᾶλλον.
ἕτεροι δὲ πεπέριδι καὶ μέλιτι τὰ αὐτὰ χρίουσι,
λίτρῳ[5] δὲ ἄλλοι καὶ κνίδης καρπῷ· σμυρνίῳ δὲ
ἤδη τινὲς ἔχρισαν καὶ λίτρῳ. ἐκ δὴ τοῦδε τοῦ
ὀδαξησμοῦ ἀκράτορες ἑαυτῶν γίνονται αἱ θήλειαι
ποῖμναι, καὶ ἐπιμαίνονται τοῖς ἄρρεσιν.

49. Τῶν κητῶν τῶν μεγίστων αἰγιαλοῖς καὶ
ἠόσι καὶ τοῖς λεπροῖς[6] καλουμένοις καὶ βραχέσι
χωρίοις προσπελάζει οὐδὲ ἕν, οἰκεῖ δὲ τὰ πελάγη.
καὶ ἔστι μέγιστα ὅ τε λέων καὶ ἡ ζύγαινα καὶ ἡ

[1] ἦρι.
[2] προσπλακέν.
[3] μελεδωνοὶ καὶ νομεῖς.
[4] νίτρου.
[5] νίτρῳ.
[6] ἐλαφροῖς.

266

their warmth in brotherly fashion. Then in the
spring they begin to swim greater distances and feed
not only upon what comes their way but on what
they have sought for and tracked down.

47. If one crushes Sea-urchins while still alive The Sea-urchin
within their shells and with their spines protruding
and then throws one bit here and another there into
the sea and leaves them, they come together again
and join up: they recognise their related fragments,
and attaching themselves grow together. And it is
by some marvellous and peculiar force of Nature that
they become whole again.

48. With a view to increasing the offspring of their Sexual stimulants for animals
animals their keepers and herdsmen at the mating
season take handfuls of salt and of sodium carbonate
and rub the genitals of their female asses and goats
and mares. These substances produce in the animals
a greater appetite for sexual intercourse. Others
rub their parts with pepper and honey; others
again with sodium carbonate and nettle-seed. And
some have in fact applied Cretan alexanders and
sodium carbonate. And from the consequent
irritation the females of a herd cannot contain
themselves but go mad after the males.

49. There is not one of the largest Cetaceans that The largest of the Cetaceans
comes near the shore or the beach or ' leprous ' (that
is, rocky) spots or into shallow water: they live in
the deeps. The largest of them are the Sea-lion,

πάρδαλις καὶ οἱ φύσαλοι καὶ ἡ πρῆστις καὶ ἡ
καλουμένη μάλθη· δυσανταγώνιστον δὲ ἄρα ⟨τὸ⟩ [1]
θηρίον τοῦτο καὶ ἄμαχον. καὶ ὁ κριὸς δεινὸν [2]
ζῷον καὶ κίνδυνον φέρον, εἰ καὶ πόρρωθεν φανείη,
τῇ τῆς θαλάττης ταράξει καὶ τῷ κλύδωνι ὃν [3]
ἐργάζεται. καὶ ὕαινα,[4] οὐκ αἴσιον ὅραμα [5] τοῖς
ναυτιλλομένοις αὕτη γε. κυνῶν δὲ περὶ διαφορᾶς
καὶ ἀλκῆς ἀνωτέρω εἶπον.

50. Αἱ καστορίδες ζῷόν εἰσι θαλάττιον, καὶ ἐπὶ
ταῖς ἀκταῖς καὶ ταῖς πέτραις ταῖς προβεβλημέναις
ἀπόφημόν τινα κωκυτὸν μεθιᾶσι, καὶ ὠρύονται
βαρύτατα. τούτου τοίνυν τοῦ ἤχου ὅστις ἂν
ἀκούσῃ,[6] ἄφυκτά οἵ ἐστι, καὶ οὐ μετὰ μακρὸν
ἀποθνήσκει. καὶ ἡ φάλλαινα δὲ τῆς θαλάττης
πρόεισι καὶ ἀλεαίνεται τῇ ἀκτῖνι. κνεφαῖαι δὲ αἱ
φῶκαι ἐξίασι μᾶλλον· ἤδη μέντοι καὶ μεσημβρίας
οὔσης [7] καθεύδουσι τῆς θαλάττης ἔξω. τοῦτό τοι
καὶ Ὅμηρος ᾔδει, καὶ ἐν Ὀδυσσείᾳ τὸν Μενέλεων
πεποίηκε τῷ Τηλεμάχῳ καὶ τῷ Πεισιστράτῳ
περιηγούμενον τὴν κοίτην αὐτῶν τήνδε, ὅτε τὰ ἐν
Φάρῳ καὶ περὶ Πρωτέως τοῦ θαλαττίου δαίμονος
αὐτοῖς ὁ Μενέλεως διεξῄει καὶ τῆς μαντείας, ἣν
ἐμαντεύσατό οἱ ὁ Πρωτεὺς ὃν εἶπον.

51. Τρίγλης πέρι ἀνωτέρω εἶπον· ὃ δὲ οὐκ
εἶπον, νῦν ἐρῶ. ἐν Ἐλευσῖνι τιμὰς ἔχει ἐκ τῶν

---

[1] ⟨τὸ⟩ add. H.
[2] ἰδεῖν ἐχθρόν.
[3] Ges : τὸ κλυδώνιον αὐτός MSS, αὐτός del. H.
[4] Schn from Opp. Hal. 1. 372 : ζύγαινα.
[5] ὁρᾷ οὐδὲ ἐργάζεται.
[6] ὑπακούσῃ.   [7] οὔσης καί.

the Hammer-headed Shark,[a] the Sea-leopard, the
great Whales,[a] the Pristis, and the fish called
Maltha. This last monster is a terrible antagonist
and invincible. The Ram-fish [b] also is a creature
to be dreaded and is dangerous, even if it emerges
at a distance, owing to the upheaval in the sea and
the wave which it creates. The Sea-hyena too is no
auspicious sight for seafarers. As to Sharks, I have
spoken above of their different kinds and of their
strength.

50. Sea-calves [c] are marine animals, and on head- The
lands and projecting rocks they utter a kind of Sea-calf
ominous cry and a very deep roar. And moreover
whoever hears this sound, for him there is no escape,
but he dies soon after.

The Whale too comes out of the sea and warms The Whale
itself in the sun. But Seals emerge for choice when
it is dark, although they do in fact sleep on shore at The Seal
midday. Homer knew this, and in the *Odyssey*
[4. 400] he has represented Menelaus explaining to
Telemachus and Pisistratus this habit they have of
resting, when he was telling them of what happened
at Pharos and of the sea-god Proteus and of the
prophecy which was uttered by the aforesaid Proteus.

51. I have spoken above [d] of the Red Mullet, but The Red
what I did not mention then I will now. At Eleusis Mullet

[a] The only animals in the list that have been certainly
identified.
[b] See below, 15. 2 n.
[c] Generally taken to mean ' Seals,' but the description that
follows points rather to the Walrus; and so Gossen (§ 215)
understands the word.
[d] See 2. 41.

μυουμένων, καὶ διπλοῦς ὁ λόγος τῆς τιμῆς [1] τῆσδε.
οἱ μέν φασιν, ἐπεὶ τρὶς τοῦ ἔτους τίκτει· οἱ δέ,
ἐπεὶ τὸν λαγὼν ἐσθίει, ὅσπερ οὖν ἐστιν ἀνθρώπῳ
θανατηφόρος. ἴσως δὲ ἐρῶ τι περὶ τρίγλης καὶ
πάλιν.

52. Πέτονται δὲ [2] ὅταν δείσωσι καὶ ἐξάλλονται
τῆς θαλάττης αἵ τε τευθίδες καὶ οἱ ἱέρακες οἱ
θαλάττιοι καὶ ἡ χελιδὼν ἡ πελαγία. καὶ αἱ μὲν
τευθίδες ἐπὶ μήκιστον ᾄττουσι τοῖς πτερυγίοις,
καὶ ἐλαφρίζουσί γε ἑαυτὰς ὑψοῦ, καὶ κατὰ ἀγέλας
ὀρνίθων δίκην φέρονται κοινῇ· αἱ δὲ χελιδόνες
χθαμαλωτέραν ποιοῦνται τὴν πτῆσιν· οἵ γε μὴν
ἱέρακες ὑπὲρ τὴν ἅλμην φέρονται ὀλίγον, ὡς μόλις
ὅτι μὴ νήχονται ἀλλὰ πέτονται καταγνῶναι.

53. Ἀλῶνται δὲ ἄρα ἰχθῦς καὶ πλανῶνται οἱ
μὲν ἀθρόοι, ὥσπερ οὖν ἀγέλαι θρεμμάτων ἢ τάξεις
ὁπλιτῶν ἰοῦσαι κατὰ ἴλας καὶ φάλαγγας, οἱ δὲ ἐν
κόσμῳ κατὰ στοῖχον ἔρχονται, οἱ δέ, φαίης ἂν
αὐτοὺς εἶναι λόχους. ἠρίθμηνται δὲ ἐς δεκάδας
ἄλλοι καὶ ταύτῃ συνέουσιν· ἤδη δὲ νήχονται καὶ
κατὰ ζεῦγός τινες. ἄλλοι δὲ οἰκουροῦσιν ἐν τοῖς
φωλεοῖς καὶ ἐνταυθοῖ καταζῶσιν.

54. Πυνθάνομαι δὲ ὅτι ἄρα [3] οἱ νομευτικὴν
δεινοὶ ὅταν ἐθέλωσιν ἐπὶ πιμελὴν τὰ ζῷα ἐπιδοῦναι,
ἀφαιροῦσιν αὐτῶν τὰ κέρατα. καὶ τοὺς τράγους
ὅταν ἐθέλωσιν ἐς μίξιν προθυμοτέρους ἐργάσασθαι,
μύρῳ χρίουσιν αὐτῶν τὰς ῥῖνας, καὶ τὰ γένεια

---

[1] αἰτίας.   [2] δὲ ἰχθύες.   [3] ἄρα ὅτι.

it is held in honour by the initiated, and of this
honour two accounts are given. Some say, it is
because it gives birth three times in a year; others,
because it eats the Sea-hare, which is deadly to man.

I shall perhaps recur to the Red Mullet.

52. Squids, Flying Gurnards,[a] and Flying-fish Flying-fish
when scared fly and leap out of the sea. Squids leap
furthest with the aid of their fins and rise high and
are borne along together in flocks like birds. Flying-
fish wing their flight at a lower level. The Flying
Gurnards however move at so little distance above
the surface of the sea, that you can hardly tell that
they are not swimming but flying.

53. It seems that Fishes roam and wander about, Fish moving
some in masses, like troops of animals or bands of in formation
hoplites marching in ranks or in lines; others advance
in an orderly column; others again you would say
were in companies. Others are numbered off by
tens and swim together in that formation; there are
even some that swim in couples, while there are
others that remain at home in their lairs and spend
their lives there.

54. I have ascertained that skilled herdsmen when Various
wishing to fatten their animals, remove their horns. treatments
And when they wish to stimulate their he-goats to for domestic
couple, they rub perfume on their nostrils; they even animals

> [a] See Thompson, *Gk. fishes*, p. 287.

μέντοι καὶ ἐκεῖνα χρίουσι τῶν αὐτῶν. πάλιν τε
τῆς ἄγαν ὀρέξεως ἀναστέλλουσιν, ἐάν τις αὐτῶν
μέσας τὰς οὐρὰς ἀποδήσῃ λίνῳ. Ἀριστοτέλης δέ
φησι τὰς ἵππους ἐκβάλλειν τὰ ἔμβρυα, ἐὰν ἐπὶ
πλέον ὀσφρήσωνται θρυαλλίδος λύχνου ἐσβεσμένης.[1]
ἀκούω δὲ ὅτι πρὸς τοὺς κύνας τοὺς οἰκουροὺς ἵνα
μὴ ἀποδιδράσκωσι τετέχνασται ἐκεῖνο. τὴν οὐρὰν
αὐτῶν καλάμῳ μετρήσαντες χρίουσι τὸν κάλαμον
βουτύρῳ, εἶτα μέντοι διδόασιν αὐτοῖς περιλιχμή-
σασθαι αὐτόν. καὶ καταμένουσί φασιν ὥσπερ οὖν
δεδεμένοι.

55. Ἴδια δὲ καὶ ἐκεῖνα κυνός. οὐχ ὑλακτοῦσιν,
εἴ τις ἔχων οὐρὰν γαλῆς σὺν ἑαυτῷ εἶτα πρόσεισι,[2]
γαλῆς δ᾽ ἣν ἐθήρασε μέν, ἀποκόψας δὲ τὴν
προειρημένην οὐρὰν εἶτα ἀφῆκε ζῶσαν αὐτήν.
ὄνος δὲ οὐ βρωμήσεται, ἐὰν αὐτοῦ τῆς οὐρᾶς λίθον
ἀπαρτήσῃς, ὥς φασιν.

56. Ἐν ὥρᾳ θερείῳ, πολλοῦ πάνυ σφόδρα τοῦ
ἡλίου ἐνακμάζοντος, οἱ ἐλέφαντες ἀλλήλους χρίουσιν ἰλύι παχείᾳ, καὶ αὐτοῖς αὕτη ψῦχος τε παρέχει
καὶ οἰκίας ὑπαντρου τινὸς ἢ δένδροις καὶ κλάδοις
ἀμφιλαφοῦς ἥδιόν ἐστι τοῖς ζῴοις τοῖς προειρημένοις. οὗτοι ῥινηλατοῦσιν ἰσχυρῶς, καὶ αἴσθησιν
ὀξυτάτην ἔχουσι. προΐασι γοῦν ἀλλήλων[3] ὁδο-
ποιοῦντες, καὶ ὅ γε πρῶτος (ἴασι γὰρ κατὰ
στοῖχον) τῆς ἐν ποσὶ πόας αἰσθόμενος καὶ ὅτι
διῆλθον ἄνθρωποι ἐκ τῆς παραψαύσεως συνεὶς
αὐτῶν, ἀνασπᾷ τὴν πόαν καὶ δίδωσι τῷ κατόπιν
ὀσφραίνεσθαι, καὶ ἐκεῖνος τῷ μετ᾽ αὐτόν· καὶ ἥδε
ἡ ἀντίδοσις ὡς ἂν εἴποις διὰ πάντων ἔρχεται. καὶ

anoint their chins as well. On the other hand they restrain an excessive appetite by tying a cord round the middle of the animals' tails. And Aristotle asserts [*HA* 604 b 30] that mares miscarry if for some length of time they smell an extinguished lamp-wick. I have heard also of this device to stop house-dogs from running away: they measure the length of their tail with a rod, smear the rod with butter, and then give it to the dog to lick. And the dogs remain at home, they say, as though they were fastened up.

55. Here is another peculiarity of Dogs. They will not bark if one approaches them holding the tail of a marten; but after cutting off the said tail of the captured marten, one must let it go alive. And a Donkey will not bray if you suspend a stone from its tail, so they say.

How to silence Dogs and Donkeys

56. In the season of summer when the sun's blaze is at its strongest Elephants smear one another with thick slime: this affords them coolness and is more agreeable to the aforesaid animals than a home beneath a cave or embowered in trees and branches. They are good at tracking by scent and have a very keen sense of smell. At any rate on the march one precedes another, and the leader (they move in single file) takes note of the grass at his feet, and when he realises from the brushing that men have passed that way, he pulls up the grass and gives it to the elephant behind him to smell, and he in turn to

The Elephant

---

¹ ἐσβεσμένην.   ² προσείει *Cobet*.   ³ ἀλλήλοις.

AELIAN

μέντοι ⟨καὶ⟩ ¹ ἐς τὸν οὐραγοῦντα ὅταν ἀφίκηται,
ὁ δὲ μέγα ἐπήχησεν, οἱ δὲ ὥσπερ οὖν σύνθημα
στρατιῶται ² λαβόντες εἶτα μέντοι ἐκτρέπονται ἐς
τὰ τῶν ὀρῶν ἄγκη καὶ δάση ἢ τῶν ἑλῶν τὰ
κοιλότερα καὶ μέντοι καὶ τῶν πεδίων ὅσα κομᾷ
τοῖς θάμνοις. πάντως δὲ ἢν ³ καταστείβουσιν
ἄνθρωποι, ταύτην ἀποδιδράσκουσιν· ὑφορῶνται
γὰρ τοῦτο τὸ ζῷον ὡς ἔχθιστον. ὅταν δὲ αὐτοὺς
αἱ νομαὶ ἐπιλίπωσιν,⁴ οἱ μὲν ⁵ τὰς ῥίζας ἐξορύττουσιν καὶ σιτοῦνται καὶ ταύτας,⁶ οἱ δὲ ἀπίασι
ζητοῦντες χιλόν. καὶ ὅ γε ἐντυχὼν τῷ θηράματι
πρῶτος αὐτῶν ὑποστρέψας καλεῖ τοὺς συννόμους
καὶ ἐπί γε τὸ ἕρμαιον αὐτοὺς ἄγει.

57. Ἐν δὲ τῷ χειμῶνι τῷ βιαιοτάτῳ, κυμαι-
νούσης μὲν τῆς θαλάττης, σκληρόν γε μὴν τῶν
ἀνέμων καὶ βίαιον καταπνεόντων, φρίττουσι τὴν
σύντροφόν τε ἅμα καὶ φίλην οἱ ἰχθύες θάλατταν.
καὶ οἱ μὲν αὐτῶν τοῖς πτερυγίοις ἐπαμῶνται τὴν
ψάμμον, καὶ ἑαυτοὺς ἐπηλυγάσαντες ὑποθάλπου-
σιν, οἱ δὲ ὑπειλοῦνταί ⁷ τινα πέτραν, ἐν σκέπῃ τε
τοῦ κρύους καὶ μάλα γε ἀσμένως ἡσυχάζουσιν·
οἱ δὲ ἐς τοὺς μυχοὺς τοῦ πελάγους καταθέοντες
εἶτα τὴν ἄνωθεν φρίκην ἐξέκλιναν κάτω καὶ ἐν
βυθῷ ⁸· οὐχ οὕτως γάρ φασιν ὥσπερ οὖν ἄνω
διοιδαίνειν ⁹ τε καὶ τύπτειν τὸ κῦμα ἀγριαῖνον.
ὑπαρχομένου δὲ τοῦ ἦρος καὶ τοῦ μὲν ἀέρος
φαιδροῦ γενομένου, τῶν δὲ φυτῶν θάλλειν ἀρχομέ-
νων καὶ τῶν λειμώνων τὰ σύντροφα κομώντων,
γαληνά τε τὰ τοῦ πελάγους καὶ ὑπεύδια αἰσθό-

¹ ⟨καὶ⟩ add. H.   ² στρατιᾶ or στρατιᾶς.
³ Jac: ῇ.   ⁴ καταλίπωσιν.

274

the one behind him. And this exchange, as you might call it, goes through the whole herd, until it comes to the one who is bringing up the rear, when he trumpets loudly. Whereupon like soldiers at a signal they turn aside to vales and thickets in the mountains or to low-lying marshes or even to level country where the bushes are dense. But at all costs they avoid land which is trodden by men, for man is a creature whom they suspect as their worst enemy. And when their feeding-grounds fail some of them dig up roots and eat them, while others go off in search of fodder. And the Elephant that is the first to find what he is seeking turns back and calls his fellows and leads them to his lucky discovery.

57. In the severest winter when the sea is stormy and the winds are blowing fierce and strong, Fish dread their native and beloved sea. And some of them heap up sand with their fins and so covered keep themselves warm, while others slip beneath some rock and are glad to rest sheltered from the cold. Others again hasten down to the recesses of the sea and there below in the depths avoid the agitation from above. For, men say, the fury of the waves does not at that depth swell and batter them as it does above. But at the beginning of spring when the sky grows bright and plants begin to put forth their leaves and the fields to wave with their natural herbage, the Fish observing that the sea is smooth and calm, mount up and leap about and

Fish in Winter

---

⁵ δέ.
⁶ αὐτάς, καὶ οἱ μὲν ἐσθίουσι καὶ ταύτας.
⁷ ὑποδύονται H.
⁸ βυσσῷ.   ⁹ διοιδάνειν H.

μένοι οἱ ἰχθύες, ἀναθέουσι καὶ πηδῶσι, καὶ
πλησίον τῆς γῆς νήχονται, ὥσπερ οὖν ἥκοντες ἐξ
ἀποδημίας.

58. Τρία δὲ ἄρα ταῦτα ἐκ βραχίστων μέγιστα
ζῷα γίνεται· τῶν μὲν ἐνύδρων ὁ κροκόδιλος, τῶν δὲ
ὑποπτέρων ἡ στρουθὸς ἡ μεγάλη, τῶν γε μὴν
τετραπόδων ὁ ἐλέφας. λέγει δὲ ὁ Ἰόβας γενέσθαι
μὲν αὐτοῦ τῷ πατρὶ πολυετῆ Λίβυν ἐλέφαντα
κατιόντα ἐκ τῶν ἄνω τοῦ γένους· καὶ Πτολεμαίῳ
δὲ τῷ Φιλαδέλφῳ Αἰθίοπα, καὶ ἐκεῖνον ἐκ πολλοῦ
βιώσαντα γενέσθαι πρᾱότατον καὶ ἡμερώτατον τὰ
μὲν ἐκ τῆς πρὸς τοὺς ἀνθρώπους συντροφίας, τὰ
δὲ [1] πωλευθέντα· Σελεύκου τε τοῦ Νικάτορος
κτῆμα ᾄδει Ἰνδὸν ἐλέφαντα, καὶ μέντοι καὶ
διαβιῶναι τοῦτον μέχρι τῆς τῶν Ἀντιόχων
ἐπικρατείας φησίν.

59. Ἰχθύες ὅσοι ποταμὸν γείτονα τῇ θαλάττῃ τῇ
συντρόφῳ κέκτηνται ἢ καὶ λίμνην τινὰ ὅταν
μέλλωσι τίκτειν, ἐκνήχονται τῆς ἅλμης, τῶν [2]
κυμάτων τὸ ἄκλυστον ὕδωρ προαιρούμενοι καὶ
ταραττόμενον ὑπὸ τῶν πνευμάτων καὶ τυπτόμενον
ἥκιστα. ἀγαθὴ γὰρ αὐτοῖς ἡ τῶν ὑδάτων εἰρήνη
⟨τὴν⟩ [3] λοχείαν ὑποδέξασθαι καὶ φυλάξαι ἀσινῆ τε
καὶ ἀνεπιβούλευτα τὰ [4] βρέφη τῇ τε ἄλλῃ καὶ
μέντοι καὶ διὰ τὴν τῶν θηρῶν ἐρημίαν καὶ σπάνιν·
φιλοῦσι δέ πως τήνδε τὴν ἐλευθερίαν ἔχειν αἵ τε
λίμναι καὶ οἱ ποταμοί. ἔνθεν τοι καὶ πολλοῖς
ἰχθύσιν εὐθενεῖται ὁ Εὔξεινος Πόντος· θηρία γὰρ
τρέφειν οὐκ ἔμαθε. φώκην δὲ εἴ που τρέφει καὶ

---

[1] τὰ δὲ ἐκ τοῦ γένους.   [2] καὶ τῶν.

swim close to the shore as though they were returning from a long journey.

58. These, it seems, are the three creatures which Longevity
from the smallest beginnings grow to the largest of the
size: among aquatic animals the Crocodile, among
birds the Ostrich, and among quadrupeds the
Elephant. And Juba relates that his father possessed
an Elephant of a great age that was descended from
remote ancestors; and that Ptolemy Philadelphus
had an Ethiopian Elephant which had lived for many
years and partly from its association with men and
partly from its training had become exceedingly
docile and gentle. He also tells of an Elephant
from India which belonged to Seleucus Nicator, and
he says moreover that it survived down to the
supremacy of the Antiochi.[a]

59. All Fish that have a river or some lake near to Sea-fish
their native sea, when they are about to spawn swim spawn in
out of the salt water, choosing in preference to the fresh water
waves water that is calm and not at all upheaved and
lashed by gales. For the tranquillity of river and
lake is well adapted to receive their offspring and to
preserve their young from harm and from attack,
both for other reasons and especially because of the
absence or paucity of savage creatures. And lakes
and rivers normally enjoy this freedom. That is the
reason why the Euxine abounds in such a quantity
of fish: it has not learnt to foster monsters. If it

[a] Seleucus Nicator reigned 312–280 B.C.; Antiochus I, 280–
261 B.C.

[3] ⟨τήν⟩ add. H.   [4] τῶν ἰχθύων τά.

δελφῖνας βραχίστους, ⟨ἀλλὰ⟩ [1] τῶν γε [2] ἄλλων
ἁπάντων οἱ τῇδε ἰχθῦς ἐν σκέπῃ εἰσίν.

60. Αἱ θαλάττιαι βελόναι λεπταὶ οὖσαι [3] καὶ
χωρητικὴν ἐμβρύων μήτραν οὐκ ἔχουσαι τὴν
αὔξην τῶν ἔνδον βρεφῶν οὐ φέρουσιν ἀλλὰ ῥήγνυνται, καὶ τοῦτον τὸν τρόπον οὐ τίκτουσιν ἀλλὰ
ἐκβάλλουσι τὰ τέκνα.

61. Λέγεται δὲ τὰ ἴχνη καὶ τὰ γνωρίσματα τῶν
τῆς ἀσπίδος δηγμάτων μὴ πάνυ τι εἶναι δῆλα καὶ
εὐσύνοπτα. καὶ τὸ αἴτιον ἐκεῖνο εἶναι πυνθάνομαι.
ὀξύτατόν ἐστι τὸ ἐξ αὐτῆς φάρμακον καὶ διαδραμεῖν
ὤκιστον. οὐκοῦν ᾗ μὲν ἐνέφυ, τὸ δὲ οὐκ ἐπιπο-
λεύει, ἀλλὰ ἐς τοὺς ἔσω πόρους κατολισθάνει, καὶ
τῆς μὲν ἐπιφανείας καὶ τοῦ χρωτὸς τοῦ ὑπὸ τὴν
ὄψιν ἀφανίζεται, ὠθεῖται δὲ ἔνδον. ἔνθεν τοι καὶ
τῆς Κλεοπάτρας ὁ θάνατος τοῖς ἀμφὶ τὸν Σεβαστὸν
οὐ πάνυ τι ῥᾳδίως ἐγνώσθη ἀλλὰ ὀψέ, δύο κεντημά-
των καὶ μάλα γε δυσθεάτων καὶ δυσθηράτων
ὀφθέντων, δι᾽ ὧν ἐφωράθη τὸ τοῦ θανάτου αἴνιγμα.
ἄλλως τε καὶ ἴχνη τοῦ τῆς ἀσπίδος σύρματος
ἐφάνη, πρόδηλα τοῖς ἔχουσι τῆς τούτων κινήσεως
τῶν ζῴων τὴν ἱστορίαν ὄντα.

62. Πομπηίου ῾Ρούφου ῾Ρωμαίοις ἀγορανομοῦν-
τος ἐν Παναθηναίοις [4] φαρμακοτρίβης ἀνὴρ καὶ
τῶν τοὺς ὄφεις ἐς τὰ θαύματα [5] τρεφόντων,

---

[1] ⟨ἀλλά⟩ add. H.   [2] τῶν δέ.
[3] οὖσαι κολπώδη.
[4] Π. οἷα εἴωθε τῇ ῾Ρωμαίων ἀγορᾷ δρᾶσθαι.
[5] Voss : τραύματα.

does breed the seal and dolphins, they are of the smallest, but from all other pests the fishes here are protected.

60. Pipefishes are slender, and having no womb The Pipefish to contain their foetus they are unable to endure the growth of their young within their bodies, but burst open; and in this way they do not give birth to, but eject, their offspring.

61. It is said that the traces and indications of The bite of the bites of the Asp are far from evident or easy to the Asp detect. And the reason for this is, I learn, as follows. The Asp's poison is exceedingly sharp and spreads very rapidly. So when the Asp fastens on a man the poison does not remain on the surface but penetrates to the inner passages of the body and disappears from view and from the skin before one's eyes, and presses inwards. That, you see, is why the manner of Cleopatra's death was by no means easily recognised by Octavian's companions, but only after a time when two punctures, hard to detect and discover, were observed, and through them was revealed the riddle of her death. Besides, marks of the Asp's trail were visible, and they were clear to persons acquainted with the movements of these creatures.

62. When Pompeius Rufus was Aedile at the Death of a Panathenaea [a] a medicine-man, one of those who snake-charmer keep snakes for show, amid a crowd of his fellow-

---

[a] Παναθήναια is used as an equivalent for the Roman *Quinquatrus*, a festival held in March. Pompeius was Consul in 88 B.C. and a colleague of Sulla.

AELIAN

ἑτέρων ὁμοτέχνων παρεστώτων πολλῶν, ἀσπίδα
κατὰ τοῦ βραχίονος προσάγει ἐς ἔλεγχον αὐτοῦ
τῆς σοφίας [1] καὶ ἐδήχθη. εἶτα τῷ στόματι
ἐξεμύζησε τὸ κακόν. ὕδωρ δὲ οὐκ ἐπιρροφήσας,
οὐ γὰρ παρῆν, καίτοι παρεσκευασμένον οἱ (ἀνετέ-
τραπτο δὲ ἐξ ἐπιβουλῆς τὸ σκεῦος), οἷα μὴ ἐκ-
κλύσας τὸν ἰὸν μηδὲ ἀπορρυψάμενος, τὸν βίον
κατέστρεψε μετὰ ἡμέραν οἶμαι δευτέραν, οὐκ
ἀλγῶν οὐδὲ ἕν, τοῦ μέντοι κακοῦ ἡσυχῇ διασήψαν-
τος αὐτοῦ τὰ οὖλα καὶ τὸ στόμα.

63. Τοῦ ἦρος ἐνακμάζοντος καὶ τῆς γῆς ἐξαν-
θούσης οἴστρου τε ἀφροδισίου τὰ ζῷα ὑποπίμπλαται καὶ μνημονεύει γάμων, καὶ ἀλλήλοις συμπλέ-
κεσθαι ὀργᾷ [2] τά τε ὄρεια καὶ ὅσα ἐνθαλαττεύει
καὶ μετεωροπορεῖ ὅσα. τῶν δὲ ἰχθύων οἱ μὲν
ταῖς [3] ψάμμοις προσαποτρίβουσι τὰ ὠὰ πυκνὰ
ὄντα καὶ ἀλλήλων ἐχόμενα, οἱ δὲ νηχόμενοι εἶτα
ἐκβάλλουσι πάμπολύ τι τῶν ὠῶν τὸ χρῆμα, καὶ οἱ
κατόπιν νέοντες τὰ πολλὰ καταπίνουσιν. ἤδη
μέντοι καὶ οἱ ἄρρενες προηγοῦνται καὶ τοῦ θοροῦ
ἀπορραίνουσιν,[4] αἱ δὲ θήλειαι ἑπόμεναι καὶ πάνυ
γε ἀπλήστως περιχανοῦσαι ἐμπίπλανται· καὶ ἥδε
ἐστὶν ἡ μίξις αὐτῶν. ὅτι δὲ ἰχθύων τινὲς καὶ
συνοικοῦσιν ὡς γαμεταῖς, καὶ φυλάττουσιν αὐτάς,
καὶ ζηλοτυπία τις καὶ ἐν ἰχθύων γένεσιν ἐξάπτεται, ἀνωτέρω εἶπον.

---

[1] σοφίας· ὁ δὲ παρέσχεν.
[2] ὀργᾷ V, ὀρέγεται other MSS.
[3] Jac : τοῖς.
[4] Jac : προσαπορραίνουσιν.

ON ANIMALS, IX. 62–63

practitioners applied an asp to his arm in order to demonstrate his skill, and was bitten. Thereupon he sucked out the poison with his mouth. He failed however to swallow some water afterwards, there being none at hand although he had got some ready (the vessel had been upset by an act of treachery), and as he had not washed off the poison and thoroughly rinsed his mouth he passed away after, I believe, two days without suffering any pain, though the poison had little by little reduced his gums and his mouth to putrescence.

63. When spring is at its height and the earth is putting forth her blossoms, animals are filled with an amorous impulse and bethink them of wedlock, and all that dwell in mountain or sea or that fly in the air desire to embrace one another. Among the Fishes there are some that rub off their eggs, massed and clinging together, on the sand; others as they swim spawn a great quantity of eggs, most of which are swallowed by those that swim in the rear.[a] In fact the males lead the way and scatter milt, and the females that follow, open-mouthed and quite insatiable, swallow it. This is their method of coupling. I have explained above how some fishes actually live with the females and look after them as though they were their wives,[b] and that even among the various kinds of fishes the fires of a sort of jealousy [c] break forth.

Fishes and their mating

[a] Cp. Hdt. 2. 93.   [b] See 1. 14.
[c] See 1. 25.

AELIAN

64. Λέγει δὲ Ἀριστοτέλης, καὶ Δημόκριτος πρὸ
ἐκείνου, Θεόφραστός τε ἐκ τρίτων καὶ αὐτός φησι,
μὴ τῷ ἁλμυρῷ ὕδατι τρέφεσθαι τοὺς ἰχθῦς, ἀλλὰ
τῷ παραμεμιγμένῳ¹ τῇ θαλάττῃ γλυκεῖ ὕδατι.
καὶ ἐπεὶ δοκεῖ πως ἄπιστον, δι' αὐτῶν τῶν ἔργων
βεβαιῶσαι βουληθεὶς τὸ λεχθὲν ὁ τοῦ Νικομάχου
λέγει εἶναί τι πότιμον ὕδωρ ἐν πάσῃ θαλάττῃ, καὶ
ἐλέγχεσθαι ταύτῃ.² εἴ τις ἀγγεῖον ἐκ κηροῦ
ποιήσας κοῖλον καὶ λεπτὸν καθείη κενὸν ἐς τὴν
θάλατταν, ἐξάψας ποθὲν ὥστε ἀνιμήσασθαι δύνασθαι, νυκτὸς³ διελθούσης καὶ ἡμέρας ἀρύεται⁴
πεπλησμένον γλυκέος τε καὶ ποτίμου ὕδατος⁵
αὐτό. καὶ Ἐμπεδοκλῆς δὲ ὁ Ἀκραγαντῖνος λέγει
τι εἶναι γλυκὺ ἐν τῇ θαλάττῃ ὕδωρ οὐ πᾶσι⁶
δῆλον, τρόφιμον δὲ τῶν ἰχθύων. καὶ τὴν αἰτίαν
τοῦδε τοῦ ἐν τῇ ἅλμῃ γλυκαινομένου λέγει φυσικήν,
ἣν ἐκεῖθεν εἴσεσθε.

65. Οἱ μυούμενοι τοῖν Θεοῖν οὐκ ἂν πάσαιντο
γαλεοῦ φασιν· οὐ γὰρ αὐτὸν εἶναι καθαρὸν ὄψον,
ἐπεὶ τῷ στόματι τίκτει. οὐ τίκτειν δὲ αὐτὸν ἔνιοι
λέγουσιν, ἀλλὰ δείσαντά τι τῶν ἐπιβουλευόντων
τὰ σκυλάκια καταπίνειν καὶ ἀποκρύπτειν, εἶτα
τοῦ φοβήσαντος παραδραμόντος ζῶντα αὖθις
ἀνεμεῖν. τῆς δὲ τρίγλης οὐκ ἂν γεύσαιντο οἱ αὐτοὶ
μύσται, οὐδὲ μὴν ἡ τῆς Ἥρας τῆς ἐν Ἄργει
ἱέρεια· καὶ τάς γε αἰτίας ἄνω που εἰπὼν οἶδα.

¹ παρακειμένῳ.   ² τοῦτο.   ³ νυκτὸς δέ.
⁴ ἀρύεται.   ⁵ ὕδατος μεστόν.   ⁶ πάνυ τι ? H.

ᵃ Aristotle.
ᵇ Demeter and Persephone, in whose honour the Eleusinian
mysteries were celebrated.

64. Aristotle [*HA* 590 a 18], and Democritus ^{Fresh water}
before him [Diels *Vorsok.*⁵ 1. 295; 2. 126], and third ^{in the sea}
in order Theophrastus [*CP* 6. 10. 2] assert that fish
are not nourished by salt water but by the fresh
water that is mingled with the sea. And since this
seems almost incredible, the son of Nicomachus,^a
wishing to confirm the statement by actual practice,
says that in every sea there is some drinkable water,
and that it can be proved in this way. If one makes
a thin, hollow vessel of wax and lets it down empty
into the sea, having attached it so that it can be
hauled up, after a night and a day it is, when drawn
up, full of fresh and drinkable water. And
Empedocles of Agrigentum asserts [*fr.* 66 Diels
PPF] that there is some fresh water in the sea, not
indeed perceptible to all, though it does nourish
fishes. And this sweetening of the water in the
brine he says is due to natural causes, which you may
learn from his writings.

65. It is said that those who have been initiated ^{Initiates}
into the Mysteries of the two goddesses ^b will not ^{abstain from}
touch Dog-fish, for (they say) it is no clean food, ^{certain fish}
since it gives birth through its mouth. Some how-
ever maintain that it does not do so, but that when
its young have been frightened by attempts on their
life, it swallows and hides them away, and that when
the scare has passed, it again ejects them alive.
And these same initiates would not taste of a Red
Mullet, nor would the priestess of Hera at Argos.
The reasons for this I know that I have explained
above somewhere.^c

^c See ch. 51.

AELIAN

66. Ἔχεως μὲν καὶ μυραίνης γάμους καὶ ὅπως
ἀλλήλοις ὁμιλοῦσιν, ἡ μὲν προϊοῦσα τῆς [1] θαλάττης,
ὁ δὲ ἐξέρπων τοῦ φωλεοῦ, ἐν τοῖς πρόσθεν εἰπὼν
οὐκ ἐπιλέλησμαι. ὃ δὲ οὐκ εἶπον νῦν ἂν εἴποιμι.
μέλλων ὁ ἔχις ὁμιλεῖν αὐτῇ, ἵνα δόξῃ πρᾶος ὡς
πρέπει [2] νυμφίῳ, τὸν ἰὸν ἀπεμεῖ καὶ ἐκβάλλει, καὶ
οὕτως ὑποσυρίσας τὴν νύμφην παρακαλεῖ, οἱονεὶ
προγάμιόν τινα ὑμέναιον ἀναμέλψας. ὅταν δὲ τὰ
τῆς ἀφροδισίου σπουδῆς τελέσωσι μετ' ἀλλήλων
ὄργια, ἡ μὲν ἐπί τε τὰ κύματα καὶ τὴν θάλατταν
ὥρμησεν, ὁ δὲ ἀναρροφήσας τὸν ἰὸν αὖθις ἐς τὰ
ἤθη τὰ οἰκεῖα ἐπάνεισιν.

1 ἐκ τῆς.   2 ὡς πρέπει] καὶ πρέπων.

284

66. I have not forgotten that I have in a previous passage [a] told of the mating of Viper and Moray and how they couple, the Moray emerging from the sea, the Viper from its den. But what I did not tell, I now will. When the Viper intends to couple with the Moray, in order to appear gentle as befits a bridegroom, he disgorges and throws up his poison, and then with a soft hissing sound, as though raising a kind of pre-nuptial wedding chant, summons his bride. And when they have together completed their amorous revels, the fish makes for the waves and the sea, while the snake gulps down his poison again and goes back to his native haunts.

Mating of Viper and Moray

[a] See 1. 50.

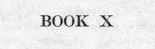

BOOK X

I

1. Σπανίως ἐλέφας ἐρᾷ, φασίν· ἔστι γὰρ
σώφρων, ὡς ἄνω μοι λέλεκται. ἀκούω δ᾽ οὖν [1]
ἐλεφάντων πάθος ἐρωτικόν, καὶ ἄξιον θαυμάσαι
αὐτό. ὃ δὲ πέπυσμαι, ἐκεῖνό ἐστιν. ἀνὴρ τῆς
τούτων ἄγρας οὐκ ἄπειρος, δύναμιν λαβὼν ἐκ
βασιλέως τοῦ Ῥωμαίων [2] καὶ σταλεὶς ἐπὶ τὴν
θήραν κατά τι ἔθος τοῖς Μαυρουσίοις ἐπιχώριον,
φησὶν ἐν συγγραφῇ νέαν μόσχον ἐλέφαντος ὡραίαν
ὡς ἐκείνοις ἰδεῖν συνελθεῖν ἐλέφαντι νέῳ καὶ
καλῷ, πρεσβύτερον δὲ ἄλλον, εἴτε ἀνὴρ ἦν εἴτε
ἐραστὴς τῆς προειρημένης, ὥσπερ οὖν ἀτιμασθέντα
ἀγανακτῆσαι· θυμῷ γὰρ βιαίῳ ἐξαφθεὶς ὥρμησε
μέν, [3] ἐλθὼν δὲ ἐπὶ τὸν νέον καὶ καλὸν καὶ συμπεσὼν
εἶτα ἐμάχετο, ὡς ὑπέρ τινος ἢ νύμφης ἢ ἐρωμένης
ἀλγῶν. καὶ ἐς τοσοῦτον ἄρα ἀλλήλοις συνήραξαν, [4]
ὡς ἀμφοτέροις ζημιωθῆναι τὰ κέρατα. ἐνίκησε
δὲ οὐδέτερος, ἀλλὰ ἀπέστρεψαν ἀπ᾽ ἀλλήλων οἱ
θηραταὶ βάλλοντες, ἐπεὶ καὶ ἀχρεῖοι [5] τὸ λοιπὸν
ἦσαν τῶν ὅπλων ἀφῃρημένοι. ἐρωτικὴ μὲν δὴ
μάχη ἐραστῶν ἐλεφάντων ἰσότιμος μέχρι τοῦ
τέλους ἐνταῦθα ἀνεπαύσατο. εἵλκετο δὲ ὁ Πάρις

---

[1] γοῦν.
[2] Ῥ. πάλαι Ἀλέξανδρος ὄνομα.
[3] μὲν καὶ ὀλίγου πάντας διέφθειρεν.
[4] συνήρραξαν.
[5] ἀχρεῖοι συμβαλεῖν.

BOOK X

An Elephant's jealousy

1. The Elephant is seldom in love, they say, for, as I have remarked earlier on,[a] it is sober. And yet I learn of Elephants experiencing the passion of love, and the tale is one to excite astonishment. And this is what I have learnt.

A man who had some knowledge of the method of hunting these animals obtained leave from the Roman Emperor and set out to hunt them in the manner of the natives of Mauretania. He tells in his narrative how he saw a young female Elephant, comely as Elephants can be, coupling with a young and beautiful male, while another older male (whether it was the husband or the lover of the aforesaid female) was furious as though it had been scorned. For inflamed with violent passion it rushed forward and coming up to the young and beautiful Elephant, fell upon it and began to fight, like a man filled with resentment over the conduct of his wife or his mistress. And the two dashed together with such force that both damaged their tusks. And neither was victorious, but the hunters separated them by hurling missiles at them, for the animals were helpless as soon as they were deprived of their weapons. So a lovers' contest between elephant lovers, equally balanced up to the end, was there brought to a close. And Paris was being dragged

[a] See 8. 17.

AELIAN

ὑπὸ τοῦ Μενέλεω καὶ ἤγχετο τοῦ ἱμάντος αὐτὸν
πιέζοντος τοῦ ὑπὸ τῷ κράνει,

καί νύ κεν εἴρυσσέν τε καὶ ἄσπετον ἤρατο κῦδος

ὁ τοῦ Ἀτρέως, εἰ μὴ ἐρράγη μὲν ὁ ἱμάς, αὐτὸν δὲ
ἐξήρπασεν ἡ Διὸς καὶ Διώνης αἰσχίστην μάχην
καὶ ἄνανδρον μεμαχημένον, καὶ ἀπελθὼν ὁ δειλὸς
ἐκάθευδε μετὰ τῆς μεμοιχευμένης.

2. Ἰχθύες δὲ ἄρα οὐ ⟨κατὰ⟩ [1] τὴν αὐτὴν ὥραν
ἐς ἀφροδίτην πρόθυμοι, ἀλλὰ οἱ μὲν ἦρος ἐπιθόρνυν-
ται, οἱ δὲ θέρειον εἰλήχασι τὴν ὁρμὴν τήνδε, ἐν [2]
τῇ ὀπώρᾳ ἄλλοι, καὶ διὰ τοῦ χειμῶνος ὑπεξάπτον-
ταί τινες ἐς τὴν προειρημένην σπουδήν. καὶ οἱ
μὲν πλεῖστοι τοῦ ἔτους ἅπαξ ὠδίναντες εἶτα
παύονται, λάβρακα δὲ ἀκούω καὶ ἐπιτίκτειν,
τρίγλην δὲ καὶ τρὶς κύειν κατηγορεῖ φασι καὶ τὸ
ὄνομα.

3. Ἡρόδοτος λέγει τὰς καμήλους ἐν τοῖς
ὄπισθεν σκέλεσιν ⟨ἔχειν⟩ [3] τέτταρας [4] μηροὺς
καὶ μέντοι καὶ γόνατα τοσαῦτα, τὰ δὲ ἄρθρα διὰ
τῶν σκελῶν τῶν κατόπιν πρὸς [5] τὴν οὐρὰν
τετράφθαι αὐταῖς.

4. Τὰς οἷς τὰς Ἀραβίων ἔχειν οὐρὰς ἀήθεις ὡς
πρὸς τὰς ἄλλας Ἡρόδοτος λέγει. γένη δὲ αὐτῶν
εἶναι διπλᾶ ὁ αὐτὸς δήπου διδάσκει, καὶ μέντοι καὶ [6]
λέγει τὰς μὲν αὐτῶν ἔχειν οὐρὰς μηκίστας, ὡς

---

[1] ⟨κατά⟩ add. H. [2] συν.
[3] ⟨ἔχειν⟩ add. Ges. [4] Ges : τε παρά.
[5] ἐς. [6] καὶ τοῦτο καί.

along by Menelaus and was being throttled by the thong that was pressing him beneath his helmet, and the son of Atreus

> ' would now have haled him away and won renown unspeakable ' [Hom. *Il.* 3. 373],

had not the thong snapped, and had not the daughter of Zeus and Dione (*i.e.* Aphrodite) snatched him away after his most shameful and unmanly fight : and he departed, the coward, and slept with the adulteress.

2. It appears that Fish are not eager for sexual intercourse at the same season, but some couple in spring, others feel the urge in summer, others in the autumn, in others again the aforesaid desire is gradually kindled during the winter. The majority after giving birth once a year, cease; though I am told that the Basse gives birth twice, whereas the very name of the Red Mullet (τρίγλη) proves, so they say, that it does so thrice. Mating season for Fishes

3. Herodotus states [3. 103] that Camels have four thigh-bones in their hind-legs, and the same number also of knees, but that their genitals between their hind-legs are turned in the direction of the tail. Anatomy of the Camel

4. Herodotus states [3. 113] that the Sheep of Arabia have tails of abnormal length compared with other sheep. And the same writer informs us that there are two kinds of Sheep, adding that one kind The Sheep of Arabia

εἶναι μετρήσαντι καὶ τριῶν πήχεων [1] οὐκ ἐλάττους.
ὥσπερ οὖν εἰ ἐφῄη τις ἐπισύρειν, ἑλκοῦσθαι ἂν
αὐτὰς πάντως παρατριβομένας πρὸς τὸ δάπεδον.
τοὺς νομέας δὲ εἶναι ἀγαθοὺς χειρουργεῖν οὐ πέρα [2]
τῶν ἁμαξίδων, αἳ [3] ὑπερείδουσι τὰς οὐρὰς τῶνδε
τῶν κτηνῶν, ὥστε ἀποστέγειν τὴν ἕλκωσιν αὐταῖς.
τὰς δὲ οἷς τὰς ἑτέρας φησὶ πλατείας φορεῖν,
πλατύνεσθαι δὲ καὶ ἐς πῆχυν αὐτάς.

5. Οἱ κοχλίαι ἴσασιν εἶναί σφισι πολεμίους
τοὺς πέρδικας καὶ τοὺς ἐρῳδιούς, καὶ αὐτοὺς
ἀποδιδράσκουσιν, οὐδ᾽ ἂν ἴδοις ἔνθα οὗτοι νέμονται
κοχλίας διέρποντας. οἱ δὲ καλούμενοι τῶν κοχλιῶν
ἀρείονες, οὗτοι μὲν καὶ ἀπατῶσι καὶ περιέρχονταί
τινι φυσικῇ σοφίᾳ τοὺς προειρημένους. τῶν γὰρ
συμφυῶν ὀστράκων προελθόντες αὐτοὶ μὲν νέμονται
κατὰ πολλὴν τὴν ἄδειαν, οἱ δὲ ὄρνιθες οὓς εἶπον
ἐπὶ τὰ κενὰ τῶν ὀστράκων ὡς ἐπ᾽ αὐτοὺς ἐκείνους
καταπέτονται, οὐδὲν δὲ εὑρόντες ἀπέρριψαν ὡς
ἀχρεῖά σφισι καὶ ἀνεχώρησαν· οἱ δὲ ἐπανελθόντες
εἶτα ἕκαστος ἐς [4] τὴν ἰδίαν οἰκίαν παρῆλθε,
κεκορεσμένος μὲν ἐκ τῆς νομῆς, σωθεὶς δὲ ἐξ ἧς
ἠπάτησε πλάνης.

6. Ποντικοὶ δὲ ἄρα κολίαι [5] τὸν Περσῶν βασιλέα
μεμίμηνται χειμάζοντα μὲν ἐν Σούσοις, θερίζοντα [6]
δὲ ἐν Ἐκβατάνοις. καὶ γὰρ οὗτοι ἐν μὲν τῇ
καλουμένῃ Προποντίδι χειμάζουσιν· ἀλεεινὴ γὰρ

---

[1] πηχῶν MSS always. [2] οὐπεράν.
[3] αἷς ? H. [4] ὡς.
[5] Gron : κοχλίαι.

has tails so long as to measure not less than three cubits. And if one were to allow the Sheep to trail their tails after them, they would be full of sores from rubbing along the ground. All that the shepherds can do is to contrive small carts which support the tails of these animals and prevent them from getting sore. But the other kind of Sheep, he says, has broad tails as much as a cubit wide.

5. Snails know that partridges and herons are their enemies; so they escape from them, and in places where these birds feed you would never see snails crawling about. But the snails which they call *Areiones* deceive and elude the aforesaid enemies by natural astuteness. Thus, they emerge from their native shells and feed without anxiety, while the birds which I mentioned swoop upon the empty shells as though they were the actual snails, but finding nothing, throw them aside as useless and go away. But the Areiones return and pass each to its own house, having eaten their fill of food and having preserved their lives by their deceptive migration.

The 'Areion' Snail

6. It seems that the Spanish Mackerel of the Euxine imitate the Persian King who spends the winter at Susa and the summer in Ecbatana. For these fish pass the winter in the Propontis as it is called, since that region is warm, but in the summer

The Spanish Mackerel

---

[6] *Spanheim* : θερίζοντα . . . χειμάζοντα.

ἥδε ἡ γῆ· θέρους δὲ πρὸς τῷ Αἰγιαλῷ διαιτῶνται. παρέχει γὰρ αὐτοῖς αὔρας μαλακὰς ἡ θάλαττα ἡ προειρημένη.

7. Πυνθάνομαι τῶν μαγείρων τοὺς τὴν τέχνην ἀκριβοῦντας ὅταν βούλωνται τῶν τριγλῶν τὰς κοιλίας ὀπτωμένων μὴ ῥήγνυσθαι, καταφιλεῖν αὐτῶν τὰ στόματα· οὗπερ οὖν γεγενημένου ὁλόκληροι διαμένουσιν, ὥς φασιν.

8. Ὁ δελφὶς ὁ θῆλυς μαζοὺς ἔχει κατὰ τὰς γυναῖκας, καὶ θηλάζει τὰ βρέφη πάνυ ἀφθόνῳ καὶ πολλῷ τῷ γάλακτι. νήχονται δὲ κοινῇ μέν, καθ᾽ ἡλικίαν δὲ διακριθέντες· καὶ τῆς μὲν πρώτης τετάχαται οἱ νέοι καὶ ἁπαλοί, ἕπονται δὲ αὐτῶν τῇ νήξει οἱ τέλειοι. φιλότεκνον γὰρ καὶ φιλόστοργον ὁ δελφὶς ζῷον, καὶ ὑπὲρ τῶν βρεφῶν ὀρρωδεῖ. καὶ φυλακῆς χάριν ὡς ἐν φάλαγγι στρατιωτικῇ οἱ μὲν τῆς πρώτης εἰσίν, οἱ δὲ τῆς δευτέρας, οἱ δὲ τῆς τρίτης· καὶ προνήχονται μὲν οἱ νέοι, ἐπινήχονται δὲ αἱ θήλειαι, καὶ οἱ ἄρρενες οὐραγοῦσιν ἐφορῶντές τε καὶ παραφυλάττοντες τὴν τῶν ἐκγόνων τε καὶ τῶν γαμετῶν νῆξιν. τί πρὸς ταῦτα ὁ Νέστωρ, ὦ καλὲ Ὅμηρε, ὅνπερ οὖν ᾄδεις τακτικώτατον τῶν ἡρώων τῶν καθ᾽ ἑαυτὸν γεγονέναι;

9. Ἔχιν ἐχίδνης οἱ μὲν τῷ γένει διαφέρειν, οὐ μέντοι τῇ φύσει φασί· τὸν μὲν γὰρ εἶναι ἄρρενα, τὴν δὲ θήλειαν. οἱ δὲ καὶ τῇ φύσει διαφέρειν οἴονται· ἀλλοῖον μὲν γὰρ τοῦτο εἶναι ζῷον, ἀλλοῖον δὲ ἐκεῖνο. ἀκούω δέ τινων λεγόντων τοὺς

they live about Aegialus,[a] because the first-named sea affords them gentle breezes.

7. I am informed that when Cooks who are masters of their art wish the stomachs of Red Mullets not to burst in the cooking, they kiss their mouths. And if this is done the fish are preserved whole, so they say.

Cooking a Red Mullet

8. The female Dolphin has breasts like a woman and suckles its young with a liberal and copious supply of milk. And they swim in a body, but separated according to age. In the front rank are ranged the young and tender, after them swim the full-grown ones. The Dolphin loves its offspring and is an affectionate creature, anxious for its children, and in order to protect them, as with soldiers in line of battle, some are with the front rank, others with the second, others with the third. The young ones swim in front, after them swim the females, and the males bring up the rear while they superintend and guard closely their offspring and their wives as they swim. What, O noble Homer, would Nestor say to this—Nestor, whom you celebrate as the best tactician among all the heroes of his day? [Cp. Hom. *Il.* 2. 555; 4. 293–309.]

The Dolphin and its young

9. Some maintain that the difference between the *Echis* and the *Echidna* is one of sex and not of kind, the former being the male viper, the latter the female. Others however consider that the difference is one of kind, and that the latter belongs to one species and the former to another. And I hear

The Viper

[a] Town on the coast of Paphlagonia.

AELIAN

μὲν ὑπὸ τοῦ ἔχεως δηχθέντας σπᾶσθαι, οὐ μὴν
τοὺς ὑπὸ τῆς ἐχίδνης. ἕτεροι δέ φασι τὸ μὲν τῆς
ἐχίδνης δῆγμα εἶναι λευκόν, τὸ δὲ τοῦ ἔχεως οὐ
τοιοῦτον, πελιδνὸν δέ. Νίκανδρος δέ φησιν ἐκ μὲν
τοῦ δήγματος ὅπερ οὖν ὁ ἔχις ἐμφύει δύο ὀδόντων
ἴχνη φαίνεσθαι· πλειόνων δέ, εἰ δάκοι ἡ ἔχιδνα.

10. Εὐθηρίας γενομένης ἐλεφάντων [1] οἷα δρῶσιν
ἐς τὸ πραῦναί τε αὐτοὺς καὶ ἡμερῶσαι εἰπεῖν
ἄξιον. πρῶτον μὲν ἐς ὕλην τινὰ ὀλίγον ἀφεστῶσαν
τῆς τάφρου ἐν ᾗ ἐθήρασαν ἄγουσιν αὐτοὺς δεδεμέ-
νους, διαλαβόντες [2] ταῖς σχοίνοις καὶ μήτε προθεῖν
ἐπιτρέποντες μήτε αὖ πάλιν ἀφίστασθαι καὶ
ἀποσπᾶν ἐς τοὐπίσω· εἶτα ἕκαστον ἐξάψαντες
μεγίστου δένδρου μεμετρημένῳ διαστήματι, ὡς
μήτε ἐς τὸ ἔμπροσθεν ἐπιπηδᾶν ἔχειν μήτε ἐπὶ
πολὺ πάλιν ἀναχωρεῖν τοῦ σκιρτᾶν καὶ ὑβρίζειν
ἐξουσίᾳ, ἀτροφίᾳ τε καὶ λιμῷ τὴν ἄγαν ἰσχὺν καὶ
ῥώμην καθαιροῦσι, καὶ μέντοι καὶ τὸν θυμὸν
αὐτῶν καὶ τῆς ψυχῆς τὸ ἄτρεπτον ἡσυχῇ κατα-
μαραίνουσιν, ὡς ἐκείνους λήθην μὲν ἴσχειν τῆς
τέως ἀμάχου ἀγριότητος, παραλύεσθαι δὲ τοῦ πρόσθεν θυμοῦ. προσιέναι τε τοὺς τῶν τοιούτων πω-
λευτὰς καὶ ἐκ χειρὸς ὀρέγειν τροφήν, τοὺς δὲ ὑπὸ
τῆς χρείας ἀναγκαζομένους λαμβάνειν καὶ μὴ
κακουργεῖν, βλέπειν τε ἤδη πρᾶόν τε καὶ κεκμηκός.
οἱ δὲ ἄγαν αὐτῶν ἰσχυροὶ καὶ τέλειοι ἀπορρήξαντες
τὰ δεσμὰ καὶ ταῖς ἀκμαῖς τῶν κεράτων καὶ ταῖς
προβοσκίσιν ἀνασπῶντες τὰ δένδρα, τὰ δὲ καὶ

[1] καὶ τῶν ἐλεφάντων ἑαλωκότων.
[2] πολὺ διαλαβόντες.

296

some say that those who have been bitten by the *Echis* are seized with convulsions, whereas victims of the *Echidna* are not. But others assert that the bite caused by the *Echidna* is white, unlike that of the *Echis* which is livid. And Nicander says [*Th.* 231] that in the bite which the *Echis* implants traces of two fangs are visible, but more if it is an *Echidna* that has bitten.

10. It is worth relating what men do after a successful Elephant-hunt to make the creatures docile and tame. First of all they lead them away bound into a wood a little distance from the trench in which they have captured them, keeping them apart by ropes and not allowing them either to run forward or to stop and pull back. Next they fasten each beast to a very large tree at a measured distance from the next one so that they can neither spring forward nor retreat backwards to any extent through being free to leap about and work mischief. And by refusing them food and by starvation they drain away their excessive strength and power, and gradually reduce their spirit and their inflexible determination, so that they forget their hitherto indomitable fierceness and abandon their former temper. The keepers of these animals go up to them and offer them food from their hands, and the Elephants under stress of need take it and do the men no harm, and already begin to wear a mild and fatigued expression. But those that are extremely powerful and full-grown, after bursting their bonds and tearing up trees with the points of their tusks and with their trunks, even smashing some by their

Taming an Elephant

297

κατάξαντες[1] ὑπὸ ῥύμης καὶ ἐμπεσόντες ἐς αὐτά,
μόγις καὶ ὀψὲ τοῦ χρόνου τὰ μὲν λιμῷ, τὰ δὲ
γλυκείᾳ τροφῇ, τὰ δὲ κέντροις ἡμερώθησαν.
τροφὴ δὲ ἡμερουμένοις τοῖσδε τοῖς ζῴοις ἄρτοι τε
οἱ μέγιστοι καὶ κριθαὶ καὶ ἰσχάδες καὶ ἀσταφίδες
καὶ κρόμμυα καὶ σκόροδα καὶ μέλι χύδην σχίνου
τε καὶ φοίνικος καὶ κιττοῦ φάκελοι καὶ πᾶν ὅσον
ἐδώδιμον ὕλης καὶ ἐκείνοις συντρόφου καὶ ἐκ
τούτου τοι καὶ φίλης.

11. Φύσεως δὲ ἰχθύων εἰσὶν ἀμαθεῖς ὅσοιπερ
οὖν τελέως ἁπάντων καταψηφίζονται σιωπὴν
αὐτῶν· ἐπεὶ καὶ συρίττουσί τινες καὶ γρυλλίζουσι.
λύρα[2] μὲν γρυλλίζει καὶ χρόμις καὶ κάπρος, ὡς
Ἀριστοτέλης φησί· χαλκεὺς[3] δὲ συρίττει, κόκκυξ
δὲ ἄρα τὸν ὁμώνυμον ὄρνιν τῇ φωνῇ μεμιμημένος
φθέγγεται παραπλήσια.

12. Ἐλέφας[4] μὲν σαρκῶν[5] ὄγκος ἐστὶν ἰδεῖν
καὶ πάνυ μέγιστος· ἐδώδιμα δὲ αὐτοῦ τὰ κρέα
οὐκ ἔστιν, ὅτι μὴ ἡ προβοσκὶς καὶ τὰ χείλη τοῦ
στόματος καὶ τῶν κεράτων ὁ μυελός. στέαρ δὲ
ἐλέφαντος ἦν ἄρα τοῖς ἰοβόλοις ἔχθιστον· εἰ γάρ
τις χρίσαιτο ἢ ἐπιθυμιάσειεν αὐτοῦ, τὰ δὲ ἀποδιδράσκει πορρωτάτω.

13. Τῶν δὲ Ἀραβίων ζῴων ἡ πολυχροιά τε καὶ
τὸ πολύμορφον πάντα γραφικὸν ἐλέγξαι δεινά, καὶ
ταῦτα οὐ μόνον τά τε ἄλκιμα καὶ γενναῖα, ἤδη δὲ

---

[1] κατεάξαντες.   [2] Ges : σαύρα.
[3] Ges : χαλκίς.   [4] Reiske : ἐλέφαντος.
[5] τῶν σαρκῶν.

onset and by assailing them, have with difficulty and only after a long while been tamed sometimes by starvation sometimes by pleasant food, at other times by means of goads. While these animals are being tamed their food consists of very large loaves of bread, barley, dried figs, raisins, onions, garlic, honey in large quantities, bundles of mastic branches and of palm-leaves and of ivy and any edible and familiar substance which is for that reason welcome to them.

11. Those who condemn all fishes without ex- Vocal Fishes ception to silence are ignorant of their nature, because there are those that whistle and those that grunt. The Gurnard grunts, so too do the Chromis and the Caprus, as Aristotle says [*HA* 535 b 17]. The John Dory whistles; the Cuckoo *a* (or ' Piper ') has a voice which resembles that of the bird whose name it bears and makes a similar sound.

12. To the eye the Elephant is a mass of flesh and The flesh of of enormous size, but his flesh is not edible, excepting the Elephant his trunk, the lips of his mouth, and the marrow of his tusks. But it seems that the fat of an Elephant is detested by poisonous creatures, for if a man rubs himself with it or burns some, they flee away to a great distance.

13. The variety of colour and of shape in the fauna The Fauna of Arabia might well put anyone skilled in painting of Arabia to the test, not only in the case of powerful and

---

a A kind of Gurnard.

καὶ τὰ ἀδοξότερα, αἵ τε ἀκρίδες καὶ οἱ ὄφεις·[1]
χρυσοειδῆ γοῦν [2] ἰνδάλματα καὶ ἐπ᾽ αὐτῶν κατέ-
στικται· οἱ δὲ ἰχθῦς ἔτι καὶ πλέον τῆς πολυκόσμου
χρόας μετειληχότες εἶτα ἰδεῖν ἐκπληκτικοί εἰσι.
καὶ τὰ ὄστρεα δὲ τὰ τῆς Ἐρυθρᾶς θαλάττης [3] τῆς
αὐτῆς ἀγλαΐας ἄμοιρα οὐκ ἔστι· ζῶναί τε γὰρ
περιέρχονται φλογώδεις αὐτά, καὶ φαίης ἂν
θεασάμενος τὴν ἶριν αὐτὰ μιμεῖσθαι τῇ κράσει τῶν
χρωμάτων,[4] γραμμαῖς παραλλήλοις ὑπὸ τῆς φύσεως
καταγραφέντα. ὁ ᾀδόμενος δὲ παρὰ τοῖς ἀνοήτοις
καὶ ἐν ταῖς γυναιξὶ θαυμαστὸς μαργαρίτης θρέμμα
μέντοι τῆς Ἐρυθρᾶς θαλάττης καὶ οὗτός ἐστι, καὶ
τίκτεσθαί γε αὐτὸν τερατολογοῦσιν ὅταν ταῖς
κόγχαις ἀνεῳγμέναις ἐπιλάμψωσιν αἱ ἀστραπαί.
θηρῶνται δὲ ἄρα αἵδε αἱ κόγχαι αἱ τῶν προειρημένων μητέρες εὐημερίας τε οὔσης καὶ τῆς θαλάττης
λείας· οἱ δὲ θηραταὶ συλλαβόντες εἶτα ἐξεῖλον
τοῦτον δὴ τὸν θέλγοντα τὰς τῶν μάχλων ψυχάς.
εὑρεθείη δ᾽ ἂν καὶ ἐν κόγχῃ μεγίστῃ μικρὸς καὶ
ἐν μικρᾷ μέγας· καὶ ἡ μὲν οὐδένα ἔχει, ἡ δὲ οὐ
πέρα ἑνός,[5] πολλαὶ δὲ καὶ πολλούς· εἰσὶ δὲ οἳ
λέγουσι καὶ εἴκοσι προσπεφυκέναι μιᾷ κόγχῃ.
καὶ ἡ μὲν κόγχη τὸ κρέας ἐστίν, ἐπιπέφυκε δὲ ἄρα
ὡς σκόλοψ ταῦτα. πρὸ καιροῦ δὲ καὶ τῆς ὠδῖνος
τῆς ἐντελοῦς εἴπερ οὖν ἀνοίξειέ τις τὰς κόγχας,
κρέας μὲν ἂν εὕροι, τῆς δὲ θήρας τὸ ἀγώνισμα
οὐχ ἕξει. λίθῳ δὲ ἄρα ὁ μαργαρίτης ἔοικε πεπωρω-
μένῳ, καὶ ἔχειν ἐν ἑαυτῷ καὶ στέγειν ὑγρὸν οὐ
πέφυκεν οὐδὲ ὀλίγον. δοκοῦσι δὲ ἄρα τοῖς τούτων

---

[1] ἀδοξότερα, . . . ὄφεις: so Gow punctuates, ἀδοξότερα. αἵτε ⟨γὰρ⟩ ἀκρίδες H.
[2] Gow: οὖν mss, del. H.

noble animals but even of the more insignificant, the
locusts and the snakes; for the markings on them
look like gold. The fish, which enjoy an even more
richly wrought colouring, are an astonishing sight.
And the oysters in the Red Sea are not without the
same glamour, for they are encircled with rings of
fiery hue, and to look at them you would say that
with the blending of their colours they were copying
the rainbow, Nature having painted parallel stripes
upon them. And the pearl, so celebrated among The Pearl
fools and admired by women, is also a nursling of
the Red Sea, and they tell a marvellous story of how
it is produced when lightning flashes upon the open
shells. So then these shells which are the mothers
of the aforesaid pearls are sought for when the
weather is fine and the sea smooth. And the
seekers collect them and extract this object which
delights the hearts of the luxurious. One may find
a small pearl even in the largest shell and a large
one in a small shell; and this one contains none, and
that not more than one, and many contain a number.
Some assert that as many as twenty have been
attached to a single shellfish. Now the shell is the
flesh, and these pearls cling to it like a thorn. But
if one were to open the shell prematurely, that is
before the birth-process is complete, one would find
the flesh indeed, but it will not contain the object of
one's quest. The pearl, it seems, is like a stone
produced by petrifaction, and it is not its nature to
contain or to admit even a drop of moisture. In the

---

³ θαλάσσης ἥπερ οὖν ἐστιν ὁ ᾿Αράβιος.

⁴ τῶν ποικίλων χ.

⁵ τοῦ ἑνός.

καπήλοις καὶ τοῖς ὠνουμένοις οἱ ἄγαν λευκοὶ καὶ
οἱ μεγάλοι κάλλιστοι καὶ τιμαλφέστατοι,[1] καὶ
πλούσιοί γε ἐξ αὐτῶν ἐγένοντο οὐ μὰ Δία ὀλίγοι
οἷς ἐντεῦθέν ἐστιν ὁ βίος. οὐκ ἀγνοῶ δὲ οὐδὲ
ἐκεῖνο, ὅτι ἄρα ἐξαιρεθέντων τῶν λίθων τῶνδε
ἀφείθησαν αὖθις αἱ κόγχαι, οἱονεὶ λύτρα δοῦσαι
τῆς ἑαυτῶν σωτηρίας τὸ σπούδασμα τὸ προειρημένον,[2] εἶτα ὑπανέφυσαν[3] αὖθις αὐτό. ἐὰν δὲ τὸ
ζῷον τὸ τρέφον αὐτὸν πρὶν ἢ ἐξαιρεθῆναι τὸν
μαργαρίτην ἀποθάνῃ, ὥς που λέγει τις λόγος, τῇ
σαρκὶ μέντοι συσσήπεται καὶ ἐκεῖνος καὶ ἀπόλλυται.
φύσει δὲ ἔχει τῆς περιφερείας τὸ λεῖον καὶ εὐπερί-
γραφον. εἰ δὲ ἐθέλοι τις τῶν πεφυκότων ἑτέρως
τινὰ σοφίας τέχνῃ περιγράψαι τε καὶ λεῖον
ἀποφῆναι τὸν λίθον, ὁ δὲ ἐλέγχει τὴν ἐπιβουλήν·
οὐ γὰρ πείθεται, τραχύτητας δὲ ὑπαναφύει, καὶ
ὅτι ἄρα ἐπιβεβούλευται ἐς κάλλος κατηγορεῖ ταύτῃ.

14. Αἰγύπτιοι τὸν ἱέρακα Ἀπόλλωνι τιμᾶν
ἐοίκασι, καὶ τὸν μὲν θεὸν Ὧρον καλοῦσι τῇ
φωνῇ τῇ σφετέρᾳ, τοὺς δὲ ὄρνιθας ἄγουσι[4]
θαυμαστούς, καὶ προσήκειν τῷ θεῷ τῷ προειρημένῳ φασὶν ὀρθῶς· οἱ[5] γὰρ ἱέρακες ὀρνίθων
μόνοι ταῖς[6] ἀκτῖσι τοῦ ἡλίου ῥᾳδίως καὶ ἀβα-
σανίστως ἀντιβλέποντες[7] καὶ δυσωπούμενοι ἥκιστα
πορείαν τε τὴν ἀνωτάτω ἴασι, καὶ αὐτοὺς ἡ θεία
φλὸξ λυπεῖ οὐδὲ ἕν. καὶ ἀνάπαλιν μέντοι πέτεσθαι
τὸν ἱέρακα οἱ ἰδόντες φασὶν ὡς ἐξ ὑπτίας νέοντα.
ἔνθεν[8] τοι καὶ πρὸς τὸν οὐρανὸν ὁρᾷ καὶ πρὸς τὸν

---

[1] τιμαλφέστατοι ὥστε ἐκείνοις κρίνειν αὐτούς.
[2] τὸ ἤδη προ-.
[3] ἐπανέφυσαν.
[4] Jac : λέγουσι.
[5] ὀρθῶς· οἱ] ὁρῶσι.

opinion of those who trade in pearls and those who buy them pearls that are pure white and large are the most beautiful and the most highly esteemed, and I can avow that many of those who make a livelihood by them have become wealthy. And I am also well aware that when these stones have been extracted and the shellfish have been released after giving up the aforesaid coveted object as ransom for their lives, they have gradually produced another one. If however the animal that fosters the pearl dies before the pearl is extracted, as is sometimes reported, both pearl and flesh rot away and perish. It has a naturally smooth and well-rounded contour, but if a man should want by artificial means to make round and smooth some stone not naturally so, the pearl confounds his design, for it declines to yield and develops roughnesses, thereby denouncing the plot that has been laid to secure its beauty.

14. The Egyptians appear to regard the Hawk as sacred to Apollo, calling the god ' Horus ' in their own language, and they regard the birds with wonder and are right in saying that they belong to the aforesaid god. For Hawks are the only birds that can face with ease and without pain the rays of the sun and are not the least dazzled; and while they fly at an immense height the divine fire does not trouble them at all. Moreover observers say that the Hawk flies upside down, like a man swimming on his back, and in this way, you see, it looks

The Hawk

---

⁶ ἀεὶ ἐν ταῖς. ⁷ βλέποντες.
⁸ ἔνθα.

AELIAN

πάντ' ἐφορῶντα ⟨Ἥλιον⟩ [1] μάλα ἐλευθέρως καὶ
ἀτρέπτως ὁ αὐτός. ὄφεων δὲ [2] καὶ δακετῶν
θηρίων ἐστὶν [3] ἔχθιστος. οὐκ ἂν γοῦν αὐτὸν
διαλάθοι οὔτε ὄφις οὔτε σκορπίος οὔτε μὴν
πονηρᾶς ὕλης ἄλλο τι ἔκτοκον. ἀκροδρύων μὲν
οὖν καὶ σπερμάτων ἄγευστος, σαρκῶν δὲ ἥδεται
βορᾷ, καὶ πίνει αἷμα, καὶ τὰ νεόττια ἐκτρέφει τοῖς
αὐτοῖς, καὶ ⟨ἐς⟩ [4] λαγνείαν ἐστὶ δριμύτατος. τὸ
δὲ αὐτοῦ τῆς κνήμης ὀστοῦν εἰ χρυσίῳ παρατεθείη,
ἕλκει τε αὐτὸ καὶ ἴυγγι ἀπορρήτῳ τινὶ πρὸς ἑαυτὸ
ἄγει καὶ ἕπεσθαι θέλγει, ὥσπερ οὖν ᾄδουσι τὸν
Ἡρακλεώτην λίθον καταγοητεύειν πως τὸν σίδη-
ρον. λέγουσι δὲ Αἰγύπτιοι καὶ ἐς πεντακόσια ἔτη
βίου προήκειν τὸν ἱέρακα, καὶ οὔπω με πείθουσιν·
ἃ δ' οὖν ἀκούω, λέγω. ἔοικε δέ φασι καὶ Ὅμηρος
ὅτι τῷ Διὸς καὶ Λητοῦς ἐστι φίλος ὑπαινίττεσθαί
πως λέγων

βῆ δὲ κατ' Ἰδαίων ὀρέων, ἴρηκι ἐοικὼς
ὠκέι φασσοφόνῳ. [5]

15. Ὁ κάνθαρος ἄθηλυ ζῷόν ἐστι, σπείρει δὲ ἐς
τὴν σφαῖραν ἣν κυλίει· ὀκτὼ δὲ καὶ εἴκοσιν
ἡμερῶν τοῦτο δράσας καὶ θάλψας αὐτήν, εἶτα
μέντοι τῇ ἐπὶ ταύταις προάγει τὸν νεοττόν.
Αἰγυπτίων δὲ οἱ μάχιμοι ἐπὶ τῶν δακτυλίων εἶχον
ἐγγεγλυμμένον κάνθαρον, αἰνιττομένου τοῦ νομοθέ-
του, δεῖν ἄρρενας εἶναι πάντως πάντη τοὺς μαχομέ-
νους ὑπὲρ τῆς χώρας, ἐπεὶ καὶ ὁ κάνθαρος θηλείας
φύσεως οὐ μετείληχεν.

[1] ⟨Ἥλιον⟩ add. H. [2] τε.
[3] ὁ αὐτός ἐστιν. [4] ⟨ἐς⟩ add. Ges.

at the sky and the all-surveying sun with complete freedom and without flinching. It is the bitter enemy of snakes and venomous creatures. At any rate no snake, no scorpion, nor indeed any other product of noxious matter would escape its notice. Fruits and seeds it will not touch; it delights to feed on flesh and drinks blood, and on these it feeds its young; it is also passionate in lechery. If the bone from its tibia is put beside gold it attracts and draws it to itself by some inexplicable fascination, persuading it to follow even as, they say, the stone of Heraclea [a] somehow bewitches iron. The Egyptians assert that the Hawk's life extends to as much as five hundred years, and they do not convince me: I merely report what I have heard. Homer, they say, seems to hint that the Hawk is beloved of the child of Zeus and Leto (i.e. Apollo) when he says [Il. 15. 237]

' And down the hills of Ida he went, like unto a swift hawk, the slayer of doves.'

15. The Scarab is a creature of which there is no The Scarab female, but it pours its semen into the heap [b] which it rolls up. After doing this and keeping the heap warm for eight-and-twenty days, on the following day it brings forth its young. Among the Egyptians the fighting class wore a Scarab engraved on their finger-rings, their ruler intimating thereby that those who fight for their country must at all costs and in every way be men, because the Scarab has in it nothing of the feminine element.

---

[a] The magnet.   [b] Of dung.

[5] φασσοφόνῳ ὁ Ἀπόλλων αὐτός.

AELIAN

16. Ἡ ὗς καὶ τῶν ἰδίων τέκνων ὑπὸ τῆς
λαιμαργίας ἀφειδῶς ἔχει, καὶ μέντοι καὶ ἀνθρώπου
σώματι ἐντυχοῦσα οὐκ ἀπέχεται, ἀλλ' ἐσθίει.
ταύτῃ τοι καὶ ἐμίσησαν Αἰγύπτιοι τὸ ζῷον ὡς
μυσαρὸν καὶ πάμβορον. φιλοῦσι δὲ οἱ φρόνιμοι
καὶ τῶν ἀλόγων τὰ πραότερα καὶ φειδοῦς ἅμα καὶ
εὐσεβείας μετειληχότα προτιμᾶν. Αἰγύπτιοι γοῦν
τοὺς πελαργοὺς καὶ προσκυνοῦσιν, ἐπεὶ τοὺς
πατέρας γηροκομοῦσιν καὶ ἄγουσι διὰ τιμῆς. οἱ
αὐτοὶ δὲ Αἰγύπτιοι καὶ χηναλώπεκας καὶ ἔποπας
τιμῶσιν, ἐπεὶ οἱ μὲν φιλότεκνοι αὐτῶν, οἱ δὲ πρὸς
τοὺς γειναμένους εὐσεβεῖς. ἀκούω δὲ καὶ Μανέ-
θωνα τὸν Αἰγύπτιον σοφίας ἐς ἄκρον ἐληλακότα
ἄνδρα εἰπεῖν ὅτι γάλακτος ὑείου ὁ γευσάμενος
ἀλφῶν ὑποπίμπλαται καὶ λέπρας· μισοῦσι δὲ ἄρα
οἱ Ἀσιανοὶ πάντες τάδε τὰ πάθη. πεπιστεύκασι
δὲ Αἰγύπτιοι τὴν ὗν καὶ ἡλίῳ καὶ σελήνῃ ἐχθίστην
εἶναι. ὅταν οὖν[1] πανηγυρίζωσι[2] τῇ σελήνῃ, θύουσιν αὐτῇ ἅπαξ τοῦ ἔτους ὗς, ἄλλοτε δὲ οὔτε ἐκείνῃ
οὔτε ἄλλῳ τῳ τῶν θεῶν τόδε τὸ ζῷον ἐθέλουσι
θύειν.[3] Ἀθηναῖοι δὲ ἐν τοῖς μυστηρίοις κατα-
θύουσι τὰς ὗς καὶ μάλα δικαίως· λυμαίνονται γὰρ
⟨τὰ⟩[4] λήια, καὶ ἐσπηδήσασαι πολλάκις τοὺς μὲν
νέους[5] τῶν ἀσταχύων καὶ οὐδέπω ὡραίους
κατακλῶσι, τοὺς δὲ ἐξορύττουσιν. Εὔδοξος δέ
φησι φειδομένους τοὺς Αἰγυπτίους τῶν ὑῶν μὴ
θύειν αὐτάς, ἐπεὶ τοῦ σίτου σπαρέντος ἐπάγουσι
τὰς ἀγέλας αὐτῶν. αἱ δὲ πατοῦσι[6] καὶ ἐς ὑγρὰν
τὴν γῆν ὠθοῦσιν, ἵνα μείνῃ ἔμβιος καὶ μὴ ὑπὸ τῶν
ὀρνίθων ἀναλωθῇ.

[1] Reiske : δέ.
[2] Αἰγύπτιοι παν-.
[3] θύειν ὡς μυσαρόν.
[4] ⟨τά⟩ add. Ges.

306

16. The Pig in sheer gluttony does not spare even **The Pig in Egypt**
its own young; moreover if it comes across a man's
body it does not refrain from eating it. That is why
the Egyptians detest the animal as polluted and
omnivorous. And sober men are accustomed to
prefer those animals which are of a gentler nature
and have some sense of restraint and reverence. At
any rate the Egyptians actually worship Storks,
because they tend and respect their parents in old
age; and these same Egyptians pay honour to
vulpansers and hoopoes, because the former are fond
of their offspring, and the latter show reverence to
their parents. And I learn that Manetho the
Egyptian, a man who attained the very summit of
knowledge, says that one who has tasted of sow's
milk becomes covered with leprosy and scaly
eruptions. And all the peoples of Asia loathe these
diseases. And the Egyptians are convinced that
the Sow is an abomination to the sun and the moon.
Accordingly when they hold the festival of the moon
they sacrifice Pigs to her once a year, but at no other
seasons are they willing to sacrifice them either to
her or to any other god. But the Athenians sacrifice
Sows at the Mysteries and very properly, for they
ruin the crops and frequently by trampling upon the
new ears of corn break some before they are ripe
and uproot others. But Eudoxus asserts that the
Egyptians refrain from sacrificing Sows, because
when the corn has been sown they drive in herds of
them, and they tread and press the seed into the soil
when moist so that it may remain fertile and not be
consumed by the birds.

---

[5] κενοῦσι. [6] πατοῦσι τοὺς πυρούς.

17. Ἀποσπώμενοι τῆς συνήθους γῆς οἱ ἐλέφαντες, καὶ ἐὰν ἡμερωθῶσι τὰ μὲν πρῶτα τοῖς δεσμοῖς
καὶ τῷ λιμῷ, τὰ δὲ ἐπὶ τούτοις ταῖς τροφαῖς καὶ
τῷ ποικίλῳ αὐτῶν, ὅμως τὸ φίλτρον τῆς θρεψαμέ-
νης χώρας οὐκ ἂν αὐτοῖς ποτε ἐξίτηλον γένοιτο.
οἱ πλεῖστοι γοῦν ὑπὸ τῆς λύπης διαφθείρονται,
ἤδη δέ τινες καὶ κλάοντες ἀστακτὶ καὶ ἀμέτροις
τοῖς δακρύοις ἐπηρώθησαν τὴν ὄψιν. ἐσάγονται
δὲ ἐς τὰς ναῦς διὰ γεφύρας, παρ' ἑκάτερα αὐτῆς
κλάδων τεθηλότων καὶ κομώντων πηγνυμένων [1]
καὶ ἄλλης ὕλης χλωρᾶς διατεινομένης ἐς ἀπάτην
τῶν θηρίων· εἰ γὰρ ταῦτα ὁρῶεν οἱ ἐλέφαντες, ἔτι
καὶ τότε διὰ τῆς γῆς ἰέναι δοκοῦσιν,[2] οὐδὲ ἐπιτρέπει ταῦτα ὁρᾶσθαι τὴν θάλατταν. βραχέα δέ ἐστι
καὶ ⟨οὐκ⟩ [3] ἀγχιβαθῆ τὰ πρὸ τῆς χώρας, ἐξ ἧς
ἀνάγκη πλεῖν αὐτούς, καὶ αἱ ναῦς ἀφεστᾶσιν αἱ
φορτίδες· καὶ διὰ ταῦτα τῆς γεφύρας δεῖ καὶ τῆς
μηχανῆς τε καὶ ἐπιβουλῆς τῆς διὰ τῶν κλάδων καὶ
τῆς ὕλης τῆς προειρημένης.

18. Ἀκούω τὸν κριὸν τὸ ζῷον ἐξ μηνῶν χειμε-
ριωτάτων κατὰ τῆς ἀριστερᾶς πλευρᾶς κεῖσθαι
καὶ καθεύδειν, ὅταν αὐτὸν αἱρῇ καὶ περιλαμβάνῃ
ὕπνος, ἀπὸ δὲ τῆς ἐαρινῆς ἰσημερίας ἔμπαλιν
ἀναπαύεσθαι, καὶ κατὰ τῆς δεξιᾶς κεῖσθαι.
οὐκοῦν καθ' ἑκάτεραν ἰσημερίαν τὴν κατάκλισιν
ἀμείβει ὁ κριός.

19. Τοὺς ἰχθῦς τοὺς φάγρους Συηνῖται μὲν [4]
ἱεροὺς νομίζουσιν, οἱ δὲ οἰκοῦντες τὴν καλουμένην

---

[1] Jac : μιγνυμένων.   [2] σφᾶς αὐτοὺς δοκοῦσιν.

17. Elephants when withdrawn from the country The
Elephant's
love of home
to which they are accustomed, though tamed at first
by captivity and hunger and after that by food and
a varied diet, nevertheless do not erase from their
memory the spell of the country that fostered them.
At any rate the majority die of grief, and some have
actually lost their sight through the floods of tears
past measuring which they have shed. And they
are brought on board ships by means of a bridge on
either side of which boughs fresh and in full leaf
have been fixed, together with other greenery that
extends the whole length in order to deceive the
beasts. For if the Elephants see these things they
imagine that they are still walking on firm ground,
and this verdure does not allow the sea to be visible.
But the water close to the shore from which they must
sail is shallow and not deep, and the cargo-vessels
are some distance out. That is why there is need
of the bridge and the device of a ruse contrived with
the boughs and greenery aforesaid.

18. I have heard that the Ram during the six The Ram
months of winter lies down upon its left side, and
sleeps so whenever sleep overtakes and constrains it.
But after the spring equinox it rests in the reverse
position and lies upon its right side. So at each
equinox the Ram changes its way of lying down.

19. The inhabitants of Syene regard the Phagrus [a] The Phagrus
and the
Maeotes
as sacred, and those who dwell in Elephantine, as it

[a] Thompson (*Gk. fishes*, p. 274) points out that φάγρος here
cannot be the *Sea-bream* of 9. 7 (i).

---

³ ⟨οὐκ⟩ add. *Ges.*   ⁴ μὲν Αἰγυπτίων.

Ἐλεφαντίνην τοὺς μαιώτας· φῦλον δὲ ἄρα καὶ
τοῦτο ἰχθύων. ἡ δὲ ἐς ἑκάτερον τὸ γένος ἐξ
ἀμφοτέρων τιμὴ τὴν γένεσιν εἴληφεν ἐντεῦθεν.
ἀνιέναι τε καὶ ἀναπλεῖν τοῦ Νείλου μέλλοντος οἱ
δὲ προθέουσί τε καὶ νήχονται, οἱονεὶ τοῦ νέου
ὕδατος ἄγγελοι, καὶ τὰς τῶν Αἰγυπτίων ἀνηρτημέ-
νας γνώμας προευφραίνουσι καλαῖς ἐλπίσι, τὴν
ἐπιδημίαν τοῦ ῥεύματος πρῶτοι συνιέντες καὶ
θαυμαστῇ τινι φύσει προμαντευόμενοι ἐκεῖνοί γε.
ἤδη δὲ καὶ τοῦτο ὑπὲρ τῆς ἐς αὐτοὺς τιμῆς φιλοῦσι
προστιθέναι οἱ προειρημένοι, λέγοντες αὐτοὺς
διαμένειν ἀλλήλων ἀγεύστους.

20. Γίνονται δὲ ἄρα ἐν τῇ Ἐρυθρᾷ θαλάττῃ
κόγχαι καὶ ἕτεραι, οὐ λεῖαι τὰ ὄστρακα, ἀλλὰ
ἔχουσαί τινας ἐντομὰς καὶ κοιλάδας. ὀξεῖαι δὲ
αὗται τὰ χείλη εἰσί, καὶ συνιοῦσαι ἐς ἀλλήλας
ἐμπίπτουσι, παραλλὰξ ἐντιθεῖσαι τὰς ἐξοχάς, ὡς
δοκεῖν δύο πριόνων [1] τοὺς κυνόδοντας ἐς ἀλλήλους
συνέρχεσθαι. οὐκοῦν τῶν ἁλιέων ὅτου ἂν νηχομέ-
νου λάβωνται καὶ δάκωσιν ὅ τι οὖν μέρος, ἀποκόπτουσιν, εἰ καὶ ὀστοῦν ὑπείη τῷ μέρει τῷ δηχθέντι,
καὶ κατὰ ἄρθρου μέντοι δακοῦσαι καὶ τοῦτο
ἀπέκοψαν, καὶ εἰκότως· τομώτατον γάρ ἐστι τὸ
δῆγμα.

21. Τοὺς κροκοδίλους Αἰγυπτίων οἱ μὲν σέβουσιν, ὡς Ὀμβῖται· καὶ οἷα ἡμεῖς τοὺς θεοὺς τοὺς
Ὀλυμπίους ἄγομεν θαυμαστούς, τοιαῦτα καὶ
ἐκείνους ἐκεῖνοι. καὶ τῶν τέκνων γε αὐτοῖς
ἐξαρπαζομένων πολλάκις οἱ δὲ ὑπερήδονται, καὶ αἵ
γε μητέρες τῶν δειλαίων γάννυνται καὶ σεμναὶ

is called, the Maeotes. (This also is a species of fish.) And the reverence which both peoples pay to either kind has its origin in this: when the Nile is about to rise and overflow, these fish come swimming in advance, as though heralding the coming water, and gladden the anxious hearts of the Egyptians with fair hopes, being the first to realise the advent of the flood and foretelling it by some marvellous natural faculty. Moreover the aforesaid peoples are accustomed to add, concerning their respect for the fish, that they never eat one another.

20. It seems that there are other Shellfish besides in the Red Sea, whose shells are not smooth but have certain grooves and hollows in them.[a] These shells have sharp lips, and when they close they fit into one another, as they make the points interlock, so that it seems as if the teeth of two saws came together.[1] And so if they catch any fisherman swimming and bite any part of him they cut it off, even though there be a bone within the bitten part; more than that, if they bite at a joint, they cut it off at once; nor is that to be wondered at, for their bite is exceedingly sharp. A Red Sea Shellfish

21. In Egypt there are some, like the people of Ombos, who venerate Crocodiles, and just as we regard the Olympian gods with awe, so do they these animals. And when, as often happens, their children are carried off by them, the people are overjoyed, while the mothers of the unfortunates The Crocodile at Ombos and Apollino-polis

---

[a] Ael. is describing the *Tridacna gigas* or its kin; see Thompson, *Gk. fishes*, s.v. κόγχη.

[1] πριόνων συνιόντων.

περιίασιν, οἷα δήπου τεκοῦσαι θεῷ βορὰν καὶ
δεῖπνον. Ἀπολλωνοπολῖται δὲ Τεντυριτῶν μοῖρα [1]
σαγηνεύουσι τοὺς κροκοδίλους, καὶ τῶν περσεῶν
(φυτὰ δέ ἐστιν ἐπιχώρια) ἐξαρτήσαντες μετεώρους
μαστιγοῦσί τε πολλὰς καὶ τὰς [2] ἐξ ἀνθρώπων
ξαίνουσι κνυζωμένους [3] καὶ δακρύοντας, εἶτα μέντοι
κατακόπτουσιν αὐτοὺς καὶ σιτοῦνται. κύει δὲ
ἄρα τὸ ζῷον τοῦτο ἐν ἑξήκοντα ἡμέραις, καὶ
τίκτει ᾠὰ ἑξήκοντα, καὶ τοσαύταις ἡμέραις θάλπει
αὐτά, σφονδύλους τε ἔχει ἐπὶ τῆς ῥάχεως τοσούτους, νεύροις τε αὐτὸν τοσούτοις φασὶ διεζῶσθαι,
λοχεία τε αὐτῷ [4] ἐς τοσοῦτον πρόεισιν ἀριθμόν,
καὶ ἔτη βιοῖ ἑξήκοντα (λέγω δὲ ταῦτα Αἰγυπτίους
φήμας τε καὶ πίστεις), πάρεστι δὲ καὶ ὀδόντας
ἑξήκοντα τοῦδε τοῦ ζῴου ἀριθμεῖν, φωλεῦον δὲ
ἄρα καθ' ἕκαστον ἔτος ἑξήκοντα ἡμερῶν ἀτρεμεῖ
τε καὶ ἀτροφεῖ. τοῖς δὲ Ὀμβίταις καὶ συνήθεις
εἰσί, καὶ μέντοι καὶ ὑπακούουσι καλούντων αὐτῶν
οἱ τρεφόμενοι ἐν ταῖς λίμναις ταῖς ὑπ' αὐτῶν
πεποιημέναις. κομίζουσι δὲ ἄρα αὐτοῖς κεφαλὰς
τῶν ζῴων τῶν θυομένων (αὐτοὶ γὰρ οὐκ ἂν
γεύσαιντο τοῦδε τοῦ μέρους) καὶ ἐμβάλλουσιν
αὐτάς, οἱ δὲ περὶ ταύταις πηδῶσιν. οἵ γε μὴν
Ἀπολλωνοπολῖται μισοῦσι κροκόδιλον, λέγοντες
τὸν Τυφῶνα ὑποδῦναι τὴν τούτου μορφήν. οἱ δὲ
οὐ ταύτην φασὶ τὴν αἰτίαν, Ψαμμύντου δὲ βασιλέως
ἀγαθοῦ καὶ δικαίου ἐς τὰ ἔσχατα ἁρπάσαι θυγατέρα
κροκόδιλον, εἶτα μέντοι μνήμῃ τοῦ τότε πάθους
μισεῖν τὸ φῦλον αὐτῶν πᾶν καὶ τοὺς κάτω τοῦ
χρόνου γεγενημένους.

---

[1] Hemst : μοῖραι.   [2] Cobet : πολλαῖς καὶ ταῖς.
[3] Cobet : κνυζομένους.   [4] αὐτῶν.

are glad and go about in pride at having, I suppose, borne food and a meal for a god. But the people of Apollinopolis, a district of Tentyra, net the Crocodiles, hang them up on persea-trees (these are indigenous), flog them severely, mangling them with all the blows in the world, while the creatures whimper and shed tears; finally they cut them up and eat them.

The Crocodile, it seems, is pregnant for sixty days, and produces sixty eggs which it broods for as many days: it has that number of vertebrae in its spine, and they say that sixty sinews girdle its body, and it bears young ones the same number of times, and it lives for sixty years (I am reporting what the people of Egypt say and believe); one may reckon the teeth of this creature as sixty in number; during sixty days of every year it remains quiet in its lair and abstains from food. The Crocodiles are accustomed to the people of Ombos, and those that are kept in the lakes made by the aforesaid people are obedient to their summons. And the people bring them the heads of the animals which they sacrifice— they themselves will never touch that part—and throw them in, and the Crocodiles come leaping round them. The inhabitants of Apollinopolis, on the contrary, detest the Crocodile, for they say that this was the shape assumed by Typho. Others however say that this is not the reason, but that a Crocodile carried off the daughter of King Psammyntus,[a] a supremely good and righteous man, and therefore in memory of that disaster even posterity abhors the whole race of Crocodiles.

[a] Psammenitus (if this is the King to whom A. is referring) was King of Egypt for six months in 526 B.C.

AELIAN

22. Βακκαῖοι¹ (γένος δὲ τοῦτο ἑσπέριον) τῶν
ἀποθνησκόντων νόσῳ τοὺς νεκροὺς ὑβρίζοντες ὡς
ἀνάνδρως καὶ μαλακῶς τεθνεώτων θάπτουσι πυρί,
τοὺς δὲ ἐν πολέμῳ τὸν βίον καταστρέψαντας ὡς
καλοὺς καὶ ἀγαθοὺς καὶ ἀρετῆς μετειληχότας γυψὶ
προβάλλουσιν, ἱερὸν τὸ ζῷον εἶναι πεπιστευκότες.
Ῥωμύλος δὲ ἄρα ἐν τῷ Παλλαντίῳ λόφῳ δώδεκα
γυψὶν οἰωνισάμενος, ὡς ἀγαθῆς τῆς μαντείας
ἔτυχε, μιμούμενος² τῶν ὀρνίθων τὸν ἀριθμόν, τῶν
Ῥωμαίων ἀρχόντων³ ἰσαρίθμους τοῖς τότε ὀφθεῖ-
σιν ὄρνισι προπορεύειν⁴ ῥάβδους ἐνομοθέτησεν.
Αἰγύπτιοι δὲ Ἥρας μὲν ἱερὸν ὄρνιν εἶναι πεπιστεύκασι τὸν γῦπα, κοσμοῦσι δὲ τὴν τῆς Ἴσιδος
κεφαλὴν γυπὸς πτεροῖς, καὶ τοῖς τῶν προπυλαίων
ὀρόφοις ἐνετόρευσαν⁵ γυπῶν πτέρυγας. εἶπον δὲ
καὶ ἀνωτέρω ὑπὲρ τοῦδε τοῦ ζῴου πολλά, ἕτερα
μέντοι.

23. Ἐν τῇ Κοπτῷ τῇ Αἰγυπτίᾳ τὴν Ἶσιν
σέβουσιν Αἰγύπτιοι ταῖς τε ἄλλαις ἱερουργίαις καὶ
μέντοι καὶ τῇ παρὰ τῶν πενθουσῶν ἢ τοὺς ἄνδρας
τοὺς σφετέρους ἢ τοὺς παῖδας ἢ τοὺς ἀδελφοὺς
λατρείᾳ τε καὶ θεραπείᾳ. ὄντων δὲ σκορπίων
ἐνταῦθα μεγέθει μὲν μεγίστων, πληγῇ δὲ ὀξυτάτων,
πείρᾳ γε μὴν σφαλερωτάτων (παίσαντες γὰρ
ἀναιροῦσι παραχρῆμα), καὶ μηχανὰς μυρίας ἐς τὴν
ἐξ αὐτῶν φυλακὴν μηχανωμένων τῶν Αἰγυπτίων,
ἀλλὰ αἵ γε πενθοῦσαι παρὰ τῇ θεῷ καὶ χαμαὶ
καθεύδουσαι καὶ ἀνυπόδητοι βαδίζουσαι καὶ μόνον

¹ Bochart : βαρκαῖοι.　　² ἀμειβόμενος.
³ τοὺς Ῥωμαίων ἄρχοντας.　　⁴ προπομπεύειν H.
⁵ ὑπετόρευσαν.

314

22. The Vaccaei [a] (they are a western people) The Vulture
insult the corpses of such as die from disease as
having died a cowardly and effeminate death, and
dispose of them by burning; whereas those who
laid down their lives in war they regard as noble,
heroic, and full of valour, and them they cast to the
Vultures, believing this bird to be sacred. And
when Romulus on the Palatine Hill, divining by the
flight of twelve Vultures, had received a favourable
augury, following the number of the birds he decreed
that the rulers of Rome should be preceded by a
number of rods [b] equal to that of the birds seen on
that occasion. And the Egyptians believe that the
Vulture is sacred to Hera, and deck the head of Isis
with Vultures' feathers, and on the roofs of the
entrances to their temples they carve the wings of
Vultures in relief.

I have earlier on said much concerning this bird,
but not to the same effect.

23. At Coptos in Egypt the natives pay homage The
to Isis in a variety of rituals but especially in the Scorpions
of Coptos
service and ministry rendered by women who are
mourning either a husband or a son or a brother.
And at Coptos there are scorpions of immense size,
possessing very sharp stings, and most dangerous in
their attack (for when they strike they kill instantly),
and the Egyptians contrive innumerable devices for
self-protection. But although the women in mourning at the temple of the goddess sleep on the floor,

---

[a] If Βακκαῖοι is correctly rendered ' Vaccaei,' they were a
tribe in the NW of Spain.

[b] Lat. *fasces*, a bundle consisting of rods and an axe, carried
by the Lictors.

οὐ πατοῦσαι τοὺς προειρημένους σκορπίους εἶτα
μέντοι ἀπαθεῖς διαμένουσι. σέβουσι δὲ ἄρα οἱ
αὐτοὶ Κοπτῖται καὶ θηλείας δορκάδας καὶ ἐκθεοῦσιν αὐτάς, τοὺς δὲ ἄρρενας καταθύουσιν. ἄθυρμα
δὲ εἶναι τὰς θηλείας τῆς Ἴσιδός φασιν.

24. Ὁ κροκόδιλος (καὶ μέντοι καὶ ταῦτα πρὸς
τοῖς ἤδη διηνυσμένοις ὑπὲρ τοῦ ζῴου ἀκήκοα)
φύσει δειλός ἐστι καὶ κακοήθης δὲ καὶ πανοῦργος
δεινῶς· καὶ ἁρπάζει μὲν καὶ ἐπιβουλεύει μάλα
ὀξέως, πέφρικε δὲ τοὺς κτύπους πάντας, δέδοικε δὲ
καὶ ἀνθρώπου βιαιοτέραν βοήν, καὶ μέντοι καὶ τοὺς
εὐθαρσέστερον ἐπιόντας ὀρρωδεῖ ἰσχυρῶς. οἱ τοίνυν καλούμενοι κατὰ τὴν Αἴγυπτον Τεντυρῖται
ἴσασι καὶ ὅθεν εὐχείρωτόν ἐστι τὸ θηρίον· μάλιστα
δ' ἂν τρωθείη ἐς δέον ὀφθαλμοὺς βληθεὶς ἢ μασχά-
λας καὶ μέντοι καὶ τὴν νηδύν. ⟨τὰ⟩ [1] νῶτα δὲ
πέφυκε καὶ τὴν οὐρὰν ἄρρηκτος· λεπίσι τε γὰρ
καὶ φολίσι πέφρακται καὶ ὡς ἂν εἴποι τις ὥπλισται,
καὶ ἐοίκασιν ὀστράκοις καρτεροῖς ἢ κόγχαις. οἱ
τοίνυν προειρημένοι [2] οὕτως εἰσὶ φιλόπονοι [3] πρὸς
τὴν αὑτῶν [4] θήραν, ὡς τὸν ἐκεῖθι ποταμὸν εἰρήνην
ἄγειν αὐτῶν βαθυτάτην. ἐνταῦθά τοι καὶ θαρροῦντες νήχονται, καὶ ἀθύρουσιν ἐν τῇ νήξει. ἐν
Ὀμβίταις [5] δὲ ἢ Κοπτίταις ἢ Ἀρσενοΐταις οὐδ'
ἀπονίψασθαι πόδας ῥάδιον, οὐδὲ ἀρύσασθαι ὕδωρ
εὔκολον· ἀλλ' οὐδὲ ταῖς ὄχθαις τοῦ ποταμοῦ ἔστιν
ἐμβαδίσαι ἐλευθέρως καὶ ἀφυλάκτως. σέβουσι δὲ
οἱ Τεντυρῖται ἱέρακας. οὐκοῦν οἱ Κοπτὸν οἰκοῦντες ὡς κροκοδίλων πολεμίους λυπεῖν προηρημένοι

---

[1] ⟨τά⟩ add. Schn.   [2] προειρημένοι θηραταὶ αὑτῶν.
[3] Jac: φιλοπόνηροι.   [4] τὴν κατ' αὑτῶν.

go about with bare feet, and all but tread on the
aforesaid scorpions, yet they remain unharmed.
And these same people of Coptos worship and deify
the female gazelle, though they sacrifice the male.
They say that the females are the pets of Isis.

24. The Crocodile (I may say that I have learned The
these facts in addition to what has already been Crocodile
recounted of this animal) is naturally timid, of an
evil disposition, and thoroughly villainous. It is
alert to seize and plan against its victims, but it
dreads all noises and is afraid even of loud shouts of
men and has a violent fear of those who boldly
attack it. Now the people of Egypt called Ten- killed at
tyrites know the best way to master the beast: the Tentyra
most effective way of wounding it is to strike it in the
eyes or the armpits and even in the belly. Its back
however, and its tail are impenetrable, for it is
fortified and, so to say, armed with scaly plates
which resemble hard earthenware or shells. Now the
aforesaid people are so assiduous in pursuit of these
creatures that the river in their district is left in
profound peace by the Crocodiles. So there they
make bold to swim and sport in their swimming.
Whereas among the people of Ombos or Coptos or worshipped
Arsinoe it is not easy even to wash one's feet nor at Coptos
can one draw water in security; why, one cannot
even walk along the river banks freely and off one's
guard. But the people of Tentyra worship Hawks.
For that reason those who live in Coptos, wishing to
annoy the Tentyrites as enemies of the Crocodiles,
often crucify Hawks. The Crocodile the people of
Coptos liken to water, that is why they worship it;

---

[5] *Ges* : "Ομβροις.

πολλάκις ἀνασταυροῦσιν ἱέρακας. εἰκάζουσι δὲ
τὸν μὲν κροκόδιλον ἐκεῖνοι ὕδατι, ἔνθεν τοι καὶ
σέβουσιν· οἱ δὲ τὸν ἱέρακα πυρί, ταύτῃ τοι καὶ
προσκυνοῦσι· μαρτύριόν τε ἐπάγουσιν † ἀπότομον
αὐτῶν εἶναι † [1] λέγοντες πῦρ καὶ ὕδωρ ἀμιγές.
ταῦτα οὖν τερατολογοῦντες [2] Αἰγύπτιοί φασιν.

25. Ὄασιν τὴν Αἰγυπτίαν διελθόντι ἀπαντᾷ
ἑπτὰ ἡμερῶν ὅλων ἐρημία βαθυτάτη. μετὰ δὲ
ταύτην Κυνοπρόσωποι νέμονται ἄνθρωποι κατὰ τὴν
ὁδὸν τὴν ἐς Αἰθιοπίαν ἄγουσαν.[3] ζῶσι δὲ ἄρα
οὗτοι θηρῶντες δορκάδας τε καὶ βουβαλίδας, ἰδεῖν
γε μὴν μέλανές εἰσι, κυνὸς δὲ ἔχουσι τὴν κεφαλὴν
καὶ τοὺς ὀδόντας. ἐπεὶ δὲ ἐοίκασι τῷδε τῷ ζῴῳ,
καὶ μάλα γε εἰκότως αὐτῶν ἐνταυθοῖ τὴν μνήμην
ἐποιησάμην. φωνῆς δ᾽ οὖν [4] ἀμοιροῦσι, τρίζουσι
δὲ ὀξύ· κάτεισι δὲ ὑπὸ τὴν ὑπήνην αὐτοῖς γένειον,
ὡς εἰκάσαι τοῖς τῶν δρακόντων αὐτό. αἱ δὲ
χεῖρες αὐτῶν ὄνυξιν ἰσχυροῖς καὶ ὀξυτάτοις εἰσὶ
τεθηγμέναι· τὸ δὲ πᾶν σῶμα δασεῖς πεφύκασι,
κατὰ τοὺς κύνας καὶ τοῦτο. ὤκιστοι δέ εἰσι καὶ
ἴσασι τὰ ἐν τοῖς τόποις δύσβατα.[5] ἐντεῦθέν [6] τοι
καὶ δυσάλωτοι δοκοῦσιν.

26. Τῷ λύκῳ ὁ τράχηλος ἐς βραχὺ συνῆκται.
οὔκουν οἷός τέ ἐστιν ἐπιστραφῆναι, ὁρᾷ δὲ ἐς τὸ
πρόσω ἀεί· εἰ δὲ βούλοιτό ποτε ἐς τοὐπίσω
θεάσασθαι, πᾶς ἐπιστρέφεται. ὀξυωπέστατον δέ

---

[1] ἀπότομον . . . εἶναι corrupt.
[2] πενθοῦντες or τερθροῦντες.
[3] Jac: εἰς Αἰ. τὴν ἄγουσαν.
[4] γοῦν or γάρ.

ON ANIMALS, X. 24-26

whereas the Tentyrites liken the Hawk to fire, hence
their adoration. And they adduce as evidence . . .[a]
maintaining that fire and water cannot mingle.

Such are the marvellous tales told by the Egyptians.

25. After traversing the Egyptian oasis one is
confronted for seven whole days with utter desert.
Beyond this live the human Dog-faces [b] along the road
that leads to Ethiopia. It seems that these creatures
live by hunting gazelles and antelopes; further,
they are black in appearance, and they have the
head and teeth of a dog. And since they resemble
this animal, it is very natural that I should mention
them here. They are however not endowed with
speech, but utter a shrill squeal. Beneath their
chin hangs down a beard; we may compare it with
the beards of dragons,[c] and strong and very sharp
nails give an edge to their hands. Their whole body
is covered with hair—another respect in which they
resemble dogs. They are very swift of foot and
know the regions that are inaccessible: that is why
they appear so hard to capture.

26. The neck of a Wolf is short and compressed;
the animal is thus incapable of turning but always
looks straight ahead. And if it wants to look back
at any time, it turns its whole body. It has the

The Dog-faces

The Wolf

---

[a] The sense required to complete the last clause appears to
be ' They account for their hostility by pointing out that, *etc.*'
[b] Gossen (§ 238) regards the Κυνοπρόσωπος as the ' Mandrill,'
a kind of baboon, native of W Africa.
[c] The δράκων in Nic. *Th.* 438 ff. is a large snake.

5 ὕδατα.   6 *Ges*: ἐνταῦθα.

319

ἐστι ζῴων, καὶ μέντοι καὶ νύκτωρ καὶ σελήνης οὐκ
οὔσης ὁ δὲ ὁρᾷ.[1] ἔνθεν τοι καὶ λυκόφως κέκληται
ὁ καιρὸς οὗτος τῆς νυκτός, ἐν ᾧ μόνος ἐκεῖνος τὸ
φῶς ὑπὸ τῆς φύσεως λαχὼν[2] ἔχει. δοκεῖ δέ μοι
καὶ Ὅμηρος λέγειν ἀμφιλύκην νύκτα καθ᾽ ἣν
δὴ[3] βλέποντες λύκοι βαδίζουσι. λέγουσι δὲ φίλον
Ἡλίῳ εἶναι αὐτόν, καὶ διὰ ταῦτα ἐς τιμὴν τὴν τοῦ
ζῴου κεκλῆσθαι καὶ τὸν ἐνιαυτὸν λυκάβαντα εἰσὶν
οἳ λέγουσι. χαίρειν δὲ αὐτῷ καὶ τὸν Ἀπόλλω
λόγος, καὶ ἡ αἰτία[4] διαρρέουσα καὶ ἐς ἐμὲ
ἀφίκετο. τὸν γάρ τοι θεὸν τοῦτον τεχθῆναί φασι
τῆς Λητοῦς μεταβαλούσης τὸ εἶδος ἐς λύκαιναν.
ταύτῃ τοι λέγει καὶ Ὅμηρος[5] λυκηγενέι κλυτο-
τόξῳ· ταύτῃ τοι καὶ ἐν Δελφοῖς ἀνακεῖσθαι λύκον
πέπυσμαι χαλκοῦν τὴν τῆς Λητοῦς ὠδῖνα αἰνιττό-
μενον. οἱ δὲ οὐ διὰ τοῦτό φασιν ἀλλὰ ἐπεὶ
κλαπέντα ἀναθήματα ἐκ τοῦ νεὼ καὶ κατορυχθέντα
ὑπὸ τῶν ἱεροσύλων λύκος[6] κατεμήνυσε. παρελθὼν
γὰρ ἐς τὸν νεὼν καὶ τῶν προφητῶν τινα τῆς
ἐσθῆτος τῆς ἱερᾶς ἑλκύσας τῷ στόματι καὶ προσ-
αγαγὼν μέχρι τοῦ τόπου ἐν ᾧ τὰ ἀναθήματα
ἐκέκρυπτο εἶτα τοῖς προσθίοις ὤρυττεν αὐτόν.

27. Κώμη Αἰγυπτία Χουσαὶ τὸ ὄνομα (τελεῖ δὲ
ἐς τὸν Ἑρμοπολίτην[7] νομόν, καὶ μικρὰ μὲν δοκεῖ,
χαρίεσσά ⟨γε⟩[8] μήν), ἐν ταύτῃ σέβουσιν Ἀφροδίτην Οὐρανίαν αὐτὴν καλοῦντες. τιμῶσι δὲ καὶ

---

[1] ὁρᾷ, καὶ ὅτε πᾶσίν ἐστι σκότος ἐκεῖνος βλέπει.
[2] λαβών.
[3] ἤδη.
[4] αἰτία ἐκείνη.
[5] Ὅ. εἰπών.
[6] λύκος τὸ ζῷον.
[7] Ἑρμουπολίτην.
[8] ⟨γε⟩ add. H.

sharpest sight of any animal, and indeed it can even
see at night when there is no moon. Hence the
name *Lycophos* (wolf's-light, *i.e.* gloaming) is applied
to that season of the night in which the Wolf alone has
light with which Nature provides him. And I think
that Homer gives the name [*Il.* 7. 433] 'twilight of
the night,' to the time during which Wolves can see to
move about. And they say that the Wolf is beloved   beloved of
of the Sun; and there are those who assert that the   Apollo
year is called *Lycabas* in honour of this animal. It is
said also that Apollo takes pleasure in the Wolf, and
the reason which is commonly reported has reached
me too. It is this: they say that the god was born
after Leto had changed herself into a she-wolf. That
is why Homer speaks of ' the wolf-born lord of the
bow '[*Il.* 4. 101]. That is why, as I learn, at Delphi
a bronze Wolf is set up, in allusion to the birth-pangs
of Leto. Others however deny this, maintaining   reveals
that it was because a Wolf gave information that   sacrilege
offerings had been stolen from the temple and had
been buried by the sacrilegious thieves. For it
made its way into the temple and with its mouth
pulled one of the priests by his sacred robe and drew
him to the spot in which the offerings had been
hidden, and then proceeded to dig the spot with its
forepaws.

27. There is a district in Egypt called Chusae (it   The Cow and
is reckoned as belonging to the province of Hermo-   Aphrodite
polis,[a] and though small in extent it possesses charm)
and there they worship Aphrodite under the title of
Urania (heavenly). They also pay homage to a cow,

---

[a] NW corner of the Nile delta.

AELIAN

θήλειαν βοῦν, καὶ τὴν αἰτίαν ἐκείνην λέγουσι. πεπιστεύκασιν αὐτὰς προσήκειν τῇδε τῇ δαίμονι· ποίαν γὰρ ἐς ἀφροδίσια ἰσχυρὰν ἔχει [1] βοῦς θῆλυς, καὶ ὀργᾷ τοῦ ἄρρενος μᾶλλον. ἀκούσασα γοῦν τοῦ μυκήματος ἐς τὴν μίξιν θερμότατα ἐξηνέμωται καὶ ἐκπέφλεκται. καὶ οἱ ταῦτά γε συνιδεῖν δεινοὶ καὶ ἀπὸ τριάκοντα σταδίων ἀκούειν ταύρου βοῦν ἐρωτικὸν σύνθημα καὶ ἀφροδίσιον μυκωμένου φασί. καὶ αὐτὴν δὲ τὴν Ἶσιν Αἰγύπτιοι βούκερων καὶ πλάττουσι καὶ γράφουσιν.

28. Σάλπιγγος ἦχον βδελύττονται Βουσιρῖται καὶ Ἄβυδος ἡ Αἰγυπτία καὶ Λύκων πόλις,[2] ἐπεί πως ἔοικεν ὄνῳ βρωμωμένῳ. ἀλλὰ καὶ ὅσοι περὶ τὴν θρησκείαν ἔχουσι τὴν τοῦ Σαράπιδος μισοῦσι τὸν ὄνον. τοῦτό τοι καὶ Ὦχος ὁ Πέρσης εἰδὼς ἀπέκτεινε μὲν τὸν Ἆπιν, ἐξεθέωσε δὲ τὸν ὄνον, ἐς τὰ ἔσχατα λυπῆσαι θέλων τοὺς Αἰγυπτίους. ἔδωκε δὲ ἄρα καὶ αὐτὸς δίκας τῷ ἱερῷ βοΐ οὐ μεμπτὰς οὐδὲ ἥττονας Καμβύσου τοῦ πρώτου τὴν θεοσυλίαν ταύτην τετολμηκότος. μισοῦσι δὲ οἱ αὐτοὶ θεραπευταὶ τοῦ Διὸς τοῦ προειρημένου καὶ τὸν ὄρυγα. τὸ δὲ αἴτιον, ἀποστραφεὶς πρὸς τὴν ἀνατολὴν τὴν τοῦ ἡλίου τὰ περιττὰ τῆς ἑαυτοῦ τροφῆς ἐκθλίβει φασὶν Αἰγύπτιοι. λέγουσι δὲ οἱ Πυθαγόρειοι ὑπὲρ[3] τοῦ ὄνου καὶ ἐκεῖνο,[4] μόνον τοῦτον τῶν ζῴων μὴ γεγονέναι κατὰ ἁρμονίαν·

[1] Jac : ἔχει ἐκεῖνος.
[2] πόλις, καὶ λέγουσι τὴν αἰτίαν.
[3] ὡς λόγος ὑπέρ.
[4] ἐκεῖνα.

322

and this, they say, is the reason: they believe that
cows are related to this goddess, because the cow
feels a strong incitement to love and is more passionate than the bull. At any rate at the sound of his
bellow the cow becomes excited and inflamed with
a burning desire to couple. And those who are
expert in these matters maintain that a cow hears a
bull as much as thirty stades *a* away when it is bellowing as a signal to love and mate. And in Egypt
sculptors and painters represent Isis herself with the
horns of a cow.

28. The people of Busiris and of Abydos in Egypt
and of Lycopolis dislike the blare of a trumpet on
the ground that it resembles the braying of an Ass.
And those who attend to the cult of Serapis also
hate the Ass. Now Ochus the Persian *b* knowing
this slew Apis and deified the Ass from a wish to pain
the Egyptians to the utmost. And so he too paid
a penalty, which all applauded, to the Sacred Bull,
no less than Cambyses *c* who was the first that dared
commit this sacrilege. And the same ministers of
the aforesaid Zeus (*i.e.* Serapis) detest the antelope
as well, and for this reason: the Egyptians maintain
that it voids its excrement after turning its back
towards the rising sun. And the followers of
Pythagoras also say this touching the Ass, that it
alone among animals was not born in tune, and

The Ass and
the Ante-
lope, hated
in Egypt

---

a Over 3 miles.
b The name of Artaxerxes III before he became King of
Persia, 359 B.C. He conquered Egypt and in 338 was poisoned
by Bagoas.
c Cambyses, King of Persia, outraged the Egyptians by his
cruelty and his insults to their religion. He died from a wound
caused by his own sword; cp. Hdt. 3. 64.

ταύτῃ τοι καὶ πρὸς τὸν ἦχον τὸν τῆς λύρας εἶναι
κωφότατον. ἤδη δὲ αὐτόν τινες καὶ τῷ Τυφῶνι
προσφιλῆ γεγονέναι φασί. ἐθέλουσι [1] δὲ καὶ
ἐκείνην αἰτίαν τῷ ὄνῳ προσάπτειν πρὸς τοῖς
προειρημένοις. πᾶν τὸ γόνιμον τετίμηται, ἐναντίως
δὲ ἄρα πρὸς ταῦτα πέφυκε τὸ ζῷον τοῦτο. δίδυμα
γοῦν ὄνον τεκοῦσαν οὐ ῥᾳδίως μέμνηταί τις λόγος.

29. Ἴδιον δὲ ἐν Αἰγυπτίοις λόγοις ἴβεως καὶ
ἐκεῖνο προσακήκοα. ὅταν ὑποκρύψηται τὴν δέρην
καὶ τὴν κεφαλὴν τοῖς ὑπὸ τῷ στέρνῳ πτεροῖς, τὸ
τῆς καρδίας σχῆμα ἀπεμάξατο. ἄλλως τε καὶ ὅτι
εἰσὶ τοῖς ἐπὶ λύμῃ καὶ ἀνθρώπων καὶ καρπῶν
ζῴοις γεγενημένοις ἔχθιστα ἤδη που καὶ ἄνω
εἶπον. [2] μίγνυνται δὲ τοῖς στόμασι, καὶ παιδοποιοῦν-
ται τὸν τρόπον τοῦτον. λέγουσι δὲ Αἰγύπτιοι
(καὶ ἐμέ γε οὐ ῥᾳδίως ἔχουσι πειθόμενον) λέγουσι
δ' οὖν [3] τοὺς ταῖς ταριχείαις τῶν ζῴων ἐφεστῶτας
καὶ δεινοὺς τήνδε τὴν σοφίαν ὁμολογεῖν τὸ τῆς
ἴβεως ἔντερον ἐξ εἶναι πήχεων καὶ ἐνενήκοντα.
διαβαίνειν δὲ κατὰ πῆχυν αὐτὴν προσακήκοα
βαδίζουσαν. σελήνης δὲ ἐκλιπούσης καταμύει,
ἔστ' ἂν ἡ θεὸς αὖθις ἀναλάμψῃ. καὶ τῷ Ἑρμῇ δέ
φασι τῷ πατρὶ τῶν λόγων φιλεῖται, ἐπεὶ ἔοικε τὸ
εἶδος τῇ φύσει τοῦ λόγου· τὰ μὲν γὰρ μέλανα
ὠκύπτερα τῷ τε σιγωμένῳ καὶ ἔνδον ἐπιστρεφομένῳ
λόγῳ παραβάλλοιτο ἄν, τὰ δὲ λευκὰ τῷ προφερο-
μένῳ τε καὶ ἀκουομένῳ ἤδη καὶ ὑπηρέτῃ τοῦ ἔνδον
καὶ ἀγγέλῳ, ὡς ἂν εἴποις. ὡς μὲν οὖν μακροβιώ-
τατόν ἐστι τὸ ζῷον καὶ δὴ εἶπον· λέγει δὲ Ἀπίων
καὶ ἐπάγεται τοὺς ἐν Ἑρμοῦ πόλει ἱερέας μάρτυρας

---

[1] λέγουσι.   [2] προεῖπον.   [3] γοῦν.

that this accounts for its being completely deaf to
the sound of the lyre. Some moreover say that it
was beloved of Typho. And in addition to the
foregoing charges they would blame the Ass for this
also: fertility in all kinds is respected, but this
animal is by nature opposed to it. At any rate it is
not easy to recall any account of a she-ass giving
birth to twins.

29. Here is another peculiarity of the Ibis which I   The Ibis
have learnt from Egyptian narratives. When it
buries its neck and head beneath its breast-feathers,
it imitates the shape of the heart. Of its special
hostility to creatures injurious to man and to crops
I think I have already spoken earlier on.[a] The birds
couple with their mouth and beget offspring in that
way. And the Egyptians say, though I for one am
not easily persuaded, yet they say that those who
see to the embalming of animals and who are experts
at it, agree that the entrails of the Ibis measure
ninety-six cubits. I have heard further that its
stride when walking measures a cubit. And when
the moon is in eclipse it closes its eyes until the goddess shines out again. It is said to be beloved of
Hermes the father of speech because its appearance
resembles the nature of speech: thus, the black
wing-feathers might be compared to speech sup-
pressed and turned inwards, the white to speech
brought out, now audible, the servant and the
messenger of what is within, so to say. Now I have
already mentioned that the bird lives to a very great
age. And Apion states that it is immortal and
adduces the priests of Hermopolis as witnesses to

[a] See 1. 38 (iv); 2. 38.

δεικνύντας οἱ ἶβιν ἀθάνατον. τοῦτο μὲν οὖν καὶ
ἐκείνῳ δοκεῖ τῆς ἀληθείας ἀφεστάναι πάμπολυ, καὶ
ἐμοὶ δὲ πάντως ἂν καταφαίνοιτο ψευδές.[1] ἔστι δὲ
τὴν φύσιν θερμότατον ἡ ἶβις, πολυβορώτατον γοῦν
ὂν καὶ κακοβορώτατον, εἴγε ὄφεις σιτεῖται καὶ
σκορπίους. ἀλλὰ τὰ μὲν πέττει ῥᾳδίως, τὰ δὲ
εὐκολώτατα ἀποκρίνει. ἴδοι δ' ἄν τις νοσοῦσαν
ἶβιν σπανιώτατα. πανταχοῦ δὲ καθιεῖσα ἶβις τὸ
ῥάμφος, τῶν ῥυπαρῶν καταφρονοῦσα καὶ ἐμβαί-
νουσα αὐτοῖς ὑπὲρ τοῦ καὶ ἐκεῖθέν τι ἀνιχνεῦσαι,
ὅμως δ' οὖν ἐς κοῖτον τρεπομένη λούει τε πρότερον
ἑαυτὴν καὶ ἐκκαθαίρει. νεοττεύει δὲ ἐπὶ τῶν
φοινίκων τοὺς αἰλούρους ἀποδιδράσκουσα· οὐ γάρ
τί που ῥᾳδίως ἐκεῖνο τὸ ζῷον ἀναρριχᾶται καὶ
ἀνέρπει κατὰ τοῦ φοίνικος, ἐκ τῶν ἐξοχῶν τῶν
ἐπὶ τοῦ πρέμνου πολλάκις ἀντικρουόμενόν τε καὶ
ἐκβαλλόμενον.

30. Καὶ ἐκεῖνα δὲ κυνοκεφάλων εἰπεῖν ἐπὶ
στόμα μοι νῦν ἀφίκετο. εἰ λάβοι κυνοκέφαλος[2]
τρωκτὰ σὺν τοῖς ὀστράκοις (ἀμυγδάλας φημὶ καὶ
τὰς τῶν δρυῶν βαλάνους καὶ κάρυα), ἐκλέπει τε
καὶ καθαίρει, καταγνὺς πρότερον πάνυ συνετῶς,
καὶ οἶδεν ὅτι ἄρα τὸ μὲν ἔνδον ἐδώδιμόν ἐστι, τὰ
δὲ ἔξω ἐκβάλλειν χρή. πίνει δὲ οἶνου, καὶ παραθέν-
των[3] ἑφθὰ κρέα καὶ ὀπτὰ ἐμπίπλαται, καὶ τοῖς
μὲν ἡδυσμένοις χαίρει, τοῖς δὲ ἀσπουδάστως
ἑφθοῖς πάνυ ἄχθεται. φείδεται δὲ καὶ ἐσθῆτος
ἐνδὺς αὐτήν, καὶ τὰ ἄλλα δρᾷ ὅσα ἀνωτέρω εἶπον.
θηλῇ δὲ γυναικὸς εἰ προσαγάγοις ἔτι νήπιον,
σπάσει τοῦ γάλακτος ὡς παιδίον.

---

[1] κατεφαίνετο ψευδὲς εἰ καὶ ἐκείνῳ δοκεῖ.

prove it. Yet even he considers that this is very far from the truth, and to me it would seem to be an absolute falsehood. The Ibis is a very hot-blooded creature, at any rate it is an exceedingly voracious and foul feeder if it really does eat snakes and and scorpions. And yet some things it digests without difficulty, while others it easily expels in its excrement. And very rarely would one see a sick Ibis, yet it thrusts its beak down in every place, caring nothing for any filth and treading upon it in the hope of tracking down something even there. And yet when it turns to rest it first of all washes itself and purges. It makes its nest in the top of date-palms in order to escape the cats, for this animal cannot easily clamber and crawl up a date-palm as it is constantly impeded and thrown off by the protuberances on the stem.

30. It occurs to me now to mention the following The Baboon additional facts relating to Baboons. If a Baboon finds some edible object with a shell on it (I mean almonds, acorns, nuts) it strips the shell off and cleans it out, after first breaking it most intelligently, and it knows that the contents are good to eat but that the outside is to be thrown away. And it will drink wine, and if boiled or cooked meat is served to it, it will eat its fill; and it likes well-seasoned food, but food boiled without any care it dislikes. If it wears clothes, it is careful of them; and it does everything else that I have described above. If you put it while still tiny to a woman's breast, it will suck the milk like a baby.

---

² τινα ἕκαστος.   ³ παραθέντος.

AELIAN

31. Τὴν δὲ θέρμουθιν ἀσπίδα, ᾗ ὄνομα ἔθεντο
Αἰγύπτιοι τοῦτο, ἱερὰν εἶναί φασι, καὶ σέβουσιν
αὐτὴν οἱ ἐκεῖθι, καὶ τῆς Ἴσιδος τὰ ἀγάλματα
ἀναδοῦσι ταύτῃ, ὥς τινι διαδήματι βασιλείῳ.
λέγουσι δὲ αὐτὴν ἐπὶ δηλήσει τῶν ἀνθρώπων καὶ
βλάβῃ μὴ φῦναι· ἐκεῖνο δὲ τερατεύονται, φείδεσθαι
μὲν αὐτὴν τῶν ἀγαθῶν, τοὺς δὲ ἀσεβοῦντας
ἀποκτιννύναι. εἰ δὲ ταῦθ᾽ οὕτως ἔχει, τοῦ παντὸς
ἂν ἡ Δίκη τιμήσαιτο τήνδε τὴν ἀσπίδα, τιμωροῦσαν
αὐτῇ καὶ ὁρῶσαν ὀξύτατα. οἱ δὲ ἐπιλέγουσιν ὅτι
ἡ Ἴσις τοῖς τὰ μέγιστα πλημμελήσασιν ἐπιπέμπει
αὐτήν. λέγουσι δὲ αὐτὴν Αἰγύπτιοι μόνην ἀσπίδων
ἀθάνατον εἶναι, ἑκκαίδεκα γένη καὶ διαφορότητας
τοῦδε τοῦ ζῴου καταλέγοντες. ἔν τε τοῖς ἱεροῖς,
ὡς ⟨οἱ⟩ [1] αὐτοί φασι, καθ᾽ ἑκάστην γωνίαν
θαλάμας τινὰς καὶ σηκώδεις ὑποδρομὰς ἐξοι-
κοδομοῦντες εἶτα μέντοι θερμούθεις ἐσῴκιζον,
στέαρ μόσχειον βορὰν παρατιθέντες ἐκ διαστημάτων.

32. Ἄκανθον τὸν ὄρνιν ἐκ τῶν τρεφουσῶν
ἀκανθῶν λαβεῖν τὸ ὄνομα οἱ σοφοὶ τὰ ὀρνίθων
φασί. φθέγγεται δὲ ἄρα ἐμμελὲς [2] καὶ εὔμουσον [3]
δεινῶς. λέγει δὲ Ἀριστοτέλης, ἐὰν τοῦδε τοῦ
ἀκάνθου καὶ μέντοι ⟨καὶ⟩ [4] τοῦ καλουμένου
αἰγίθου [5] τὸ αἷμα ἐς ταὐτὸν ἀγγεῖόν τις ἀναμίξῃ
καὶ κεράσαι θελήσῃ, τὰ δὲ μὴ συνιέναι μηδ᾽
ἑνοῦσθαι ἐς μίαν κρᾶσιν. ἱερόν τε εἶναι τὸν ἄκανθον
τῶν δαιμόνων τῶν κατὰ τὴν ὁδὸν πομπευόντων καὶ
ἀγόντων τοὺς ἀνθρώπους φασί.

---

[1] ⟨οἱ⟩ add. H.   [2] Jac: ἐκμελές.

31. They say that the asp to which the Egyptians have given the name *Thermuthis* is sacred, and the people of the country worship it, and bind it, as though it was a royal headdress, about the statues of Isis. And they deny that it was born to destroy or injure man, but when they maintain that it does not touch virtuous people but kills evildoers they are romancing. If however this is so, then Justice would value this asp beyond all things, for taking vengeance on her behalf and for its piercing sight. Others add that Isis sends it against the worst transgressors. And the Egyptians assert that the Thermuthis alone among asps is immortal, and they reckon sixteen different species and varieties. And in their temples, as they say, they build dens and burrows like shrines in every corner and make homes for the Thermuthes, and at intervals they provide them with calves' fat to eat.

The 'Thermuthis' asp

32. Those who know about birds say that the bird Acanthus [a] derives its name from the acanthus which provides it with food. And its voice is wonderfully harmonious and tuneful. And Aristotle says [*HA* 610 a 6] that if one pours the blood of the Acanthus and of the Aegithus, as it is called, into the same vessel and wants to mix them, the two kinds will not mix and unite into a single compound. They say that the Acanthus is sacred to the gods who escort and conduct men on a journey.

The Linnet

[a] Linnet or perhaps Siskin; identical with the *Acanthis* of Arist. *HA l.c.*; and *Aegithus* has been taken to mean the same, though ' Blue Tit ' is more probable.

---

3 *Ges* : ἄμουσον.   4 ⟨καί⟩ add. H.
5 *Ges* : αἰγιθάλου.

33. Ὅτι σώφρων ἐστὶν ἡ τρυγὼν καὶ πλὴν τοῦ
συννόμου, ὅτῳ καὶ συνῆλθεν ἐξ ἀρχῆς, μὴ ἄν ποτε
ὁμιλήσειεν [1] ἀσπαζομένη λέχος ὀθνεῖόν τε καὶ
ἀλλότριον, ἄνω μοι λέλεκται. ἀκούω δὲ τῶν
ἀκριβούντων τὴν ὑπὲρ τῶν τοιούτων ἱστορίαν καὶ
λευκὰς τρυγόνας φανῆναι πολλάκις. λέγουσι δὲ
αὐτὰς [2] ἱερὰς εἶναι Ἀφροδίτης τε καὶ Δήμητρος,
Μοιρῶν δὲ [3] καὶ Ἐρινύων τὰς ἄλλας. [4]

34. Ὤφθησάν ποτε καὶ χελιδόνες λευκαί, ὡς
Ἀλέξανδρος ὁ Μύνδιός φησιν. ἐν δὲ τῇ Ἀλεξάνδρου τοῦ Πύρρου παιδὸς σκηνῇ χελιδὼν νεοττεύουσα
εἶτα μέντοι ἀτελῆ τὴν πρᾶξιν αὐτῷ ἐφ᾽ ἥνπερ οὖν
ὡρμᾶτο ὑπεσήμηνεν οὐ πάνυ τι οὖσαν ἀγαθήν. καὶ
Ἀντιόχῳ δὲ . . . νεοττεύουσα ἐν . . . [5] αὐτοῦ τὰ
μέλλοντά οἱ ἀπαντήσεσθαι ὑπηνίξατο· ἀνελθὼν
γὰρ ἐς τοὺς Μήδους εἶτα μέντοι οὐκ ἐπανῆλθεν ἐς
τοὺς Σύρους, ἀλλ᾽ ἑαυτὸν κατά τινος ἔωσε κρημ-
νοῦ.[6] ὡρμᾶτο δὲ ἄρα ἐπὶ πρᾶξιν οὐ χρηστὴν καὶ
οὗτος. ἡνίκα δὲ ἐξέλιπε τὴν ἀκρόπολιν τὸ πρότερον Διονύσιος, συνανήχθησάν οἱ καὶ αἱ νεοττεύουσαι
χελιδόνες ἐκεῖθι, καὶ ἐμαντεύοντο τὴν ἐπάνοδον.
τιμᾶται δὲ ἡ χελιδὼν θεοῖς μυχίοις καὶ Ἀφροδίτῃ,
μυχίᾳ μέντοι καὶ ταύτῃ.

[1] ὁμιλήσῃ.
[2] αὐτοί.
[3] καὶ Μοιρῶν.
[4] αὗται.
[5] To fill the lacunae H suggests καὶ ᾽Α. δὲ ⟨χελιδὼν⟩ ν. ἐν ⟨τῇ
σκηνῇ⟩ αὐτοῦ.
[6] τινα . . . κρημνόν.

---

[a] The *Ring*-dove is so described in 3. 44.

33. I have stated earlier on that the Turtle-dove The Turtle-
dove
is continent [a] and does not, from a desire for some
strange and alien bed, consort with any other mate
than the one it originally joined.   And I learn from
those who enquire minutely into such matters that
white Turtle-doves are often to be seen.   These,
they say, are sacred to Aphrodite and Demeter,
while the other kind is sacred to the Fates and the
Erinyes.

34. Even white Swallows have been seen at times, The Swallow
as omen
according to Alexander of Myndus.   A Swallow
made its nest in the tent of Alexander the son
of Pyrrhus [b] and then indicated that, whatever
the somewhat discreditable expedition on which he
was setting out, it would be ineffectual.   And ⟨a
Swallow⟩ which made its nest ⟨in the tent⟩ of Antiochus [c] hinted obscurely at the future in store for
him.   For he went up against the Medes and never
returned to Syria but threw himself over a precipice.
He too therefore embarked on no prosperous affair.
And when Dionysius [d] first left his citadel, the
Swallows which had their nests there withdrew at the
same time and foretold his return.   The Swallow is
held sacred to the Gods of the Household and to
Aphrodite, for she also is one of them.

  [b] Alexander II became King of Epirus, 272 B.C.; he expelled
Antigonus Gonatas from Macedonia, but was in turn expelled
from Macedonia and Epirus by the son of Antigonus.
  [c] Antiochus VII, King of Syria, defeated by the Parthians
(τοὺς Μήδους), 128 B.C.
  [d] Dionysius the elder, tyrant of Syracuse, who lived c. 430–
367 B.C.; he made Ortygia into a fortress where he took refuge
during a revolt which he subsequently quelled.

35. Οἱ πέρδικες ὅταν ἐπῳάζωσι, προβάλλονταί τινας θάμνους καὶ δάση ἕτερα[1] ὑπὲρ τοῦ καὶ δρόσους καὶ ὄμβρους[2] καὶ πᾶν ὅ τι ἂν ᾖ[3] νοτερὸν ἀποστέγειν αὐτῶν. εἰ γὰρ διάβροχά πως γένοιτο, ἐὰν μὴ πάλιν ἡ τεκοῦσα ὑποθάλψῃ[4] αὐτὰ ἐπελθοῦσα ταχέως, γίνεται ἄγονα. ἀθρόα δὲ καὶ πεντεκαίδεκα ᾠὰ ἀποτίκτει. Παφλαγόνων δὲ ἄρα περδίκων διπλῆν ὁρᾶσθαι καρδίαν[5] Θεόφραστός πού φησιν. ἄθυρμα δὲ ὁ πέρδιξ τῆς[6] Διὸς καὶ Λητοῦς ὥς ἐστιν ἄλλοι λέγουσιν.

36. Εἶπον μὲν καὶ ἀνωτέρω περὶ τῶν κύκνων, εἰρήσεται δὲ ἄρα καὶ νῦν ὅσα οὐ πρότερον εἶπον. Ἀριστοτέλης λέγει ἐν τῇ θαλάττῃ τῇ Λιβύων φανῆναί ποτε κύκνων ἀγέλην, καὶ ἀκουσθῆναί τι μέλος αὐτῶν ὡς ἐκ χοροῦ τινος ὁμοφώνου, πάνυ μὲν ἡδύ, γοερόν γε μήν, καὶ οἷον ἐς οἶκτον ἐπικλάσαι τοὺς ἀκούοντας. καί τινας ἐπὶ τῷ μέλει φησὶ φανῆναι τεθνεῶτας αὐτῶν. φίλος δὲ ἦν ἄρα ὁ κύκνος πηγαῖς τε καὶ τενάγεσι καὶ λίμναις καὶ ταῖς ὅσαι πεφύκασιν ὑδάτων σύρροιαί[7] τε καὶ ἀφθονίαι. ἐνταῦθα γοῦν καὶ τὰς ἑαυτοῦ μούσας αὐτὸν φιλοσοφεῖν οἱ σοφοὶ τούτων φασί.

37. Ἡ γλαῦξ ἐπί τινα σπουδὴν ὡρμημένῳ ἀνδρὶ συνοῦσα καὶ ἐπιστᾶσα οὐκ ἀγαθὸν σύμβολόν φασι. μαρτύριον δέ,[8] ὁ Ἠπειρώτης Πύρρος νύκτωρ εὐθὺ τοῦ Ἄργους ᾔει, καὶ αὐτῷ ἐντυγχάνει

---

[1] ἕτερα ἀποκρύπτουσαι.
[2] ὄμβρον or ὄμβρος.
[3] εἴη.
[4] ἐπιθάλψῃ.
[5] *Reiske* : καὶ καρδίαν.
[6] τῷ (*sc.* Apollo) *Oud, cp.* 10. 14, 49 ; 11. 10.

35. When Partridges are sitting on their eggs they The
screen them with branches and other thick leafage Partridge
in order to keep out the dews and showers and every
kind of damp. For if their eggs get soaked, unless
the mother bird is quickly on the spot to warm them
again, they become sterile. Partridges lay as many
as fifteen eggs at a sitting. Theophrastus says
somewhere [fr. 182] that a double heart is to be seen
in the Partridges of Paphlagonia. Other sources
tells us that the Partridge is the darling of the
daughter of Zeus and Leto.[a]

36. I have indeed spoken earlier on about Swans, The Swan
but I shall now relate what I did not mention then.
Aristotle says [HA 615 b 4] that a flock of Swans was
once seen in the Libyan Sea, and that a melody was
heard proceeding from them as from a choir singing
in unison; and very sweet it was, although mournful
and calculated to move the hearers to pity. And
some of the birds, he says, when the music was
ended were seen to have died. It seems that the
Swan is devoted to springs and pools and meres and
to all spots where waters meet and abound. At any
rate that is where those learned in these things say
that the bird meditates its music.

37. If an Owl accompanies and stays beside a man The Owl,
who has set out on some business, they say it is no an evil omen
good omen. Witness the case of Pyrrhus of Epirus
who set out for Argos by night: this bird met him

---

[a] Artemis.

---

[7] ἐπίρροιαι.   [8] δὲ καὶ ἐκεῖνο ἦν.

AELIAN

ἥδε ἡ ὄρνις καθημένῳ μὲν ἐπὶ τοῦ ἵππου, φέροντί
⟨γε⟩[1] μὴν τὸ δόρυ ὀρθόν. εἶτα ἐπὶ τούτου
ἑαυτὴν ἐκάθισεν, οὐδὲ ἀπέστη, δορυφοροῦσα οὐ
χρηστὴν τὴν δορυφορίαν ἡ ὄρνις ἡ προειρημένη
τήνδε. παρῆλθε γοῦν ὁ Πύρρος ἐς τὸ Ἄργος, καὶ
ἀκλεέστατα ἀνθρώπων ἀπέθανεν. ἔνθεν μοι δοκεῖ
καὶ Ὅμηρος εἰδὼς καλῶς τῆς ὄρνιθος τὸ οὐδαμῇ
εὐσύμβολον ἐρῳδιὸν μὲν τὸν ἐκ τῶν ποταμῶν
ἀνεῖναι τοῖς ἀμφὶ τὸν Διομήδην τὴν Ἀθηνᾶν
φάναι, ὅτε ἀπῄεσαν κατασκεψόμενοι τὰ τῶν
Τρώων, μὴ μέντοι τὴν γλαῦκα, εἰ καὶ δοκεῖ φίλη
εἶναι αὐτῇ. ὅτι δὲ ἡ Ἰλιὰς γῆ ἔνδροσός τε καὶ
κατάρρυτός ἐστιν, Ὅμηρος τεκμηριῶσαι ἱκανὸς ἐν
τοῖς πρὸ τῆς τειχομαχίας.

38. Φοβερός ἐστι[2] τῷ καράβῳ ὁ πολύπους.
ἐὰν γοῦν ἁλῶσί ποτε δικτύῳ ἑνί, οἱ κάραβοι
τεθνήκασι παραχρῆμα. Λουσίας δὲ ποταμὸς ἐν
Θουρίοις ὀνομάζεται, ὅσπερ οὖν ἔχει μὲν λευκότα-
τον ὑδάτων αὐτὸς καὶ ῥεῖ διειδέστατα, τίκτει δὲ
ἰχθῦς μέλανας ἰσχυρῶς.

39. Ἄμπελον ὁμώνυμον τῷ φυτῷ πάρδαλίν τινα
οὕτω καλεῖσθαί φασι φύσεως ἰδίας παρὰ τὰς
λοιπὰς μετειληχυῖαν, καὶ οὐρὰν οὐκ ἔχειν ἀκούω
αὐτήν. ἥπερ οὖν εἰ ὀφθείη[3] γυναιξίν, ἐς νόσον
ἐμβάλλει ἀδόκητον αὐτάς.

40. Ἐν τῇ Σκυθίᾳ γῇ γίνονται ὄνοι κερασφόροι,
καὶ στέγει τὰ κέρατα ἐκεῖνα τὸ ὕδωρ τὸ Ἀρκαδικὸν

---

[1] ⟨γε⟩ add. H. [2] ἐστι καὶ ἰσχυρός.
[3] Schn: ὀφθῇ.

334

as he was on horseback and bearing his lance erect.
Whereupon it perched upon the lance and would not
leave him: it was no safe lancer-guard that the bird I
named afforded him. At any rate Pyrrhus reached
Argos and met the most inglorious death in the
world.[a] That is why I think that Homer knowing
full well that the Owl was nowhere a favourable
omen, says [*Il.* 10. 274] that Athena sent a heron
from the rivers to the comrades [b] of Diomedes when
they went off to spy upon the Trojans' camp—a
heron, not an owl, even though it appears to be her
favourite. And that the country about Troy is moist
and well-watered Homer can bear witness in the lines
that precede the Battle at the Wall [*Il.* 12. 18].

38 (i). The Octopus is the terror of the Crayfish.
At any rate if they chance to be caught in one and
the same net, the Crayfish dies on the spot. Octopus and Crayfish

(ii). There is a river at Thurii called the Lusias, of
which the water is of the purest and is absolutely
transparent in its flow, and yet it produces fish of
a deep black hue. Black fish

39. They say that there is a leopard called the
Ampelus, like the plant (*i.e.* grape-vine), and that its
nature is peculiar compared with other leopards; and
I have heard that it has no tail. If it is seen by
women it afflicts them with an unexpected ailment. The 'Ampelus' leopard

40. In Scythia there are Asses with horns, and
these horns hold water from the river of Arcadia The Horned Ass of Scythia

[a] He was struck by a tile thrown by a woman from a house-
top.
[b] Odysseus alone accompanied D.

τὸ καλούμενον τῆς Στυγός· τὰ δὲ ἄλλα ἀγγεῖα
διακόπτει πάντα, κἂν ᾖ σιδήρου πεποιημένα.
τούτων τοι τῶν κεράτων ⟨ἕν⟩ [1] ὑπὸ Σωπάτρου
κομισθῆναί φασιν Ἀλεξάνδρῳ τῷ Μακεδόνι, καὶ
ἐκεῖνον πυνθάνομαι θαυμάσαντα ἐς Δελφοὺς ἀνάθημα ἀναθεῖναι τῷ Πυθίῳ τὸ κέρας, καὶ ὑπογράψαι
ταῦτα

σοὶ τόδ᾽ Ἀλέξανδρος Μακεδὼν κέρας ἄνθετο,
          Παιάν,
κάνθωνος Σκυθικοῦ, χρῆμά τι [2] δαιμόνιον,
ὃ Στυγὸς ἀχράντῳ [3] Λουσηΐδος [4] οὐκ ἐδαμάσθη
ῥεύματι, βάσταξεν δ᾽ ὕδατος ἠνορέην.

Δημήτηρ δὲ ἄρα τὸ ὕδωρ ἀνῆκε τοῦτο πλησίον Φε-
νεοῦ, τὴν δὲ αἰτίαν εἶπον ἀλλαχόθι.

41. Εὐπόλιδι τῷ τῆς κωμῳδίας ποιητῇ δίδωσι
δῶρον Αὐγέας ὁ Ἐλευσίνιος σκύλακα ἰδεῖν ὡραῖον,
Μολοττὸν τὸ γένος, καὶ καλεῖ τοῦτον ὁ Εὔπολις
ὁμωνύμως τῷ δωρησαμένῳ αὐτόν. κολακευθεὶς
οὖν ταῖς τροφαῖς, καὶ ἐκ τῆς συνηθείας ὑπαχθεὶς
τῆς μακροτέρας, ἐφίλει τὸν δεσπότην ὁ Αὐγέας ὁ
κύων. καί ποτε ὁμόδουλος αὐτῷ νεανίας, ὄνομα
Ἐφιάλτης, ὑφαιρεῖται δράματά τινα τοῦ Εὐπόλιδος,
καὶ [5] οὐκ ἔλαθε κλέπτων, ἀλλὰ εἶδεν αὐτὸν ὁ κύων,
καὶ ἐμπεσὼν ἀφειδέστατα δάκνων ἀπέκτεινεν.
χρόνῳ δὲ ὕστερον ἐν Αἰγίνῃ τὸν βίον ὁ Εὔπολις
κατέστρεψε, καὶ ἐτάφη ἐνταῦθα· ὁ δὲ κύων
ὠρυόμενός τε καὶ θρηνῶν τὸν τῶν κυνῶν θρῆνον,

---

[1] ⟨ἕν⟩ add. H.   [2] Kühn: σχῆμα τό.
[3] Brunck: ἀχράντου.   [4] Reinesius: Λουσηθίδος.
[5] ἅ.

known as the Styx; all other vessels the water cuts through, even though they be made of iron. Now one of these horns, they say, was brought by Sopater [a] to Alexander of Macedon, and I learn that he in his admiration set up the horn as a votive offering to the Pythian god at Delphi, with this inscription beneath it:

'In thine honour, O God of Healing, Alexander of Macedon set up this horn from a Scythian ass, a marvellous piece, which was not subdued by the untainted stream of the Lusean [b] Styx but withstood the strength of its water.'

It was Demeter who caused this water to well up in the neighbourhood of Pheneus, and the reason for it I have stated elsewhere.[c]

41. Augeas of Eleusis gave Eupolis, the writer of comedies, a hound of fine appearance, a Molossian, which Eupolis named after the donor. Now Augeas the hound, pampered in its feeding and influenced by long association with its master, came to love him. On one occasion a young fellow-slave of the name of Ephialtes stole some plays of Eupolis, and the theft did not pass unnoticed, for the hound saw him, fell upon him, and, biting him mercilessly, killed him. Some time afterwards Eupolis ended his days in Aegina and was buried there, and the hound, howling and lamenting after the manner of dogs, let

Eupolis and his dog

---

[a] Writer of burlesques, lived at Alexandria.
[b] Lusi, a town in northern Arcadia.
[c] In no surviving work.

εἶτα μέντοι λύπῃ καὶ λιμῷ ἑαυτὸν ἐκτήξας ἀπέθανεν ἐπὶ τῷ τροφεῖ καὶ δεσπότῃ, μισήσας τὸν βίον
ὁ κύων. καὶ ὅ γε τόπος καλεῖται μνήμῃ τοῦ τότε
πάθους Κυνὸς Θρῆνος.

42. Μύρμηκος εἶδος θανατηφόρου φασὶν εἶναί
τι, καὶ λαέρτην ὄνομα ἔχειν τόνδε τὸν μύρμηκα.
καὶ σφῆκας δέ τινας ἐκάλουν λαέρτας. λέγει
ταῦτα Τήλεφος ὁ κριτικὸς ὁ ἐκ τοῦ Μυσίου
Περγάμου.

43. Διὰ τοῦ θέρους τοῦ πυρωδεστάτου τὰ τῶν
Αἰγυπτίων πεδία ὁ Νεῖλος ἐπικλύσας ὄψιν μὲν
αὐτοῖς θαλάττης ἡπλωμένης καὶ λείας δίδωσι, καὶ
ἁλιεύουσι κατὰ τῆς τέως ἀρουμένης Αἰγύπτιοι,
καὶ πλέουσι σκάφαις ἐς ταύτην τὴν ὥραν καὶ τήνδε
τὴν ἐπιδημίαν τοῦ ποταμοῦ πεποιημέναις. εἶτα ὁ
μὲν ὑπονοστεῖ καὶ ἐς τὰ ἑαυτοῦ μέτρα ὑποστρέφει
τὰ ἐκ τῆς φύσεως οἱ[1] νενομισμένα, ἰχθῦς δὲ[2]
χῆροι τοῦ πατρὸς καὶ ἔρημοι τοῦ ἀναπλεύσαντος
ὕδατος ὑπολείπονται,[3] ἐν ἰλύι παχείᾳ[4] τρεφόμενοι
γεωργοῖς δεῖπνον. καὶ βιαιότερον μὲν εἰρήσεται,
ἄμητος δ' οὖν ἰχθύων οὗτός ἐστιν Αἰγύπτιος.

44. Γένη δὲ ἄρα καὶ τεττίγων οὐκ ὀλίγα ἦν,
καὶ αὐτὰ οἱ δεινοὶ ⟨ταῦτα⟩[5] εἰδέναι ἀριθμοῦσι,
καὶ ὀνόματα ᾄδουσιν αὐτῶν. ὁ μὲν γὰρ τεφρὰς ἐκ
τῆς χρόας ὀνομάζεται. ὁ δὲ ἄρα μέμβραξ ὁπόθεν[6]
οὐκ οἶδα, καὶ λακέτας δὲ ἦν ἄρα τέττιγος ὄνομα,
καὶ κερκώπην[7] ἀκοῇ παρεδεξάμην καὶ ἀχέταν

---

[1] οἱ αὐτῆς ἀεί.   [2] δὲ ἐκεῖνοι.

himself pine away through grief and starvation and, disgusted with life, died soon after on the grave of the master that had fed it. And in memory of the sad event the place is named *Hound's Dirge*.

42. They say that there is a species of deadly Ant, and that it goes by the name of *Laertes*. The name has also been applied to certain kinds of Wasp. This is what Telephus the grammarian from Pergamum in Mysia says.

The 'Laertes' ant and wasp

43. All through the hottest summer the Nile in flood gives the fields of Egypt the appearance of a calm stretch of open sea, and over what was till then ploughland there the Egyptians fish and sail in boats manufactured against that season and against this visitation by the river. Later the river retreats and returns to within its naturally proper limits, while the fish bereft of their sire and abandoned by the flood-water are left behind, nurtured in the thick slime to provide a meal for the farmers. This then, though the expression is somewhat violent, is the Egyptian fish-harvest.

Fish in the Nile mud

44. There are, it seems, many species of Cicada, and those who are skilled in these matters enumerate them and report their names. Thus, the *Ashen* one is so called from its colour; whence the *Membrax* got its name I do not know; and *Chirper*, it appears, is the name for a Cicada; and I have heard tell of the *Long-tail* and the *Shriller* and the *Prickly* one. Well,

The Cicada: various kinds

---

³ ὑπαπολείπονται.

⁵ ⟨ταῦτα⟩ add. H.

⁷ W Dindorf: κερκώπαν.

⁴ τραχεία.

⁶ τοῦτο ὁπόθεν.

τινὰ καὶ ἀκάνθιον. ἐγὼ μὲν ⟨οὖν⟩[1] τοσαῦτα
τεττίγων ἀκούσας γένη ·μέμνημαι· ὅτῳ δὲ καὶ
πλείω τῶν προειρημένων ἐς γνῶσιν ἀφίκετο,
λεγέτω ἐκεῖνος.

45. Καὶ ταῦτα μέντοι κυνῶν προσακήκοα. τὰ
σκυλάκια τυφλὰ τίκτεται, καὶ οὐχ ὁρᾷ τῆς μητρῴας
ὠδῖνος προελθόντα. καὶ τρισκαίδεκα ἡμερῶν τῶν
πρώτων κατείληπται τῷ πάθει τῷδε, ὅσων καὶ ἡ
σελήνη οὐ φαίνει νύκτωρ, εἶτα μέντοι ζώων
ὀξυωπέστατος γίνεται ὁ κύων. τιμῶσι δὲ αὐτὸν
Αἰγύπτιοι, καὶ νομόν τινα ἐκάλεσαν ἐξ αὐτοῦ, καὶ
τῆς γε τιμῆς διπλῆν εἶναι τὴν αἰτίαν φασί, τὴν
⟨μὲν⟩[2] λέγουσαν ὅτι ἄρα τῆς Ἴσιδος ζητούσης
πανταχόσε τὸν Ὄσιριν κύνες προηγούμενοι τὰ μὲν
ἐπειρῶντο συνανιχνεύειν αὐτῇ τὸν παῖδα, τὰ δὲ
ἀναστέλλειν τὰ θηρία· ἡ δὲ ἑτέρα, ὅτι ἄρα ἤδη
μὲν ἀνατέλλει τὸ ἄστρον ὁ κύων, ὃν Ὠρίωνος ἡ
φήμη γενέσθαι λέγει, συνανίσχει δὲ αὐτῷ τρόπον
τινὰ καὶ ὁ Νεῖλος ἐπιὼν ἐς τὴν ἀρδείαν τῆς γῆς
τῆς Αἰγυπτίας, καὶ ἀναχεῖται περὶ τὰς ἀρούρας.
ὡς οὖν ἄγοντα τόδε τὸ γόνιμον ὕδωρ καὶ παρακα-
λοῦντα τιμῶσιν Αἰγύπτιοι.

46. Ὀξύρυγχος οὕτως[3] ἰχθὺς κέκληται, καὶ
ἔοικεν ἐκ τοῦ προσώπου λαβεῖν τὸ ὄνομα καὶ τοῦ
σχήματος τοῦ κατ' αὐτό. τρέφει δὲ ἄρα τὸν

---

[1] ⟨οὖν⟩ add. H.   [2] ⟨μέν⟩ add. Schn.
[3] Schn : οὗτος.

---

[a] Cynopolis, close to Oxyrhynchus.

these are all the kinds of Cicada of which I remember having heard the names, but if anyone has got to know more than those that I have mentioned, he must tell them.

45. Here are further facts relating to Dogs which I have heard. Puppies are born blind, and when they emerge from their dam's womb they cannot see. For the first fortnight they are afflicted in this way, that is for as many nights as the moon does not appear, but after that the Dog has the sharpest sight of any animal. And it is held in honour by the Egyptians, for they have named a district *a* after it, and they assert that the reason for this is twofold : first, when Isis was seeking everywhere for Osiris,*b* Dogs led the way and tried both to help her to trace his son and also to keep off the wild beasts. And the second reason is this, that at the same time that the Dog-star rises (the story goes that it was the dog of Orion), the Nile also in a sense rises, coming up to water the land of Egypt, and pours over the plough-lands. And so the Egyptians pay honour to the Dog for bringing and summoning this fertilising water.

The Dog honoured in Egypt

46. There is a fish that goes by the name of *Oxyrhynchus,c* and it appears to derive its name from its face and from the shape of it. The Nile breeds

The 'Oxy-rhynchus' fish

---

b Osiris was the husband of Isis ; he was murdered by his brother Typhon. Plut. *de Is. et Os.* 356F tells how Osiris, mistaking Nephthys for her sister Isis, begat upon her Anubis. Isis sought for ' his son ' to help in her search for the body of Osiris.

c That is, ' sharp-snouted.'

AELIAN

προειρημένον ὁ Νεῖλος, καὶ μέντοι καὶ ἐξ αὐτοῦ
κέκληται νομὸς ἔνθα δήπου καὶ τιμὰς ὁ ἰχθὺς ἔχει
ὁ αὐτός. ἀγκίστρῳ δὲ θηραθέντα ἰχθὺν οὐκ ἄν
ποτε φάγοιεν οἶδε οἱ ἄνδρες, δεδιότες μή ποτε ἄρα
αὐτῷ περιπαρεὶς ἔτυχεν ὁ παρὰ σφίσιν ἱερὸς καὶ
θαυμαστὸς ἰχθὺς ὃν εἶπον. ἐὰν δὲ καὶ δικτύοις
ἁλῶσί ποτε ἰχθύες, ἀνιχνεύουσι ταῦτα, μὴ παρα-
λαθὼν ἐνέπεσεν ἐκεῖνος, καὶ προτιμῶσιν ἀθηρίαν
ἢ ἁλόντος ἐκείνου τὴν μάλιστα εὐθηρίαν. λέγουσι
δὲ αὐτὸν οἱ περίχωροι ἐκ τῶν Ὀσίριδος τραυμάτων
γεγονέναι· νοοῦσι δὲ τὸν Ὄσιριν ἄρα τὸν αὐτὸν
τῷ Νείλῳ εἶναι.

47. Ὁ δὲ ἰχνεύμων ὁ αὐτὸς ἄρα καὶ ἄρρην καὶ
θῆλυς ἦν, μετειληχὼς καὶ τῆσδε τῆς φύσεως καὶ
τῆσδε, καὶ σπείρειν τε καὶ τίκτειν τῷ αὐτῷ[1] ἡ
φύσις δέδωκεν. ἀποκρίνονται δὲ ἐς τὸ ἀτιμότερον
γένος οἱ ἡττηθέντες κατὰ μάχην[2]· οἱ γὰρ κρατή-
σαντες ἀναβαίνουσι τοὺς ἡττημένους καὶ ἐς αὐτοὺς
σπείρουσιν. οἱ δὲ ἆθλον τῆς ἥττης φέρονται
ὠδῖνάς τε ὑπομεῖναι καὶ ἀντὶ πατέρων[3] γενέσθαι
μητέρες. τοῖς γε μὴν πολεμιωτάτοις ἀνθρώπῳ
ζῴοις, ἀσπίδι καὶ κροκοδίλῳ, ἔχθιστον ὁ ἰχνεύμων,
καὶ τόν γε πόλεμον αὐτῶν ἀνωτέρω εἶπον. λέγονται δὲ οἱ ἰχνεύμονες ἱεροὶ εἶναι Λητοῦς καὶ
Εἰλειθυιῶν· σέβουσι δὲ αὐτοὺς Ἡρακλεοπολῖται,[4]
ὥς φασιν.

[1] τοῖς αὐτοῖς.
[2] μετὰ τὴν μ.
[3] καὶ ὑπὲρ τοῦ τέως ἀντὶ π.
[4] αὐτοὺς Αἰγύπτιοι Ἡ.

the aforesaid fish; and after it too [a] a district is named, where, I believe, this same fish is held in veneration. Should the inhabitants catch a fish on a hook they will never eat it for fear lest the aforesaid fish, which they regard as sacred and to be worshipped, may have chanced to impale itself on the hook. And whenever fish are netted, they search the nets in case this famous fish has fallen in without their noticing it. And they would rather catch nothing at all than have the largest catch which included this fish. And the people who live round about maintain that it was born from the wounds of Osiris. They identify Osiris with the Nile.

47. The Ichneumon is both male and female in the same individual, partaking of both sexes, and Nature has enabled each single same animal both to procreate and to give birth. Those that are worsted in a fight are degraded into the less honoured class, for the victors mount the vanquished and inseminate them. And the latter carry with them as prize of their defeat endurance of birth-pangs and motherhood for fatherhood. The Ichneumon is most hateful to man's deadliest enemies, the asp and the crocodile: I have earlier on described how they war with each other.[b] Ichneumons are said to be sacred to Leto and the Goddesses of Birth, and the people of Heracleopolis worship them, so they say.

The Ichneumon

[a] That is, like the Dog in ch. 45. Oxyrhynchus lay on the W side of the Nile between lat. 28 and 29; Cynopolis lay on the opposite bank.
[b] See 3. 22 and 8. 25.

48. Λυκάονι τῷ βασιλεῖ τῆς Ἠμαθίας γίνεται
παῖς, ὄνομα Μακεδών, ἐξ οὗ καὶ ἡ χώρα κέκληται
μετὰ ταῦτα τὸ ἀρχαῖον ὄνομα οὐκέτι φυλάξασα.
τούτῳ δὲ ἄρα παῖς ἀνδρεῖος ἦν καὶ κάλλει διαπρε-
πής, Πίνδος ὄνομα· ἦσαν δέ οἱ καὶ ἄλλοι παῖδες,
ἀνόητοι δὲ οὗτοι τὴν ψυχὴν καὶ τὸ σῶμα οὐ
ῥωμαλέοι, οἵπερ οὖν χρόνῳ ὕστερον βασκήναντες
τἀδελφῷ τῆς τε ἀρετῆς καὶ τῆς ἄλλης εὐδαιμονίας
ἐκεῖνον μὲν διέφθειραν, ἑαυτοὺς δὲ ἐπαπώλεσαν
διδόντες δίκην κατὰ τὴν Δίκην. αἰσθόμενος γὰρ
ὅδε ὁ Πίνδος τὴν ἐκ τῶν ἀδελφῶν ἐς ἑαυτὸν
ἐπιβουλήν, τὴν πατρῴαν ἀρχὴν ἀπέλιπεν, ᾤκει δὲ
ἐν χώρῳ, καὶ ἦν τῇ τε ἄλλῃ ῥωμαλέος, καὶ οὖν
καὶ κυνηγετικὸς ἦν. καί ποτε ἐθήρα νεβρούς. καὶ
οἱ μὲν ἔθεον ᾗ ποδῶν εἶχον, ὁ δὲ μεταδιώκων εἶτα
μέντοι τὸν ἵππον ἀνὰ κράτος ἤλαυνε, καὶ τῶν μὲν
συνθηρατῶν ἀποσπᾷ πολύ, οἱ νεβροὶ δὲ ἐς φάραγγα
κοίλην τε καὶ βαθυτάτην ἐσδύντες καὶ ἑαυτοὺς
σώσαντες ἀπὸ τῆς ὄψεως τοῦ διώκοντος ἠφανίσθη-
σαν. οὐκοῦν ἀποπηδήσας τοῦ ἵππου ὁ Πίνδος
ἐκεῖνον μὲν τοῦ ῥυμοῦ πρός τι τῶν παρεστώτων
ἐξῆψε δένδρων, αὐτὸς δὲ οἷος ἦν τὴν φάραγγα
διερευνᾶν καὶ μαστεύειν τοὺς προειρημένους.
εἶτα ἀκούει φωνῆς,[1] καὶ ἔλεγεν αὕτη 'τῶν νεβρῶν
μὴ ἅψαι'. οὐκοῦν ἐπεὶ πολλὰ περιβλέψας οὐδὲν
ἐθεάσατο, ἔδεισε τὸ φώνημα ὡς ἔκ τινος αἰτίας
κρείττονος προσπεσόν. καὶ τότε μὲν ᾤχετο ἀπιὼν
καὶ τὸν ἵππον ἀπάγων, τῇ δὲ ὑστεραίᾳ μόνος
ἀφικνεῖται, καὶ πάρεισι μὲν ἐς τὴν φάραγγα οὐδαμῶς
μνήμῃ τε τῆς φωνῆς τῆς προσπεσούσης αὐτοῦ ταῖς
ἀκοαῖς καὶ δέει. στρέφοντι δ' ἐν ἑαυτῷ βουλὴν
καὶ διαποροῦντι τίς ἦν ἄρα ὁ τῇ προτεραίᾳ ἀναστεί-

48. To Lycaon King of Emathia was born a son of the name of Macedon, after whom the country has thenceforward been called, no longer preserving its ancient name. Now his son was a vigorous youth of remarkable beauty and his name was Pindus. Other sons he had besides, but they were foolish in spirit and not robust of body, and so in course of time growing jealous of the valour and the general good fortune of their brother, they slew him; but it was to their own undoing, and they paid the penalty as was right. For Pindus realising that his brothers were plotting against him, left his father's kingdom and lived in the country. And besides being vigorous in other respects he was also a great hunter. And on one occasion he was pursuing some fawns, and they fled as fast as their legs could carry them, while he rode at full speed in pursuit, leaving his fellow huntsmen far behind. But the fawns entered a hollow and very deep ravine, escaped out of their pursuer's sight, and disappeared. Accordingly Pindus leapt from his horse and fastened it by the rein to one of the trees hard by and was just about to investigate the ravine and to search for the fawns, when he heard a voice which said 'Touch not the fawns!' And so after looking all round and seeing nothing, he was in fear of the voice,[1] thinking that it proceeded from some mightier agency. And then he departed taking his horse with him. But on the following day he came unaccompanied, but remembering the voice that had fallen on his ears and being afraid, he did not enter the ravine. And while he was taking council with himself and was perplexed

---

[1] βοῆς.

λας αὐτὸν τῆς ἐπὶ τὴν ἄγραν ὁρμῆς, καὶ περιβλέ-
ποντι οἷα εἰκὸς ἢ νομέας ὀρείους [1] ἢ θηρατὰς
ἑτέρους, μέγα τι χρῆμα ὁρᾶται δράκων τὸ μὲν
πλεῖστον τοῦ σώματος ἐπισύρων, ὀλίγην [2] δὲ
ἀνατείνας ὡς πρὸς αὐτὸ [3] τὴν δέρην [4] (καὶ ἦν ἡ δέρη
σὺν τῇ κεφαλῇ ὑπὲρ τέλειον ἄνδρα τὸ μέγεθος),
εἶτα ὀφθεὶς ἐξέπληξεν. οὐ μὴν ἐς φυγὴν ὁ Πίνδος
ἐξώρμησεν, ἀλλ᾽ ἀθροίσας ἑαυτὸν σοφίᾳ περιέρχεται
τὸν θῆρα· τῶν γὰρ ὀρνίθων ὧν θηράσας [5] ἔτυχε
προσήγαγε, καὶ προύτεινέν οἱ ξένια ταῦτα καὶ
ἑαυτοῦ ζωάγρια. ὁ δέ, οἷα δήπου τοῖς δώροις μει-
λιχθεὶς καὶ καταγοητευθεὶς ὡς ἂν εἴποις, ᾤχετο
ἀπιών. ταῦτα τὸν νεανίαν ἦσε, καὶ τὸ ἐντεῦθεν
ἀπέφερε τῷ δράκοντι μισθὸν σωτηρίας, ὡς ἄνθρωπος ἀγαθός, ὧν εἶχε θηραμάτων ἀπαρχὰς κεχαρι-
σμένας ἢ τῆς ἄγρας τῆς ὀρείου ἢ τῆς πτηνῆς. καὶ
τῷ μὲν τὰ τῆς προειρημένης δωροφορίας ἐνεργό-
τατα ἦν, ὑπήρχετο δὲ καὶ τὰ ἐκ τοῦ δαίμονος
εὐθενεῖσθαι τῷ Πίνδῳ, καὶ ὁσημέραι χωρεῖν ἐς
τὸ σοβαρώτερον· θηρῶντι γὰρ ἀπήντων εὐθηρίαι,
ὅσαι τε τῶν ἐν ταῖς ὕλαις ζῴων, τῶν τε
ὀρνίθων ὅσαι. ἦν οὖν αὐτῷ καὶ περιβολή, καὶ
διεῖρπε μέντοι καὶ κλέος ὡς ὁμόσε τοῖς θηρίοις
ἰόντος καὶ ἀτρέπτως αἱροῦντος αὐτά· ἦν δὲ καὶ
ἰδεῖν μέγας καὶ οἷος ἐκπλῆξαι τῷ τε ὄγκῳ τοῦ
σώματος καὶ τῇ εὐεξίᾳ προσέτι, τῇ δὲ ὥρᾳ τὸ
θῆλυ πᾶν ἀναφλέγων [6] καὶ ἐς ἑαυτὸν ἐξάπτων ἦν
δῆλος. καὶ ἐφοίτων ἐπὶ θύρας τὰς ἐκείνου οἷα
δήπου βεβακχευμέναι ὅσον μὲν τῶν [7] γυναικῶν ἦν

[1] ὀρείους ἢ αὔλιον.
[2] ὀλίγον.
[3] ἑαυτόν.
[4] δέρην αὐτήν.
[5] Jac : θύσας.
[6] ἀνέφλεγεν.

as to who it was that the day before had checked his pursuit of the quarry, and while he was looking about, as was natural, for shepherds on the hills or other hunters, he beheld a monstrous serpent trailing most of its body behind but with the neck, which was small compared with the rest of the body, held aloft. (Neck and head together exceeded in size that of a full-grown man.) The sight filled him with terror. Pindus however did not take to flight, but pulled himself together and by his adroitness tricked the serpent, for he brought forward the birds which he happened to have caught and offered them as friendly gifts and as a ransom for his own life. And the serpent mollified presumably and bewitched, as you might say, by the gifts, departed. This pleased the youth and thereafter, being a good man, he used to bring payment for the saving of his life to the serpent, giving freely the firstfruits of the chase, whether beast or bird from the hills. And this bestowal of gifts had the most fruitful results for Pindus, and his fortune began to prosper and grew every day more impressive, for whether it was[7] beasts of the forest or whether it was birds, with all of them his hunting was successful. Accordingly he enjoyed abundance; moreover his fame spread abroad, of how he fearlessly attacked and captured wild beasts. His figure was tall and such as to cause astonishment by reason of the bulk of his body and of his splendid condition also. And it was clear that his beauty inflamed and kindled the hearts of all women with desire for him: all who were widowed would throng his doors like people crazed, while

---

[7] μὲν τῶν] μέντοι.

χῆρον, αἵ γε μὴν συνοικοῦσαι τοῖς γεγαμηκόσι
φρουρούμεναι μὲν τῷ νόμῳ, τῷ κλέει δὲ τοῦ
κάλλους τοῦ κατὰ τὸν Πίνδον δεδουλωμέναι
προὐτίμων συνοικεῖν ἐκείνῳ ἢ θεαὶ γεγονέναι.
καὶ ἐτεθήπεσάν γε αὐτὸν οἱ πολλοὶ τῶν ἀνδρῶν
καὶ ἐφίλουν, ἐχθροὶ δὲ οἱ ἀδελφοὶ μόνοι ἦσαν. καὶ
ποτε θηρῶντα ἐλλοχῶσι μόνον, καὶ ποταμοῦ γε ἦν
ἡ θήρα πλησίον, εἶτα ἐρήμῳ συμμάχων οἱ τρεῖς
ἐπιστάντες ἔπαιον τοῖς ξίφεσιν αὐτόν, ὁ δὲ ἐβόα.
ἀκούει ταῦτα ὁ ἑταῖρος αὐτοῦ δράκων· ὀξυήκοον
δὲ καὶ ὀξυωπέστατον τὸ ζῷόν ἐστιν. οὐκοῦν
πρόεισι τῆς ἑαυτοῦ κοίτης, καὶ τοῖς ἀνοσίοις περι-
πλακεὶς ἀπέκτεινεν αὐτοὺς ἐς πνῖγμα ἄγχων· αὐτὸς δὲ οὐ κατέλυσε τὴν φυλακήν, ἔστε [1] οἱ προσήκοντες τῷ νεανίᾳ ποθοῦντες αὐτὸν ἀφίκοντο, καὶ
ἐνέτυχον κειμένῳ. καὶ ὠλοφύροντο μέν, προσελ-
θεῖν δὲ ἐπὶ κηδεύσει τοῦ νεκροῦ οὐκ ἐτόλμων
δέει τοῦ φρουροῦ. ὁ δὲ συνεὶς φύσει τινὶ
ἀπορρήτῳ ὅτι ἀναστέλλει αὐτούς, ᾤχετο ἀπιὼν
κατὰ πολλὴν τὴν σχολήν, ἀπολιπὼν ἐκεῖνον τῆς
τελευταίας ἐκ τῶν προσηκόντων χάριτος τυχεῖν.
οὐκοῦν καὶ ἐτάφη μεγαλοπρεπῶς καὶ ὁ γείτων τῷ
φόνῳ ποταμὸς ἐκλήθη Πίνδος ἐκ τοῦ νεκροῦ καὶ
τοῦ κατ᾽ αὐτὸν τάφου. ἴδιον μὲν δὴ τῶν ζῴων
ἐκτίνειν χάριτας τοῖς εὐεργέταις, ᾗπερ οὖν καὶ
ἄνω λέλεκται, καὶ νῦν δὲ οὐχ ἥκιστα.

49. Ἰδίᾳ δὲ καὶ ἐν τῇ Κλάρῳ τὸν Διὸς καὶ
Λητοῦς τιμῶσιν οἱ Κλάριοι καὶ πᾶν τὸ Ἑλληνικόν.
οὐκοῦν ὁ ἐνταῦθα χῶρος τοῖς ἰοβόλοις θηρίοις
ἀστιβής τε ἅμα καὶ ἔχθιστός ἐστι, τοῦτο μὲν καὶ

[1] ἐστ᾽ ἄν.

those who were married to husbands and whom
custom confined indoors were enslaved by the fame
of Pindus's beauty and would rather have been his
wife than become goddesses. As to the men, most
of them admired and loved him; only his brothers
hated him. And once when he was hunting by
himself they lay in wait for him, and the hunting-
ground was near a river, and the three set upon him
as he had none to help him and smote him with their
swords. Whereupon he cried aloud. His cry was
heard by his companion the serpent. (This creature
is keen of hearing and has very sharp eyes.) And so
it emerged from its lair and coiling round the miscreants killed them by choking them to death. But
the snake continued to mount guard until the youth's
relations, who were anxious for him, arrived and
found him lying dead. But though they made lamen-
tation for him they did not dare to attend to the dead
body for fear of its guardian. The serpent however
realising by some mysterious instinct that it was
keeping them away, departed at a very leisurely pace,
leaving Pindus to receive the last kind service from
his kin. And so he was buried with great pomp,
and the river which was close by the scene of murder
was called Pindus after the dead man and the tomb
over him. It is then a characteristic of animals to
render thanks to their benefactors, as I have stated
earlier on, and especially on this occasion.

49. Particularly in Clarus do the inhabitants and
all Greeks pay honour to the son of Zeus and Leto.[a]
And so the land there is untrodden by poisonous
creatures and is also highly obnoxious to them.

Clarus free
from
noxious
creatures

[a] Apollo.

AELIAN

τῇ τοῦ δαίμονος βουλῇ, πάντως δὲ καὶ πεφρικότων
τῶν θηρίων αὐτὸν ἅτε καὶ αὐτὸν σῴζειν [1] εἰδότα
καὶ μέντοι καὶ τὸν σωτῆρα καὶ νόσων ἀντίπαλον
Ἀσκληπιὸν [2] φύσαντα. ἀλλὰ [3] καὶ Νίκανδρος οἷς
λέγω μάρτυς. λέγει δὲ Νίκανδρος

> οὐκ ἔχις οὐδὲ φάλαγγες ἀπεχθέες οὐδὲ βαθυπλὴξ
>   ἄλσεσιν ἐνζώει [4] σκορπίος ἐν Κλαρίοις,
> Φοῖβος ἐπεί ῥ᾽ αὐλῶνα βαθὺν μελίῃσι [5] καλύψας
>   ποιηρὸν δάπεδον θῆκεν ἑκὰς δακετῶν.

50. Ἀκούω λεγόντων ἐν Ἔρυκι, ἔνθα δήπου καὶ
ὁ τῆς Ἀφροδίτης ὑμνούμενος νεώς ἐστιν, οὗπερ
οὖν καὶ ἀνωτέρω μνήμην [6] ἐποιησάμην τῶν ἐκεῖθι
περιστερῶν εἰπὼν τὰ ἴδια, εἶναι μὲν καὶ χρυσὸν
πολὺν καὶ ἄργυρον πάμπλειστον καὶ ὅρμους καὶ
δακτυλίους μέγα τιμίους, ἄσυλα δὲ εἶναι καὶ
ἄψαυστα ταῦτα τῷ τῆς θεοῦ δέει, καὶ ἀεὶ τοὺς
ἄνω τοῦ χρόνου δι᾽ αἰδοῦς ἄγειν θαυμαστῆς καὶ
τὴν δαίμονα τὴν προειρημένην καὶ τὰ κειμήλια
ἐκείνης. Ἀμίλκαν δὲ πυνθάνομαι τὸν Λίβυν
συλήσαντα αὐτὰ καὶ ἐργασάμενον χρυσίον καὶ
ἀργύριον εἶτα μέντοι διανεῖμαι τῇ στρατιᾷ πονη-
ρὰν νομήν, καὶ ὑπὲρ τούτων αὐτὸν ἀλγεινότατα καὶ
βαρύτατα αἰκισθέντα ἅμα καὶ κολασθέντα κρεμα-
σθῆναι, πᾶν δὲ ὅσον αὐτῷ τῆς πράξεως καὶ τῆς
ἐκδίκου θεοσυλίας ἐγένετο μέτοχον βιαίοις τε καὶ
δεινοῖς χρήσασθαι θανάτοις, ἥ τε πατρὶς αὐτοῦ
εὐδαιμονιζομένη τέως [7] καὶ ζηλωτὴ δοκοῦσα ἐν
ὀλίγαις τῶν ἱερῶν χρημάτων ἐσκομισθέντων ἡ δὲ
δούλη ἦν. καὶ ταῦτα μὲν πρὸς τὴν παροῦσάν μοι

---

[1] αὐτὸν σῴζειν] σῴζειν θεόν.   [2] τὸν Ἀ.

350

The god wills it so, and the creatures in any case dread him, since the god can not only save life but is also the begetter of Asclepius, man's saviour and champion against diseases. Moreover Nicander also bears witness to what I say, and his words are:

'No viper, nor harmful spiders, nor deep-wounding scorpion dwell in the groves of Clarus, for Apollo veiled its deep grotto with ash-trees and purged its grassy floor of noxious creatures' [Nic. *fr.* 31].

50. I have heard it said that in Eryx, where of course the famous temple of Aphrodite is (the pigeons there and their peculiarities I mentioned earlier on),[a] there is a store of gold, an immense store of silver, necklaces, and finger-rings of great price; and that dread of the goddess renders them safe from robbers and untouched; and that men in ancient times always regarded the aforesaid goddess and her treasures with veneration and awe. But I learn that Hamilcar the Carthaginian [b] looted these objects, melted down the silver and gold, and then distributed an infamous largesse to his troops. And for these deeds he suffered the most painful and grievous torments and was punished with crucifixion, while all his accomplices and partners in that unholy sacrilege died violent and terrible deaths. And his native land which till then was so prosperous and

The worship of Aphrodite at Eryx

---

[a] See 4. 2.
[b] Defeated at Himera and killed, 480 B.C.; see Hdt. 7. 165-7.

---

³ ἀλλά γε.   ⁴ *Bernhardy* : ἐν ζῴοις.
⁵ *OSchn* : μελίαισι.   ⁶ τὴν μνήμην.
⁷ αὐτοῦ εὖ. τέως] ὡς εὖ. τέως αὐτῷ MSS, ὡς *del. Jac.*

χρείαν σεμνὰ ὄντα ὅμως οὐχ ὁρᾷ πω, τὸ δὲ τοῖς
λόγοις ⟨τοῖσδε⟩ ¹ συμμελὲς ² τοῦτο εἰρήσεται.
ἀνὰ πᾶν ἔτος καὶ ἡμέραν πᾶσαν θύουσι τῇ ³ θεῷ
καὶ οἱ ἐπιχώριοι καὶ οἱ ξένοι. καὶ ὁ μὲν βωμὸς
ὑπὸ τῷ οὐρανῷ ὁ μέγιστός ἐστι, πολλῶν δὲ ἐπ᾽
αὐτοῦ ⁴ καθαγιζομένων θυμάτων ὁ δὲ πανημέριος
καὶ ἐς νύκτα ἐξάπτεται. ἕως ⁵ δὲ ⁶ ὑπολάμπει,
καὶ ἐκεῖνος οὐκ ἀνθρακιάν, οὐ σποδόν, οὐχ ἡμι-
καύτων ⁷ τρύφη δαλῶν ⁸ ὑποφαίνει, δρόσου δὲ
ἀνάπλεώς ἐστι καὶ πόας νεαρᾶς, ἥπερ οὖν ἀναφύεται
ὅσαι νύκτες. τά γε μὴν ἱερεῖα ἑκάστης ἀγέλης
αὐτόματα φοιτᾷ καὶ τῷ βωμῷ παρέστηκεν,
ἄγει δὲ ἄρα αὐτὰ πρώτη μὲν ἡ θεός, εἶτα ἡ
δύναμίς τε καὶ ἡ τοῦ θύοντος βούλησις. εἰ γοῦν
ἐθέλοις θῦσαι οἶν, ἰδού σοι τῷ βωμῷ παρέστηκεν
οἷς, καὶ δεῖ χέρνιβα κατάρξασθαι ⁹· εἰ δὲ εἴης τῶν
ἁδροτέρων καὶ ἐθέλοις ⟨θῦσαι⟩ ¹⁰ βοῦν θήλειαν ἢ ¹¹
καὶ ἔτι πλείους, εἶτα ὑπὲρ τῆς τιμῆς οὔτε ⟨σὲ⟩ ¹²
ὁ νομεὺς ἐπιτιμῶν ζημιώσει οὔτε σὺ λυπήσεις
ἐκεῖνον· τὸ γὰρ δίκαιον τῆς πράσεως ἡ θεὸς
ἐφορᾷ. καὶ εὖ καταθεὶς ἵλεων ἕξεις αὐτήν· εἰ δὲ
ἐθέλοις τοῦ δέοντος πρίασθαι εὐτελέστερον, ¹³ σὺ
μὲν κατέθηκας τὸ ἀργύριον ἄλλως, τὸ δὲ ζῷον
ἀπέρχεται, καὶ θῦσαι οὐκ ἔχεις. ἴδιον μὲν δὴ
Ἐρυκίνων ζῴων εἰρήσθω καὶ τοῦτο ἡμῖν ἐπὶ τοῖς
ἄνω.

---

¹ ⟨τοῖσδε⟩ add. H.   ² συγγενές.
³ Ges : τῷ.   ⁴ Jac : αὐτόν.
⁵ Jac : ὡς.   ⁶ δὲ καί.
⁷ ἡμικαύστων.   ⁸ Reiske : δάδων.
⁹ χέρνιβος καὶ κατάρξασθαι εἴτε αἶγα εἴτε ἔριφον.
¹⁰ ⟨θῦσαι⟩ add H.   ¹¹ Ges : εἰ.
¹² ⟨σὲ⟩ add. H.   ¹³ Ges : λυσιτελέστερον.

which was reputed enviable above most lands, after these sacred objects had been imported, was reduced to slavery.[a] But impressive though these facts are they have no bearing on my present object, but what is relevant to this discourse shall now be told.

On every day throughout the whole year the people of Eryx and strangers too sacrifice to the goddess. And the largest of the altars is in the open air, and upon it many sacrifices are offered, and all day long and into the night the fire is kept burning. The dawn begins to brighten, and still the altar shows no trace of embers, no ashes, no fragments of half-burnt logs, but is covered with dew and fresh grass which comes up again every night. And the sacrificial victims from every herd come up and stand beside the altar of their own accord; it is the goddess in the first place that leads them on, and in the second place it is the ability to pay, and the wish, on the part of the sacrificer. At any rate should you desire to sacrifice a sheep, lo and behold, there is a sheep standing at the altar, and you must begin the ceremonial washing. But if you are a man of substance and wish to sacrifice one cow or even more than one, then the herdsman will not mulct you by charging too much, nor will you disappoint him,[b] for the goddess sees that the sale-prices are just, and if you pay fairly you will win her favour. If however you want to buy at a cheaper rate than is proper, you will pay down your money in vain—the animal departs and you are unable to sacrifice.

So much then for this peculiarity of animals at Eryx in addition to those which I have mentioned earlier on.

[a] This is entirely false.   [b] By attempting to bargain.

353

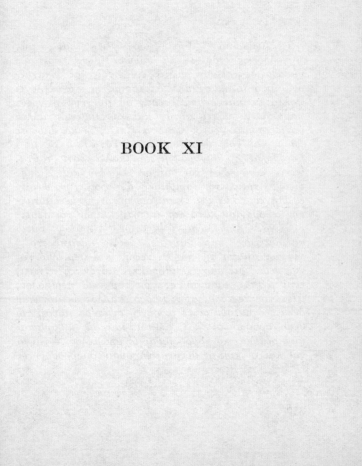

BOOK XI

ΙΑ

1. Ἀνθρώπων Ὑπερβορέων γένος καὶ τιμὰς Ἀπόλλωνος τὰς ἐκεῖθι ᾄδουσι μὲν[1] ποιηταί, ὑμνοῦσι δὲ καὶ συγγραφεῖς, ἐν δὲ τοῖς καὶ Ἑκαταῖος, οὐχ ὁ Μιλήσιος, ἀλλ' ὁ Ἀβδηρίτης. ἃ δὲ λέγει πολλά τε καὶ σεμνὰ ἕτερα, οὔ μοι νῦν ἡ χρεία παρακαλεῖν δοκεῖ αὐτά, καὶ οὖν καὶ ἐς ἄλλον ὑπερθήσομαι χρόνον ἕκαστα εἰπεῖν, ἡνίκα ἐμοί τε ἥδιον καὶ τοῖς ἀκούουσι λῷον ἔσται· ἃ δέ με[2] μόνα ᾔδε ἡ συγγραφὴ παρακαλεῖ ἔστι ταῦτα. ἱερεῖς εἰσι τῷδε τῷ δαίμονι Βορέου καὶ Χιόνης υἱεῖς,[3] τρεῖς τὸν ἀριθμόν, ἀδελφοὶ τὴν φύσιν, ἑξαπήχεις τὸ μῆκος. ὅταν οὖν οὗτοι τὴν νενομισμένην ἱερουργίαν κατὰ τὸν συνήθη καιρὸν τῷ προειρημένῳ[4] ἐπιτελῶσιν, ἐκ τῶν Ῥιπαίων οὕτω καλουμένων παρ' αὐτοῖς ὁρῶν καταπέτεται[5] κύκνων ἄμαχα τῷ πλήθει νέφη,[6] καὶ περιελθόντες τὸν νεὼν καὶ οἱονεὶ καθήραντες αὐτὸν τῇ πτήσει, εἶτα μέντοι κατίασιν ἐς τὸν τοῦ νεὼ περίβολον, μέγιστόν τε τὸ μέγεθος καὶ τὸ κάλλος ὡραιότατον ὄντα. ὅταν οὖν οἵ τε ᾠδοὶ τῇ σφετέρᾳ μούσῃ τῷ θεῷ προσᾴδωσι καὶ μέντοι καὶ οἱ κιθαρισταὶ συγκρέκωσι τῷ χορῷ παναρμόνιον μέλος, ἐνταῦθά τοι καὶ οἱ κύκνοι συναναμέλπουσιν ὁμορροθοῦντες

---

[1] μὲν καὶ.
[2] μοι.
[3] υἱέες.
[4] τὸν προειρημένον.

356

BOOK XI

1. The race of the Hyperboreans and the honours Swans and the worship of Apollo there paid to Apollo are sung of by poets and are celebrated by historians, among whom is Hecataeus, not of Miletus but of Abdera. The many other matters of importance which he narrates I think there is no need for me to bring in now, and in fact I shall postpone the full recital to some other occasion, when it will be pleasanter for me and more convenient for my hearers. The only facts which this narrative invites me to relate are as follows. This god has as priests the sons of Boreas and Chione, three [a] in number, brothers by birth, and six cubits in height. So when at the customary time they perform the established ritual of the aforesaid god there swoop down from what are called the Rhipaean mountains [b] Swans in clouds, past numbering, and after they have circled round the temple as though they were purifying it by their flight, they descend into the precinct of the temple, an area of immense size and of surpassing beauty. Now whenever the singers sing their hymns to the god and the harpers accompany the chorus with their harmonious music, thereupon the Swans also with one accord join in the chant and

[a] Or rather two, Calais and Zetes.
[b] A fabulous range of mountains from which the N wind was supposed to issue; beyond them lived the Hyperboreans.

---

[5] καταπέτονται. [6] ἀμήχανα . . . τὰ νέφη.

357

καὶ οὐδαμῶς οὐδαμῇ ἀπηχὲς καὶ ἀπῳδὸν ἐκεῖνοι
μελῳδοῦντες, ἀλλὰ ὥσπερ οὖν ἐκ τοῦ χορολέκτου
τὸ ἐνδόσιμον λαβόντες καὶ τοῖς σοφισταῖς τῶν
ἱερῶν μελῶν τοῖς ἐπιχωρίοις συνάσαντες. εἶτα
τοῦ ὕμνου τελεσθέντος οἱ δὲ ἀναχωροῦσι τῇ πρὸς
τὸν δαίμονα τιμῇ τὰ εἰθισμένα λατρεύσαντες καὶ
τὸν θεὸν ἀνὰ πᾶσαν τὴν ἡμέραν οἱ προειρημένοι ὡς
εἰπεῖν χορευταὶ πτηνοὶ μέλψαντές [1] τε ἅμα καὶ
ᾄσαντες.

2. Θύουσι δὲ [2] καὶ ἄλλως οἱ Ἠπειρῶται τῷ
Ἀπόλλωνι καὶ αὐτοὶ καὶ πᾶν ὅσον τῶν ξένων
ἐπίδημόν ἐστι,[3] καὶ τούτῳ δὴ [4] τὴν μεγίστην
ἑορτὴν ἄγουσι μιᾶς ἡμέρας τοῦ ἔτους σεμνήν τε καὶ
μεγαλοπρεπῆ. ἔστι δὲ ἄνετον τῷ θεῷ ἄλσος, καὶ
ἔχει κύκλῳ περίβολον, καὶ ἔνδον εἰσὶ [5] δράκοντες,
καὶ τοῦ θεοῦ ἄθυρμα οὗτοί γε. ἡ τοίνυν ἱέρεια,
γυνὴ [6] παρθένος, πάρεισι μόνη, καὶ τροφὴν τοῖς
δράκουσι κομίζει. λέγονται δὲ ἄρα ὑπὸ τῶν
Ἠπειρωτῶν ἔκγονοι τοῦ ἐν Δελφοῖς Πυθῶνος
εἶναι. ἐὰν μὲν οὖν οὗτοι παρελθοῦσαν τὴν ἱέρειαν
προσηνῶς θεάσωνται καὶ τὰς τροφὰς προθύμως
λάβωσιν, εὐθενίαν [7] τε ὑποδηλοῦν ὁμολογοῦνται
καὶ ἔτος ἄνοσον· ἐὰν δὲ ἐκπλήξωσι μὲν αὐτήν, μὴ
λάβωσι δὲ ὅσα ὀρέγει μειλίγματα, τἀναντία τῶν
προειρημένων οἱ μὲν μαντεύονται, οἱ δὲ ἐλπίζουσιν.

3. Ἐν Αἴτνῃ δὲ ἄρα τῇ Σικελικῇ Ἡφαίστου
τιμᾶται νεώς, καὶ ἔστι περίβολος καὶ δένδρα ἱερὰ
καὶ πῦρ ἄσβεστόν τε καὶ ἀκοίμητον. εἰσὶ δὲ

---

[1] τέρψαντες.   [2] τε.
[3] ἐστι θύουσι.   [4] ἤδη.

never once do they sing a discordant note or out of tune, but as though they had been given the key by the conductor they chant in unison with the natives who are skilled in the sacred melodies. Then when the hymn is finished the aforesaid winged choristers, so to call them, after their customary service in honour of the god and after singing and celebrating his praises all through the day, depart.

2. The people of Epirus and all strangers sojourning there, beside any other sacrifice to Apollo, on one day in the year hold their chief festival in his honour with solemnity and great pomp. There is a grove dedicated to the god, and round about it a precinct, and in the enclosure are Serpents, and these self-same Serpents are the pets of the god. Now the priestess, who is a virgin, enters unaccompanied, bringing food for the Serpents. And the people of Epirus maintain that the Serpents are sprung from the Python at Delphi. If, as the priestess approaches, they look graciously upon her and take the food with eagerness, it is agreed that they are indicating a year of prosperity and of freedom from sickness. If however they scare her and refuse the pleasant food she offers, then the Serpents are foretelling the reverse of the above, and that is what the people of Epirus expect.

Serpents sacred to Apollo in Epirus

3. At Etna in Sicily honour is paid to a temple of Hephaestus, and there are a precinct, sacred trees, and a fire that is never extinguished, never sleeps.

Dogs sacred to Hephaestus at Etna

AELIAN

κύνες περί τε τὸν νεὼν καὶ τὸ ἄλσος ἱεροί, καὶ
τοὺς μὲν σωφρόνως καὶ ὡς πρέπει τε ἅμα καὶ
χρὴ παριόντας ἐς τὸν νεὼν καὶ τὸ ἄλσος οἵδε
σαίνουσι καὶ αἰκάλλουσιν, οἷα [1] φιλοφρονούμενοί [2]
τε καὶ γνωρίζοντες δήπου· ἐὰν δέ τις ᾖ ⟨τὰς⟩ [3]
χεῖρας ἐναγής, τοῦτον μὲν καὶ δάκνουσι καὶ
ἀμύσσουσι, τοὺς δὲ ἄλλως ἔκ τινος ὁμιλίας
ἥκοντας ἀκολάστου μόνον διώκουσιν.

4. Τὴν Δήμητρα Ἑρμιονεῖς σέβουσι, καὶ θύουσιν
αὐτῇ μεγαλοπρεπῶς τε καὶ σοβαρῶς, καὶ τὴν
ἑορτὴν Χθόνια [4] καλοῦσι. μεγίστας [5] γοῦν [6] ἀκούω
βοῦς ὑπὸ τῆς ἱερείας τῆς Δήμητρος ἄγεσθαί τε
πρὸς τὸν βωμὸν ἐκ τῆς ἀγέλης καὶ θύειν ἑαυτὰς
παρέχειν. καὶ οἷς λέγω μάρτυς Ἀριστοκλῆς, ὅς
πού φησι

Δάματερ πολύκαρπε, σὺ κὴν Σικελοῖσιν ἐναργὴς
  καὶ παρ' Ἐρεχθείδαις. ἐν δέ τι ⟨τοῦτο⟩ [7]
  μέγα
κρίνετ' ἐν Ἑρμιονεῦσι· τὸν ἐξ ἀγέλης γὰρ
  ἀφειδῆ
  ταῦρον, ὃν οὐ χειροῦντ' [8] ἀνέρες οὐδὲ δέκα,
τοῦτον γραῦς στείχουσα μόνα μόνον οὔατος ἕλκει
  τόνδ' ἐπὶ βωμόν, ὁ δ' ὡς ματέρι παῖς ἕπεται.
σὸν τόδε, Δάματερ, σὸν τὸ σθένος· ἵλαος εἴης,
  καὶ πάντως [9] θάλλοι κλᾶρος ἐν Ἑρμιόνῃ.

[1] οἷα δή.
[2] Ges : φιλόφροσιν ἐνούμενοι.
[3] ⟨τάς⟩ add. H.
[4] Meursius : χθονίαν.
[5] Ges : μεγίστους.
[6] οὖν.
[7] ⟨τοῦτο⟩ add. Ges.
[8] οὐχ αἱροῦσιν.
[9] πάντων.

And about the temple and the grove there are sacred Hounds which greet and fawn upon such as pass into the temple and the grove with honest hearts in seemly fashion as is their duty, as though the animals had a kindness for them and presumably recognised them. If however a man has his hands stained with crime, they bite and tear him, whereas those who only come from the bed of debauchery they simply chase away.

4. The people of Hermione [a] worship Demeter and sacrifice to her in splendid and impressive style; and they call her festival the 'Festival of the Earth.' At any rate I have heard that the largest cattle allow themselves to be led from the herd by the priestess to the altar of Demeter and be sacrificed. And Aristocles bears witness to my statement when he says somewhere

The worship of Demeter at Hermione

'Demeter, goddess of abundance, thou dost manifest thyself both to the people of Sicily and to the sons of Erechtheus, but this among the dwellers in Hermione is judged a mighty feat: the bull of surpassing size from a herd, which not even ten men can master, this bull an aged woman, coming by herself, leads by the ear alone to this altar, and he follows as a child after its mother. Thine, even thine, Demeter, is the power. Show us thy favour and grant that every farm in Hermione may thrive exceedingly.'

[a] Town on the SE coast of Argolis.

5. Ἐν ⟨τῇ⟩ ¹ γῇ τῇ Δαυνίᾳ ² νεὼν μὲν εἶναι
τῆς Ἀθηνᾶς τῆς Ἰλιάδος ᾄδουσι· τοὺς δὲ ἐνταυθοῖ
κύνας τρεφομένους ὑμνοῦσι τῶν μὲν Ἑλλήνων τοὺς
ἀφικνουμένους σαίνειν, ὑλακτεῖν δὲ τοὺς βαρβάρους.

6. Ἐν Ἀρκαδίᾳ δὲ χώρᾳ ἐστὶν ἱερὸν Πανός·
Αὐλὴ τῷ χώρῳ τὸ ὄνομα. οὐκοῦν ὅσα ἂν ἐνταυθοῖ
τῶν ζῴων καταφύγῃ ὥσπερ οὖν ἱκέτας ὁ θεὸς δι'
αἰδοῦς ἄγων ³ εἶτα μέντοι σῴζει τὴν μεγίστην
σωτηρίαν αὐτά· οἱ γάρ τοι λύκοι οἱ διώκοντες
παρελθεῖν ἔσω πεφρίκασι καὶ ἀναστέλλονται μόνον
θεασάμενοι οἳ ⁴ κατέφυγεν. ἴδια δὴ ⁵ καὶ τούτων
τῶν ζῴων ἔοικε ⁶ πρὸς σωτηρίαν ἀγαθά.

7. Ἐν Κουριάδι ⁷ αἱ ἔλαφοι (πλῆθος δὲ ἄρα
τούτων τῶν θηρίων ἐνταῦθά ἐστι, καὶ πολλοὶ
θηραταὶ περὶ τὴν ἄγραν αὐτῶν ἠνέμωνται) ὅταν
καταφύγωσιν ἐς τὸ τοῦ Ἀπόλλωνος ἱερὸν τὸ
ἐνταυθοῖ ⁸ (ἔστι δὲ ἄλσος μέγιστον), ὑλακτοῦσι
μὲν οἱ κύνες, πλησίον δὲ ἐλθεῖν οὐχ ὑπομένουσιν·
αἱ δὲ συστᾶσαι ⁹ νέμονται ἄτρεπτον ¹⁰ καὶ ἀδεᾶ ¹¹
τὴν νομήν, ἀπορρήτῳ τινὶ φύσει τὴν ὑπὲρ ἑαυτῶν
σωτηρίαν τῷ θεῷ πιστεύουσαι αἱ ἔλαφοι.

8. Ἄνω που λέλεκταί μοι τὰς μυίας τὰς ἐν
Ὀλυμπίᾳ τῇ τῆς πανηγύρεως ἐπιδημίᾳ ἑκούσας
ἀφίστασθαι καὶ ὡς ἂν εἴποι τις μετὰ τῶν γυναικῶν
ἐπὶ τὴν ἀντιπέρας ὄχθην τοῦ Ἀλφειοῦ ἀπιέναι.

---

¹ ⟨τῇ⟩ add. H.　　　　² Gron : Δαυλίᾳ.
³ ἄγων τὰ ζῷα.　　　　⁴ οὗ.
⁵ Schn : δέ.　　　　　⁶ ἔθηκαι or ἔφην.
⁷ Schn : κουριδίῳ.　　⁸ ἐνταυθοῖ τιμήσιον.

362

5. In the country of the Daunii [a] there is a temple to Athena of Ilium which is celebrated. And they say that the Hounds that are kept there fawn upon any Greeks that arrive but bark at foreigners.

Dogs sacred to Athena in Daunia

6. And in Arcadian territory there is a shrine of Pan; Aule is the name of the place. Now any animals that take refuge there the god respects as suppliants and protects in complete safety. For wolves in pursuit are afraid to enter it and are checked at the mere sight of the place of refuge. So there is private property for these animals too to enable them to survive.

A refuge for hunted animals in Arcadia

7. On Curias [b] when the Deer (of which there are a great number and many hunters keen in pursuit of them) take refuge in the temple of Apollo there (the precinct is of very wide extent), the hounds bay at them but do not dare to approach. But the Deer in a body graze undeterred and without fear and by some mysterious instinct trust to the god for their safety.

A refuge for hunted Deer in Cyprus

8. I have mentioned somewhere earlier on [c] how on the occasion of the national assembly at Olympia the flies absent themselves of their own free will and, so to speak, depart along with the women to the opposite bank of the Alpheus. And in the island of

Flies avoid the festival of Apollo

[a] A people in the NW of Apulia.
[b] Promontory on the S coast of Cyprus.
[c] See 5. 17.

---

[9] ἐνστᾶσαι.   [10] ἄτρεστον.
[11] ἀδεῆ.

ἐν δὲ τῇ Λευκάδι ἄκρα μέν ἐστιν ὑψηλή, νεὼς δὲ
Ἀπόλλωνι ἵδρυται, καὶ Ἄκτιόν γε αὐτὸν οἱ
τιμῶντες ὀνομάζουσιν. οὐκοῦν τῆς πανηγύρεως
ἐπιδημεῖν μελλούσης, καθ᾽ ἣν καὶ τὸ [1] πήδημα
πηδῶσι τῷ θεῷ, θύουσι βοῦν ταῖς μυίαις, αἱ δὲ
ἐμπλησθεῖσαι τοῦ αἵματος ἀφανίζονται. δεκασθεῖ-
σαι μὲν οὖν ἀπαλλάττονται αὗται, αἱ δὲ Πισαῖαι
ἀδέκαστοι. κρείττους ἄρα ἐκεῖναι, αἰδοῖ τοῦ θεοῦ,
ἀλλὰ μὴ μισθοῦ τὰ δέοντα πράττουσαι.

9. Ἴκαρός ἐστι νῆσος, καὶ τῇ γε Ἐρυθρᾷ
θαλάττῃ ἔγκειται. ἐνταῦθα τοίνυν νεώς ἐστιν
Ἀρτέμιδος, καὶ πλήθη αἰγῶν τε ἀγρίων καὶ
δορκάδων εὖ μάλα εὐτραφῶν καὶ λαγῶν μέντοι.
τούτων οὖν ἐάν τις αἰτήσας λαβεῖν παρὰ τῆς θεοῦ
εἶτα ἐπιχειρήσῃ θηρᾶν ὅσα ἂν ἔχῃ καλῶς, οὐ
διαμαρτάνει τῆς σπουδῆς, ἀλλὰ καὶ λαμβάνει καὶ
τῷ δώρῳ χαίρει· ἐὰν δὲ μὴ αἰτήσῃ, οὔτε αἱρεῖ
καὶ δίδωσι δίκας, ἃς ἄλλοι λέγουσιν.

10. Εἶτα κύκνων μὲν τῶν Ῥιπαίων τῶν ἐν τοῖς
Ὑπερβορέοις, ἐπεὶ λατρεύουσι τῷ Διὸς καὶ
Λητοῦς ὁσημέραι φιλοπόνως, ἐποιησάμην μνήμην,
ταύρου δὲ ἱεροῦ, ὅνπερ οὖν ἐκθεοῦσιν Αἰγύπτιοι,
οὐκ ἐροῦμεν τὰ ἴδια; καὶ πῶς ἡμᾶς οὐκ ἂν μέμψαι-
το καὶ ἡ συγγραφὴ καὶ ἡ φύσις, ἧς καὶ τοῦτο
ἔργον τε καὶ δῶρον; ἀλλὰ † μήσιός † [2] γε οὐδὲ

[1] Jac: τι.   [2] μήσιος corrupt.

---

[a] Strabo (10. 452) relates that at the annual festival a
criminal, to whom a number of live birds were attached in

364

Leucas there is a high promontory on which a temple
of Apollo has been built, and worshippers style him
Apollo of Actium. Now when the festival is about to
be held there in which they make the Leap[a] in honour
of the god, men sacrifice an ox to the flies, and when
the latter have sated themselves with the blood they
disappear. Yes, but they are bribed to depart,
whereas the flies at Pisa need no bribe. So the
latter are superior because they do what is required
out of reverence for the god and not for a reward.

9. Icarus is an island and lies in the Red Sea.[b] Hunting on
Ichara
Now there is a temple of Artemis there and quantities
of wild goats and plump gazelles and hares also.
If a man ask leave of the goddess to take them and
then starts to hunt whatever is allowed, he does
not fail in his object but succeeds and is glad of her
gift. But should he fail to ask, he takes nothing
and is punished in a way that others describe.

10. And now, when I have mentioned the swans Apis, the
sacred bull
of the
Egyptians
from the Rhipaean mountains in the country of the
Hyperboreans on account of their daily and assiduous
service of the son of Zeus and Leto, shall I refrain
from telling of the special characteristics of the
sacred Bull which the Egyptians deify? How then
could I avoid being censured by history and by
Nature, who made and gave this gift also to man?
But ⟨no one shall accuse me of negligence on this

order to break his fall (or 'leap'), was thrown into the sea,
was then picked up by boatmen and taken from the
country.
  [b] Or rather at the northern end of the Persian Gulf. The
more usual spelling is 'Ichara.'

ταύτῃ ῥᾴθυμον, καὶ εἰκότως εἰρήσεται καὶ ἡ
θεολογία ἥδε. θεὸς Αἰγυπτίοις [1] ἐναργέστατος ὁ
Ἆπις εἶναι πιστεύεται. γίνεται δὲ ἐκ βοός, ἐς
ἣν οὐράνιον σέλας ἐμπεσὸν σπορᾶς αἴτιόν ἐστι τῷ
προειρημένῳ. ⟨καὶ⟩ Ἕλληνες ⟨μὲν⟩ [2] αὐτὸν καλοῦσιν Ἔπαφον, καὶ γενεαλογοῦσίν οἱ μητέρα Ἰὼ
τὴν Ἀργείαν τὴν Ἰνάχου· Αἰγύπτιοι δὲ ἐκβάλλουσι τὸν λόγον ὡς ψευδῆ, καὶ χρῶνται τῷ χρόνῳ
μάρτυρι. φασὶ γὰρ Ἔπαφον μὲν ὀψὲ καὶ κάτω
γενέσθαι, τὸν δὲ Ἆπιν τὸν πρῶτον μυριάδας ἐτῶν
παμπόλλας τὴν ἐς ἀνθρώπους ἐπιδημίαν προειληφέ-
ναι. σημεῖα δὲ αὐτοῦ καὶ γνωρίσματα λέγει μὲν
καὶ Ἡρόδοτος καὶ Ἀρισταγόρας, οὐχ ὁμολογοῦσι
δὲ αὐτοῖς Αἰγύπτιοι· ἐννέα καὶ εἴκοσι γὰρ αὐτὰ
εἶναί φασι καὶ ἐμπρέπειν τῷδε τῷ ἱερῷ βοΐ. τίνα
δὲ ταῦτά ἐστι καὶ ὅπως διέσπαρται κατὰ τοῦ
σώματος τοῦ ζῴου, καὶ ὄντινα τρόπον οἱονεὶ
διήνθισται αὐτοῖς, ἀλλαχόθεν εἴσεσθε· ὅτου δὲ
τῶν ἀστέρων ἕκαστον σημεῖον διὰ συμβόλων
αἰνίττεται τὴν φύσιν, Αἰγύπτιοι τεκμηριῶσαι
ἱκανοί. καὶ γάρ τοι καὶ τὴν ἄνοδον τὴν τοῦ
Νείλου ὑποδηλοῦν σημεῖά [3] φασι καὶ τὸ τοῦ
κόσμου σχῆμα· ἀλλ᾽ ὄψει τι καὶ σύμβολον, ὡς
ἐκεῖνοι λέγουσιν, ὅπερ οὖν αἰνίττεται τοῦ φωτὸς
εἶναι τὸ σκότος πρεσβύτερον. καὶ τὸ μηνοειδὲς
τῆς σελήνης κατηγορεῖ σχῆμα ⟨τῷ⟩ [4] συνιέντι [5]
σημεῖον [6] ἄλλο, καὶ ἄλλα δὲ ἐπὶ τούτοις ἄλλων
αἰνίγματα βεβήλοις τε καὶ ἀμαθέσιν ἱστορίας
θεοπρεποῦς οὐκ εὐσύμβολα ταῦτα ὀφθαλμοῖς ὄντα.

---

[1] Ges : Αἰγύπτιος.
[2] ⟨καὶ⟩ . . . ⟨μέν⟩ add. H.
[3] σημεῖον.
[4] ⟨τῷ⟩ add. H.

point?⟩,[a] and I will describe also, as is reasonable, this system of religion.

Among the Egyptians Apis is believed to be the god whose presence is most manifest. He is born of a cow on which a flash of light from heaven has fallen and caused his engendering. The Greeks call him Epaphus and trace his descent from his mother the Argive Io, daughter of Inachus. The Egyptians however reject the story as false, and appeal to time as their witness, for they maintain that Epaphus was born late down the ages, whereas the first Apis visited mankind many, many thousands of years earlier. Herodotus [3. 28] and Aristagoras [Müller *FHG* 2. 98] adduce evidence and tokens of this; but the Egyptians do not acknowledge them, for they assert that there are nine-and-twenty marks clearly to be seen on this sacred bull. But what these marks are, and how they are distributed over the body of the animal, and in what fashion the bull is, as it were, adorned with them, you may learn from another source. And the Egyptians are able to explain which of the stars each mark symbolises. And they say further that the marks indicate when the Nile will rise and the shape of the universe. But you will also see a mark (so the Egyptians assert) which suggests that darkness is older than light. And another mark explains the shape of the crescent moon to him who understands; there are besides, other mysterious signs of different import which to the eyes of the profane and those uninstructed in divine history are hard to interpret. And whenever

[a] The text is defective and the translation conjectural.

---

[5] *Ges*: συνιόν τι.   [6] μέρος.

ὅταν δὲ διαρρεύσῃ ἡ φήμη τὸν θεὸν Αἰγυπτίοις
τετέχθαι λέγουσα, τῶν γραμματέων τῶν ἱερῶν
τινες, οἷσπερ [1] οὖν μάθημα παιδὶ ἐκ πατρὸς
παραδοθὲν ἀκριβοῦν [2] τὸν ὑπὲρ τῶν σημείων
ἔλεγχον, ἥκουσιν ἐνταῦθα, οὗ τῆς θεοφιλοῦς [3] βοὸς
τὸ βρέφος ἐτέχθη, καὶ κατά γε τὴν ὑφήγησιν τὴν
Ἑρμοῦ τὴν πρεσβυτάτην οἰκίαν [4] ἐγείρουσιν, ἔνθα
δήπου καὶ διαιτήσεται τήν γε πρώτην, ἐς ἡλίου
μὲν ἀνατολὰς ὁρῶσαν, τροφοὺς [5] δὲ τὰς τοῦ
βρέφους ὑποδέξασθαι καὶ μάλα γε ἱκανήν· τετ-
τάρων γὰρ δεῖ μηνῶν ἐν γάλαξι τόνδε εἶναι τὸν
μόσχον. ἐπὰν δὲ γένηται ἐκτραφείς,[6] ἐνταῦθά τοι
ὑπανισχούσης σελήνης νέας ἀπαντῶσι γραμματεῖς
ἱεροὶ καὶ προφῆται, καὶ μέντοι καὶ ναῦν ἀνὰ ἔτος
ἐς τοῦτο τῷδε [7] τῷ δαίμονι ἱερὰν κοσμοῦσι, καὶ
ταύτῃ πορθμεύουσιν αὐτὸν ἐς Μέμφιν, ἔνθα φίλτατα
ἤθη αὐτῷ καὶ διατριβαὶ [8] κεχαρισμέναι καὶ ἐνηβητή-
ρια καὶ δρόμοι καὶ κονίστραι καὶ γυμνάσια καὶ
θηλειῶν βοῶν ὡραίων οἶκοι [9] καὶ φρέαρ καὶ κρήνη
ποτίμου νάματος· οὐ [10] γάρ οἵ φασιν οἱ θεραπευταί
τε καὶ ἱερεῖς λυσιτελεῖν ἀεὶ Νειλῴου πίνειν· καὶ
γὰρ πιαίνεσθαι [11] γλυκέος τούτου τοῦ ῥεύματος
καὶ ἐς ὄγκον σαρκῶν ὄντος ἀγαθοῦ. πομπὰς δὲ
ἃς πέμπουσι, καὶ ἱερουργίας ⟨ἃς⟩ [12] ἐπιτελοῦσι
τοῦ νέου [13] δαίμονος τὰ θεοφάνια θύοντες Αἰγύπτιοι,
καὶ χορείας ⟨ἃς⟩ [14] χορεύουσι, καὶ θαλίας καὶ

---

[1] Jac : ὥσπερ.
[2] Jac : ἀκριβοῖ.
[3] θεοῦ φασιν Αἰγυπτίου.
[4] οἰκίαν τε.
[5] Röhl : τροφάς MSS, H.
[6] τραφείς.
[7] καὶ τῷδε.
[8] τριβαί.
[9] οἶκοι, οἱονεὶ θάλαμοι, ὅτε ἐθέλοι καὶ ἢν ἐρᾷ θυμὸς ἀναβαίνειν αὐτόν.
[10] τοῦτο.

the report gets abroad which tells the Egyptians that the god has been born, some of the sacred scribes to whom there has been handed down from father to son the science whereby they verify these marks, come to the spot where the calf has been born to the heifer beloved of god, and in accordance with the immemorial precepts of Hermes erect a house where the calf will live at any rate for the time being; it faces the rising sun and is quite large enough to take in the nurses[a] of the calf, for it is essential that the calf should be at the udder for four months. And when it has been weaned, then at the rising of the new moon the sacred scribes and priests go out to meet it and moreover year by year make ready a sacred vessel for this god and transport him on board to Memphis, where he finds abodes after his heart and delightful spots to linger in and places where he may amuse himself, where he may run and roll in the dust and exercise himself, and the homes of beautiful cows, and a well and a spring that yield water for drinking, for his ministers and priests say that it is not good for him always to drink of the Nile. Moreover he is said to grow fat on this sweet water which helps to build up a mass of flesh. As for the processions which they hold and the sacred offices which they perform when the Egyptians celebrate the revelation of the new god, the dances

[a] The 'nurses' are the cows which supply the Apis-calf with milk.

---

[11] καὶ γὰρ π.] καταπιαίνεσθαι.
[12] ⟨ἃς⟩ add. Schn.
[13] Reiske: νέου καὶ ὕδατος.
[14] ⟨ἃς⟩ add. Reiske.

AELIAN

πανηγύρεις ἃς ἐπιτελοῦσι, καὶ ὅπως αὐτοῖς καὶ
πόλις ἅπασα καὶ κώμη δι' εὐφροσύνης ἔρχεται,
μακρὰ ἂν εἴη λέγειν. ἐκεῖνος δέ,[1] ἐν ὅτου τῇ
ἀγέλῃ τόδε τὸ θεῖον ἐγένετο ζῷον, δοκεῖ τε
εὐδαίμων καὶ ἔστιν, ἄγουσί τε Αἰγύπτιοι θαυμαστόν
γε αὐτόν. μάντις δὲ[2] ἦν ἄρα ἀγαθὸς ὁ Ἆπις, οὐ
καθίζων μὰ Δία κόρας ἢ πρεσβυτέρας γυναῖκας
ἐπί τινων τριπόδων, οὐδὲ μὴν πώματος ἱεροῦ
ἐμπιπλάς, ἀλλ' ὁ μέν τις εὔχεται τῷ θεῷ τῷδε,[3]
παῖδες δὲ ἀθύροντες ἔξω καὶ πρὸς αὐλοὺς[4]
σκιρτῶντες, ἐπίπνοοι γενόμενοι σὺν τῷ ῥυθμῷ
αὐτὰ ἕκαστα προλέγουσιν, ὡς εἶναι ⟨ἀληθέστερα
τῶν ἐπὶ⟩ Σάγρᾳ[5] τὰ λεχθέντα. εἰκάζουσι δὲ
ἄρα καὶ τῷ Ὥρῳ αὐτὸν Αἰγύπτιοι, ὅνπερ οὖν
πεπιστεύκασι φορᾶς καρπῶν[6] καὶ εὐετηρίας αἰτιώ-
τατον ἁπάσης. ἔνθεν τοι καὶ ὑπὲρ τῆς πολυχροίας
αὐτοῦ φιλοσοφοῦσι, τὸ ποικίλον τῶν καρπῶν
ὑπαινιττόμενοι διὰ συμβόλων. λέγει δέ τις τῶν
προφητῶν λόγος οὐ πᾶσιν ἔκπυστος ὅτι ἄρα
⟨Μῆνις⟩[7] ὁ τῶν Αἰγυπτίων βασιλεὺς ἐπενόησε
ζῷον ὥστε σέβειν ἔμψυχον, εἶτα μέντοι προείλετο
ταῦρον, ἁπάντων ὡραιότατον εἶναι αὐτὸν πεπιστευ-
κὼς καὶ τῆς γε Ὁμήρου κρίσεως τῆς ὑπὲρ τούτων
κατ' ἴχνια ἰὼν ὁ Μῆνίς φασιν. εἰπεῖν γὰρ καὶ
Ὅμηρον ἐν Ἰλιάδι

ἠΰτε βοῦς ἀγέληφι μέγ' ἔξοχος ἔπλετο πάντων
ταῦρος· ὁ γάρ τε βόεσσι μεταπρέπει ἀγρομένῃσι.

---

[1] Schn : λέγει δὲ ἐκεῖνος.　　[2] τε.
[3] τῷδε, καὶ μαθεῖν ἐθέλει αὐτοῦ.　　[4] ἀλλήλους.
[5] ⟨ἀληθέστερα τῶν ἐπὶ⟩ Σάγρᾳ Anon.: εἶναι Σάγραν.
[6] Ges : τρόπων.
[7] ⟨Μῆνις⟩ add. H.

which they execute, the feasts and the assemblies which they organise, and how every town and village is filled with joy—all this would make a long story. But the man in whose herd this divine animal was born is counted fortunate and is so, and the Egyptians regard him with admiration.

Apis, it seems, is in effect a good prophet: he to be sure never sets girls or elderly women on tripods, never fills them with some sanctified draught, but a man prays to this god, and children without, who are playing and dancing to the music of pipes, become inspired and proclaim in time with the music the actual response of the god, so that what they say is more true than what occurred by the Sagras.[a]

The Egyptians liken Apis to Horus whom they believe to be the prime cause of the fertility of their crops and of every good season. That is how they come to reason about his varied colouring, seeing in it a hidden symbolical reference to the variety of the crops. And there is a story of the priests not known to all, that Menis the King of Egypt, thinking of some living animal that he might worship, elected a bull, believing it to be the finest of all animals, and at any rate following Homer in his judgment on these matters, so they say. For Homer too in his *Iliad* [2. 480] says

'Even as a bull standeth out far foremost in the herd, for he is conspicuous amid the pasturing kine.'

[a] A river (no longer identifiable) in Bruttium which was the scene of a battle between the Locrians and the people of Croton at some date during the 6th cent. B.C. The Locrians with the aid of the Dioscuri defeated a force more than ten times their number. The news of the victory reached Sparta on the very same day, and was received with incredulity. See Suidas, ἀληθέστερα κτλ., Smith, *Dict. Geogr.* 2. 873.

οἷα δὲ ἐς μυθολογίαν ὑπὲρ τοῦδε τοῦ ζῴου ἐκτρέ-
πουσιν Αἰγύπτιοι[1] οἱ τὰ περὶ τῶν ζῴων γράψαντες,
οὔ μοι δοκεῖ φίλα εἶναι.

11. Ἀλλ' ἄγε δὴ μετάβηθι φαίη ἂν ὁ λόγος,
καὶ οὐχ ἵππον μὰ Δία οὐδὲ μὴν τὸν ἐν αὐτῷ
λόχον ᾄδε, ἀλλὰ Μνεῦιν βοῦν. καὶ τοῦτον Αἰγύπτιοι Ἡλίου φασὶν ἱερόν, ἐπεὶ τόν γε Ἆπιν ἀνάθημα
εἶναι Σελήνῃ λέγουσιν. ἴδιον δὲ ἄρα καὶ τούτῳ
γνώρισμα ἐς τὸ εἶναι μήτε κίβδηλον μήτε μὴν
νόθον ἀλλὰ φίλον τῷ θεῷ τῷ προειρημένῳ φασὶν
Αἰγύπτιοι. καὶ ὑπὲρ τούτων μὲν[2] ἐρεῖ ἄλλος,
ὃν δὲ ἤκουσα λόγον Αἰγύπτιον ἐς βάσανον τοῦδε
τοῦ ταύρου καὶ ἔλεγχον, εἴτε σπορᾶς ἐστι κρείττονος εἴτε μή, τοῦτον εἰπεῖν ἐθέλω. Βόκχορις ὁ
τῶν Αἰγυπτίων βασιλεύς, κλέος[3] ψευδὲς καὶ
φήμην οὐδὲν λέγουσαν ὑγιὲς οὐκ οἶδα ὅπως
ἁρπάσας, δίκαιός τε ἐν ταῖς κρίσεσιν ἐδόκει καὶ
πρὸς τὸ ὅσιον[4] τὴν ψυχὴν κεκοσμημένος· ἦν δὲ
ἄρα ἔμπαλιν πεφυκὼς ἐκεῖνος. καὶ τὰ μὲν πλείω
ἐῶ νῦν, τὸν δὲ Μνεῦιν, λυπῆσαι θέλων Αἰγυπτίους,
οἷα ἐργάζεται γοῦν αὐτόν. ταῦρον ἄγριον ἐπάγει
οἱ ἀντίπαλον. οὐκοῦν μυκᾶται μὲν ὁ Μνεῦις,
ἀντεμυκήσατο δὲ ὁ ἔπηλυς. εἶτα ὑπὸ ⟨τοῦ⟩[5]
θυμοῦ φέρεται μὲν ἐμπεσεῖν τῷ ταύρῳ τῷ θεοφιλεῖ
ὁ ξένος ἐθέλων, καὶ σφάλλεται, καὶ ἐς περσέας
ἐμπεσὼν πρέμνον ἐπεσχέθη τὸ κέρας, ὁ δὲ κατὰ
τῆς πλευρᾶς[6] ὁ Μνεῦις τιτρώσκων[7] ἀπέκτεινεν

[1] Αἰγύπτιοι τῇδε τῇ περὶ τῶν ζῴων ἀληθείᾳ καὶ ἰδιότητι.
[2] δέ.
[3] κατὰ κλέος.
[4] θεῖον.
[5] ⟨τοῦ⟩ add. H.
[6] κατὰ πλεύραν.

But the facts which Egyptian writers on zoology distort into legends about this animal are not to my taste.

11. 'Nay, but change the theme' [Hom. *Od.* 8. 492], as the phrase might go, and sing not of the Horse[a] nor yet of the ambush within, but of the bull Mneuis. And he, say the Egyptians, is sacred to the Sun, whereas Apis, they say, is dedicated to the Moon. And according to the Egyptians he also bears a special mark to show that he is no counterfeit, no bastard, but beloved of the aforesaid god. On these topics another shall speak, but what I wish to tell is the Egyptians' account of the test and the proof to which they put this bull to see whether he is of superior birth or not.

Bocchoris the King of Egypt[b] acquired—I do not know how—a false reputation and a fictitious renown and appeared to be just in his judgments and to have his heart set on righteousness. But by nature, it seems, he was the reverse. Most of his actions I pass over at present, but this is how, from a desire to cause pain to the people of Egypt, he treated Mneuis. He set a wild bull against him. So Mneuis began to bellow and the newcomer bellowed in answer. And then the stranger rushed forward in anger intending to fall upon the bull beloved of the god, but tripped and falling against the stem of a persea-tree, broke his horn, whereupon Mneuis

Mneuis, the sacred bull of the Egyptians

and King Bocchoris

[a] The Wooden Horse whereby the Greeks gained entry into Troy. See Verg. *Aen.* 2. 13-267.

[b] Perh. 9th cent. B.C.

[7] τιτρώσκων τῇ κεφαλῇ.

αὐτόν. αἰδεῖται Βόκχορις, καὶ μισοῦσιν αὐτὸν
Αἰγύπτιοι. εἰ δέ τις αἴσχιστον [1] οἴεται ἐκ τῶν
φυσικῶν λόγων ἐς μύθους ἐμπεσεῖν, μωρός ἐστι.
λέγω γὰρ ὅσα τε δρᾶται ἐπὶ τοῖσδε τοῖς ταύροις
καὶ ὅσα ἐπράχθη καὶ ἀκούω λεγόντων Αἰγυπτίων.
οὐκ ἦν δὲ ἄρα . . . [2] οὕτω τὸ ψεῦδος ἐκείνοις
ἔχθιστόν ἐστι.

12. Οἱ δελφῖνες, τὸ μὲν φιλόμουσον αὐτῶν καὶ
περὶ τὴν ᾠδὴν σπουδαῖόν [3] τε καὶ φιλόπονον
κεκήρυκταί τε καὶ ἐς πολλοὺς ἐξεφοίτησε, καὶ ὣς
εἰσι φιλάνθρωποι ἄλλοι τε εἶπον καὶ ἡμεῖς ἄνω
που διεξήλθομεν τῷ λόγῳ· ἐνταυθοῖ δὲ εἰπεῖν
ὑπὲρ τῆς συνέσεως αὐτῶν οὐ χεῖρόν ἐστιν. ὅταν
γοῦν δικτύῳ περιπέσῃ δελφίς, τὰ μὲν πρῶτα
ἡσυχάζει καὶ φυγῆς οὐδέν τι μέμνηται, εὐωχεῖται
δὲ τῶν συνεαλωκότων ἰχθύων, καὶ ὥσπερ ἐπὶ
δαῖτα ἥκων κλητὸς εἶτα ἐμφορεῖται αὐτῶν· ὅταν
δὲ αἴσθηται ἐπισυρόμενος ὅτι γίνεται τῆς γῆς
πλησίον, ἐνταῦθά τοι τὸ δίκτυον διατραγὼν
ἀπαλλάττεται καὶ ἔστιν ἐλεύθερος. ἐὰν δέ ποτε
ἁλῷ, οἱ χαριέστεροι τῶν ἁλιέων ὁλόσχοινον
αὐτοῦ διείραντες τῶν ῥινῶν ἀφῆκαν αὐτόν [4]· ὁ
δὲ οἷα τὸν ἔλεγχον αἰδούμενος οὐκέτι πλησιάζει
σαγήνῃ τὸ ἐντεῦθεν. λέγει δὲ Ἀριστοτέλης ὅτι
κἂν ἁλῷ καὶ δεθῇ καὶ ἐν τῇ σκάφῃ ᾖ, πολλοὶ
περινήχονται δελφῖνες τὴν ἁλιάδα, καὶ ἐς τοσοῦτον
πηδῶσί τε καὶ σκιρτῶσι δίκην ἱκετῶν, ἔστε

---

[1] ἔχθιστος.   [2] Lacuna.
[3] ᾠδὴν σπ.] φωνὴν ᾠδικόν.
[4] ἐπαφῆκαν αὐτὸν γνώρισμα τοῦτο εἰ ἐμπέσοι ἄρα τοῦ καὶ
πρόσθεν ἁλῶναι αὐτὸν περιφέρει.

wounded him in the flank and killed him. Boccharis
was put to shame and the Egyptians loathed him.

But if anyone considers it highly undignified to
drop from natural history into legend, he is a fool.
For I am stating what the practice is with these bulls,
and what then occurred, and what I hear Egyptians
say . . .[a] a lie to them is an abomination.

12. The Dolphins' love of music and their eager The Dolphin
pursuit of song have been noised abroad and spread
to many quarters, and others have told of their
friendliness to man, and we ourselves have discoursed
upon it earlier on,[b] I think. But here I shall do well
to speak of their intelligence. At any rate whenever a Dolphin is enclosed in a net he keeps quiet to
begin with and does not think of escaping, but feasts
upon the fish that have been caught with him and,
as though invited to a banquet, takes his fill of them.
But as soon as he realises, while being drawn along,
that he is nearing the shore, he thereupon bites
through the net, escapes, and is free. If however
he is caught, the more kindly fishermen pass a
rush through his nostrils and let him go; and the
Dolphin, as though he were ashamed of the evidence
of his capture, never comes near a drag-net again.
And Aristotle says [*HA* 631 a 11] that whenever
one is caught and made fast and is in the fish-box,[c]
Dolphins swim round the boat in numbers and leap
so high and writhe like suppliants, until the fisher-

[a] The text is defective. The sense of the missing words
was perhaps ' This is no mere idle tale, for, *etc*.'
[b] See 2. 6.
[c] Or ' tub ' into which the caught fish are thrown.

AELIAN

παθεῖν τι τοὺς ἁλιέας, καὶ οἰκτεῖραι μὲν τὸν
δεσμώτην, εἶξαι δὲ τοῖς δεομένοις καὶ ἀπολῦσαι
αὐτοῖς τὸν ᾑρημένον.

13. Δάφνιδος τοῦ βουκόλου τοῦ Συρακοσίου
παθόντος ὑπὸ τῆς νύμφης ταῦτα δήπου τὰ ὑμνούμενα, πέντε τροφίμους κύνας, τὸν Σάννον [1] καὶ
τὸν Πόδαργον καὶ τὴν Λαμπάδα καὶ τὸν Ἄλκιμον
καὶ τὸν Θέοντα, θεασαμένους τοῦ δεσπότου τὰς
πάθας ἐπ' αὐτῷ θάνατον ἑλέσθαι φασί, καὶ πολλὰ
μὲν ὀδύρασθαι [2] πρότερον, κλαῦσαι δὲ πάμπολλα.

14. Διάφορα μὲν καὶ ποικίλα τῆς τῶν ἐλεφάντων
ἰδιότητος ἄνω μοι λέλεκται· νῦν δὲ εἰρήσεται ὅτι
καὶ μνήμην ἀγαθόν ἐστι τὸ ζῷον τοῦτο, καὶ
ἐντολὰς φυλάξαι οἶδε καὶ μὴ ψεύσασθαι τὴν τῶν
παρακαταθεμένων ὅ τι οὖν αὐτῷ προσδοκίαν τε
καὶ ἐλπίδα. ὅτε γοῦν Ἀντίγονος ἐπολιόρκει
Μεγαρέας, ἑνὶ τῶν ἐλεφάντων τῶν πολεμικῶν
συνετρέφετο καὶ θῆλυς, ὄνομα Νίκαια. ταύτῃ
τοίνυν ἡ τοῦ τρέφοντος αὐτὸν γυνὴ παιδίον, ⟨ὃ⟩ [3]
ἔτυχε τεκοῦσα πρὸ ἡμερῶν τριάκοντα, παρακατέ-
θετο [4] φωνῇ τῇ Ἰνδῶν, ἧς ἀκούουσιν ἐλέφαντες.
ὁ δὲ καὶ ἐφίλει τὸ παιδίον καὶ ἐφύλαττε, καὶ
κειμένου πλησίον ἥδετο, καὶ κνυζωμένου [5] παρέ-
βλεπε, καὶ καθεύδοντος τῇ προβοσκίδι τὰς μυίας
ἀπεσόβει [6] καλάμου κλαδὶ τοῦ παραβαλλομένου ἐς

[1] Σάνον.
[2] Ges : ὀδύρεσθαι.
[3] ⟨ὃ⟩ add. Jac.
[4] Jac : καὶ παρα-.
[5] κνυζομένου.
[6] Schn : ἀνεσόβει.

376

men feel a touch of sympathy and take pity on the prisoner and yield to the entreating creatures and release the captive to them.

13. They say that the five hounds, Sannus, Podargus, Lampas, Alcimus, and Theon, kept by Daphnis the neatherd of Syracuse who suffered his well-known punishment [a] at the hands of the Nymph, at the sight of their master's misfortune chose to die after he died, having previously bewailed him deeply and shed tears in abundance.

The Hounds of Daphnis

14. I have earlier on spoken of the differences and the varieties in the character of Elephants, and I shall now tell what a good memory too this animal has, how it can remember orders and not belie the expectation and the hope of those who entrust it with whatever it may be. For instance when Antigonus [b] was besieging Megara a female elephant of the name of Nicaea was being kept along with one of the war-elephants. Now to this animal the wife of the keeper entrusted a baby which she happened to have borne a month before, speaking the Indian language, which Elephants understand. And the Elephant grew fond of the child and used to look after it, and liked to have it lying near, and would glance at it when it whimpered; and when it slept the Elephant would scare away the flies, holding in her trunk a spray from the reeds which were thrown

The Elephant as nurse

---

[a] See Ael. *VH* 10. 18. D. was beloved by a Nymph and vowed to be faithful to her or to lose his sight. He was seduced by a King's daughter and suffered the penalty.

[b] A. Gonatas, vice-gerent of Demetrius II, King of Macedon, fought against Pyrrhus, besieged and recovered Megara, perh. in 270 B.C. See W. W. Tarn, *Ant. Gon.* 286.

τροφήν· εἰ δὲ ¹ μὴ παρῆν τὸ βρέφος, τότε καὶ
τροφὴν ἀνεστέλλετο. οὐκοῦν ἔδει τὴν τρέφουσαν
αὐτὸ ἐμπλῆσαι μὲν τοῦ γάλακτος, παραθεῖναι δὲ
τῷ κηδεμόνι, ἢ πάντως ἀγανακτῶν ἡ Νίκαια ἦν
δῆλος καὶ τεθυμωμένος καί τι καὶ δρασείων τῶν
δεινῶν. πολλάκις δὲ καὶ ἀνακλαύσαντος αὐτοῦ
εἶτα τὴν σκάφην ᾗ ἐνέκειτο διέσεισε, παραμυθούμε-
νος τῷ σεισμῷ τὸ βρέφος, οἷα δήπου φιλοῦσι καὶ
αἱ τροφοὶ καὶ αἱ τίτθαι δρᾶν ποιῶν, ὦ ἄνθρωποι, ὁ
ἐλέφας.

15. Ζηλοτυπίαν ζῴων ἐνεργοτάτην διαφόρων ἐν
καιρῷ οἶδα εἰπών, πορφυρίωνος καὶ κυνὸς καὶ
μέντοι καὶ πελαργοῦ νὴ Δία ἐκ τρίτου· νῦν δὲ ἔοικα
λέξειν ἐλέφαντος ὀργὴν ἐς γάμον ἀδικούμενον.
μοιχευομένην γὰρ τῆς τοῦ πωλεύσαντος αὐτὸν καὶ
τρέφοντος γυναῖκα ἐπ᾽ αὐτοφώρῳ καταλαβών, δι᾽
ἀμφοτέρων θάτερον διεὶς ² τοῖν κεράτοιν, ἀπέκτεινε
καὶ τὸν μοιχὸν καὶ τὴν μοιχευομένην, καὶ εἴασε
κεῖσθαι κατὰ τῶν στρωμάτων ⟨τῶν⟩ ³ ὑβρισμένων
καὶ τῆς εὐνῆς τῆς πεπατημένης, ὡς ἐλθόντα τὸν
πωλευτὴν καταγνῶναι καὶ τὸ ἀδίκημα καὶ τὸν
τιμωρήσαντα αὐτῷ γνωρίσαι. καὶ τοῦτο μὲν
Ἰνδικὸν τὸ ἔργον, ἐκεῖθεν δὲ ἐξεφοίτησε δεῦρο
ἀκούω δὲ καὶ ἐπὶ Τίτου ⁴ ἀνδρὸς καλοῦ καὶ ἀγαθοῦ
ἐν ⁵ τῇ Ῥώμῃ ταὐτὸν γεγονέναι· προστιθέασι δὲ
ὅτι ἄρα ὁ ἐνθάδε ἐλέφας ἀπέκτεινεν ⁶ ἀμφοτέρους,
καὶ ἱματίῳ ⁷ κατεκάλυψε, καὶ ἐλθόντι τῷ τροφεῖ
ἀποβαλὼν τὸ ἱμάτιον κειμένους ἀλλήλων πλησίον

¹ Ges : εἴ γε.　　　² πείρας.
³ ⟨τῶν⟩ add. H.

378

beside her as her fodder.   And if the child was not
there she would actually put her own food aside.
And so the mother was obliged to give the child its
fill of milk and then place it beside its guardian,
otherwise Nicaea gave unmistakable signs of being
annoyed and angered and even of threatening mischief.   And often, if the baby started to cry, she
rocked the cradle in which it lay, comforting it as
nurses are in the habit of doing by the swaying—
and this, my fellow-men, was an Elephant.

15. I know that I have spoken appropriately of   An Elephant
the very violent jealousy on the part of different   punishes
animals, viz the coot, the dog, and in the third place   adultery
the stork.   But now I intend to speak of the anger
of an Elephant over an outraged marriage.   Having
detected the wife of its trainer and keeper in the
very act of adultery, it drove one tusk through the
woman and one through her lover and killed them
both and left them lying amid the dishonoured
coverings on the desecrated bed, so that when the
trainer came he might note their sin and recognise
his avenger.   This happened in India, but the deed
travelled from there to these shores, and I learn
that in the reign of Titus, that good and noble
man, the same thing occurred in Rome, but they
add that the Elephant there killed both the offenders
and covered them with a cloak which on the arrival
of its keeper it threw off and revealed the two lying

---

⁴ *Ges*: τόπου Ῥωμαίων βασιλεύοντος MSS, Ῥωμ. βασ. *del. H.*
⁵ καὶ ἐν.   ⁶ ἀπέκτεινε μέν.
⁷ ἐν ἱματίῳ.

AELIAN

ἀπέδειξε, καὶ τὸ κέρας δέ, ὧπερ οὖν διέπειρεν
αὐτούς, καὶ τοῦτο ἡμαγμένον ἑωρᾶτο.

16. Ἴδιον δὲ ἦν ἄρα τῶν δρακόντων καὶ ἡ
μαντική. ἐν γοῦν Λαουινίῳ[1] τῷ[2] πολίσματι,
ὅπερ τῆς Λατίνων χώρας ἐστί (κέκληται δὲ ἀπὸ
τῆς Λατίνου θυγατρὸς Λαουινίας,[3] ἡνίκα Λατῖνος
Αἰνείᾳ συνεμάχησε κατὰ τῶν καλουμένων Ῥου-
τουλῶν, εἶτα ἐνίκησεν αὐτούς· ἔκτισε[4] δὲ
Αἰνείας ὁ Ἀγχίσου ὁ Τρὼς τὴν πόλιν τὴν προειρημένην, εἴη δ᾽ ἂν τῆς Ῥώμης μητρομήτωρ, ὡς
ἂν εἴποι τις· ἐντεῦθεν γὰρ ὁρμηθεὶς Ἀσκάνιος ὁ
Αἰνείου καὶ Κρεούσης τῆς Τρωάδος ᾤκισε τὴν
Ἄλβαν, Ἀλβανῶν δὲ ἡ Ῥώμη ἄποικος)· οὐκοῦν
ἐν τῷ Λαουινίῳ[5] ἄλσος τιμᾶται μέγα καὶ δασύ,
καὶ ἔχει πλησίον νεὼν Ἥρας Ἀργολίδος. ἐν δὲ
τῷ ἄλσει φωλεός ἐστι μέγας καὶ βαθύς, καὶ ἔστι
κοίτη δράκοντος. παρθένοι τε ἱεραὶ νενομισμέναις
ἡμέραις παρίασιν ἐς τὸ ἄλσος ἐν τοῖν χεροῖν
φέρουσαι μᾶζαν καὶ τοὺς ὀφθαλμοὺς τελαμῶσι
κατειλημέναι[6]· ἄγει δὲ αὐτὰς εὐθύωρον ἐπὶ τὴν
κοίτην τοῦ δράκοντος πνεῦμα θεῖον, καὶ ἀπταίστως
προΐασι βάδην καὶ ἡσυχῆ, ὥσπερ οὖν ἀκαλύπτοις
ὁρῶσαι τοῖς ὀφθαλμοῖς. καὶ ἐὰν μὲν παρθένοι
ὦσι, προσίεται τὰς τροφὰς ⟨ἅτε⟩[7] ἁγνὰς ὁ
δράκων καὶ πρεπούσας ζῴῳ θεοφιλεῖ· εἰ δὲ μή,
ἄπαστοι[8] μένουσι, προειδότος αὐτοῦ τὴν φθορὰν

[1] Schn: Λαουαινείῳ.   [2] τῷδε τῷ.
[3] Schn: Λαουινείας.
[4] Freinsheim: ἐνίκησε MSS, ᾤκισε H.
[5] Schn: Λαουινείῳ.
[6] Cobet: κατειλημμέναι MSS, H.

380

side by side, while the tusk with which it had pierced them was seen to be stained with blood.

16. It seems that one peculiarity of snakes is their faculty of divination. At any rate in the town of Lavinium,[a] which is in Latium—it is so named after Lavinia the daughter of Latinus at the time when he fought as an ally of Aeneas against the people called Rutulians and overcame them. And Aeneas of Troy, son of Anchises, founded the aforesaid town; and it might be, in a manner of speaking, the grandmother of Rome, because it was from Rome that Ascanius, the son of Aeneas and Creüsa the Trojan, set out to found Alba, and Rome was a colony of Alba.—Well, there is a sacred grove in Lavinium of wide area and thickly planted, and near by is a shrine to Hera of Argolis. And in the grove there is a vast and deep cavern, and it is the lair of a Serpent. And on certain fixed days holy maidens enter the grove bearing a barley-cake in their hands and with their eyes bandaged. And divine inspiration leads them straight to the Serpent's resting-place, and they move forward without stumbling and at a gentle pace just as if they saw with their eyes unveiled. And if they are virgins, the Serpent accepts the food as sacred and as fit for a creature beloved of god. Otherwise the food remains untasted, because the Serpent already knows and has divined their impurity. And ants crumble the cake of the deflowered maid

The Serpent of Lavinium

---

[a] A. has confused 'Lavinium' and 'Lanuvium'; see Prop. 4. 8. 5 ff.

---

[7] ⟨ἅτε⟩ add. Jac.   [8] ἄψαυστοι W, H.

καὶ μεμαντευμένου. μύρμηκες δὲ τὴν μάζαν τὴν
τῆς διακορηθείσης ἐς μικρὰ καταθρύψαντες, ὡς
ἂν εὔφορα αὐτοῖς εἴη, εἶτα ἐκφέρουσιν ἔξω τοῦ
ἄλσους, καθαίροντες τὸν τόπον. γνωρίζεταί τε
ὑπὸ τῶν ἐπιχωρίων τὸ πραχθέν, καὶ αἱ παρελθοῦσαι
ἐλέγχονται, καὶ ἥ γε τὴν παρθενίαν αἰσχύνασα ταῖς
ἐκ τοῦ νόμου κολάζεται τιμωρίαις. μαντικὴν μὲν
δὴ δρακόντων ἂν ἀποφήναιμι τὸν τρόπον τοῦτον.

17. Λέγει μὲν οὖν Ὅμηρος χαλεποὶ δὲ θεοὶ
φαίνεσθαι ἐναργεῖς. οὐκοῦν ἔχει τι καὶ δράκων
ὁ ἐν ταῖς ἁγιωτάταις τιμαῖς θειότερον, καὶ ἰδεῖν [1]
οὐ λυσιτελὲς αὐτόν. καὶ ὅ γε λέγω τοιοῦτόν ἐστιν.
ἐν Μετήλει [2] τῆς Αἰγύπτου δράκων ἐστὶν ἱερὸς ἐν
πύργῳ, καὶ τετίμηται καὶ ἔχει θεραπευτὰς καὶ
ὑπηρέτας, καὶ κεῖταί οἱ τράπεζα καὶ κρατήρ. ἐς
τοῦτον οὖν ἀνὰ πᾶσαν [3] ἡμέραν ἄλφιτα ἀναδεύσαν-
τες μελικράτῳ εἶτα ἀπίασι, καὶ τῇ ὑστεραίᾳ
ὑποστρέψαντες κενὸν τὸν κρατῆρα εὑρίσκουσιν.
οὐκοῦν ὁ πρεσβύτατος τῶνδε τῶν ὑπηρετῶν
ἵμερον δριμύτατον ἔσχε θεάσασθαι τὸν δράκοντα,
καὶ παρελθὼν μόνος καὶ ποιήσας τὰ εἰθισμένα ὑπ-
απέστη [4]. ὁ δὲ ἀνελθὼν ἐπὶ τὴν τράπεζαν ὁ
δράκων εἱστιᾶτο. καὶ τὰς θύρας ὁ πολυπράγμων
ἀνοίξας (ἔτυχε γὰρ κατὰ τὰ εἰθισμένα ἐπικλείσας)
ψόφον εἰργάσατο ἰσχυρόν. ὁ δράκων δὲ ἠγανάκ-
τησε καὶ ἀνεχώρησεν, ὁ δὲ ἰδὼν ὃν ἐπόθει σὺν
τῷ ἑαυτοῦ κακῷ, γίνεται μὲν ἔκφρων, εἰπὼν δὲ
ὅσα εἶδε καὶ ὡς ἠσέβησεν ὁμολογήσας, ἦν ἄφωνος,
εἶτα οὐ μετὰ μακρὸν πεσὼν ἀπέθανεν.

[1] Schn : εἰδέναι. [2] Wesseling : Μελίτη.
[3] πᾶσαν τήν. [4] ἀπέστη H.

into small pieces so that they can be carried easily, and transport them without the grove, cleansing the spot. And the inhabitants get to know what has occurred and the maidens who came in are examined, and the one who has shamed her virginity is punished in accordance with the law.

This is the way in which I would demonstrate the faculty of divination in serpents.

17. Now Homer says [*Il.* 20. 131] 'but gods are hard to endure when seen clear to view.' And so even a serpent which is honoured by the most sacred rites has in it something of the divine, and to look upon it is not profitable. And what I mean is this. In Metelis,[a] a town of Egypt, there is a sacred Serpent in a tower, and it receives honours and has ministers and servants, and before it are set a table and a bowl. So every day they pour barley into this bowl and soak it in honey and milk and then depart, returning on the following day to find the bowl empty. Now the eldest servant felt a keen desire to set eyes upon the Serpent, and coming by himself performed the usual duties and withdrew. And the Serpent mounted on the table and feasted. And this busybody in opening the doors (he had closed them as was the custom) made a loud noise. The Serpent was indignant and retired, while the man who had seen the creature whom he wished to see, to his own undoing, went out of his mind, told what he had witnessed, and confessed his impious deed, became dumb, and shortly afterwards fell down dead.

A sacred Serpent and the penalty of inquisitiveness

[a] Town in the NW of the Delta.

18. Ἴδια δὲ ἄρα τῶν ζῴων καὶ ταῦτά ἐστιν. ὁ
ταὼς ὑπὲρ τοῦ μὴ βασκανθῆναι λίνου ῥίζαν οἱονεὶ
περίαπτόν τι φυσικὸν ἀναζητήσας, ὑπὸ τῇ ἑτέρᾳ
πτέρυγι βύσας περιφέρει. λέγεται δὲ καὶ ἵππος [1]
τὰ οὖρα εἰ ἐπισχεθείη,[2] παρθένος [3] λύσασα ἣν
φορεῖ ζώνην ἐὰν αὐτὸν παίσῃ [4] κατὰ τοῦ προσώπου
τῇ ζώνῃ, παραχρῆμα ἐξουρεῖν ἀθρόως καὶ τῆς
ὀδύνης παύεσθαι. θήλειαν δὲ ἵππον ἐς ἀφροδίσια
λυττήσασαν πάνυ σφόδρα παῦσαι ῥᾳδίως ἔστιν, ὡς
Ἀριστοτέλης λέγει, εἴ τις αὐτῆς ἀποκείρειε [5] τὰς
κατὰ τοῦ τένοντος τρίχας· αἰδεῖται γάρ, καὶ οὐκ
ἀτακτεῖ, καὶ παύεται τῆς ὕβρεως καὶ τοῦ σκιρτήμα-
τος τοῦ πολλοῦ, κατηφήσασα ἐπὶ τῇ αἰσχύνῃ.
τοῦτό τοι καὶ Σοφοκλῆς αἰνίττεται ἐν τῇ Τυροῖ
τῷ δράματι· πεποίηται δέ οἱ αὕτη λέγουσα, καὶ
ἃ λέγει ταῦτά ἐστιν

κόμης δὲ πένθος λαγχάνω πώλου δίκην,
ἥτις συναρπασθεῖσα βουκόλων ὕπο
μάνδραις ἐν ἱππείαισιν ἀγρίᾳ χερὶ
θέρος θερισθῇ ξανθὸν αὐχένων ἄπο,
σπασθεῖσα [6] δ' ἐς λειμῶνα [7] ποταμίων ποτῶν
ἴδῃ σκιᾶς εἴδωλον ἀνταυγὲς τύπῳ [8]
κουραῖς ἀτίμως διατετιλμένης φόβην.[9]
φεῦ, κἂν ἀνοικτίρμων τις οἰκτείρειέ νιν
πτήσσουσαν αἰσχύνησιν, οἷα μαίνεται
πενθοῦσα καὶ κλαίουσα τὴν πάρος φόβην.

19. Μελλούσης δὲ οἰκίας καταφέρεσθαι αἰσθητικῶς ἔχουσιν οἵ τε ἐν αὐτῇ μύες καὶ μέντοι καὶ

---

[1] Jac: ἵππου.   [2] ἐπισχεθῇ.
[3] Jac: παρθένου, -ον.   [4] παίῃ.

18. Here are further peculiarities of animals. The Peacock in order to escape the influence of the evil eye seeks out a root of flax as a kind of natural amulet and carries it about packed under one wing. And it is said that if a horse suffers from a retention of urine, and a maiden strikes him across the face with the girdle she is wearing, he immediately stales copiously and is relieved of his pain. And when a mare shows an altogether frenzied desire to go a-horsing it is easy to arrest her, according to Aristotle [*HA* 572 b 7], if one clips the mane on her neck. For she feels shame and is no longer skittish and drops her wantonness and her constant frisking and is downcast at her disgrace. And Sophocles, you remember, in his drama of *Tyro* hints at this. Tyro is represented as speaking, and this is what she says [*fr.* 659 P]:

'But it is my lot to grieve for my hair, even as a filly which seized by neatherds in the stables has had the yellow harvest reaped from her neck with ruthless hand; and haled to the meadow to drink of the stream, beholds the mirrored image of her reflexion with the hair cropped beneath the shears to her dishonour. Alas! even a pitiless heart would pity her, cowering in her shame, to see how wild are her grief and her tears for her lost hair.'

19. When a house is on the verge of ruin the mice in it, and the martens also, forestall its collapse and

Safeguards and remedies for animals

Animals give warning of impending disaster

---

⁵ ἀποκείρει.   ⁶ σπάσουσα H after GHermann.
⁷ Pearson : ἐν λειμῶνι MSS, H.
⁸ Pearson : αὐγασθεῖσ' ὑπό MSS, αἰκισθεῖσ' Haupt, H.
⁹ Brunck : φόβης MSS, H.

385

AELIAN

⟨αἱ⟩ [1] γαλαῖ, καὶ φθάνουσι τὴν καταφορὰν καὶ
ἐξοικίζονται. τοῦτό τοί φασι καὶ ἐν Ἑλίκῃ
γενέσθαι. ἐπειδὴ γὰρ ἠσέβησαν ἐς τοὺς Ἴωνας
τοὺς ἀφικομένους οἱ Ἑλικήσιοι, καὶ ἐπὶ βωμοῦ
ἀπέσφαξαν αὐτούς, ἐνταῦθα δήπου (τὸ Ὁμηρικὸν
τοῦτο) τοῖσιν δὲ θεοὶ τέραα προὔφαινον· πρὸ
πέντε γὰρ ἡμερῶν τοῦ ἀφανισθῆναι τὴν Ἑλίκην,
ὅσοι μύες ἐν αὐτῇ ἦσαν καὶ γαλαῖ καὶ ὄφεις καὶ
σκολόπενδραι καὶ σφονδύλαι καὶ τὰ λοιπὰ ὅσα ἦν
τοιαῦτα, ἀθρόα ὑπεξῄει τῇ ὁδῷ τῇ ἐς Κερύνειαν [2]
ἐκφερούσῃ. οἱ δὲ Ἑλικήσιοι ὁρῶντες τὰ [3] πρατ-
τόμενα ἐθαύμαζον μέν, οὐκ εἶχον δὲ τὴν αἰτίαν
συμβαλεῖν. ἐπεὶ δὲ ἀνεχώρησε τὰ προειρημένα
ζῷα, νύκτωρ γίνεται σεισμός, καὶ συνιζάνει ἡ
πόλις, καὶ ἐπικλύσαντος πολλοῦ κύματος ἡ Ἑλίκη
ἠφανίσθη, καὶ κατὰ τύχην Λακεδαιμονίων ὑφορμοῦσαι [4] δέκα νῆες συναπώλοντο τῇ προειρημένῃ.[5]
χρῆται δὲ ἅμα ἐς τιμωρίαν τῶν ἀσεβῶν ἀνδρῶν
ὑπηρέταις τοῖς ζῴοις ἡ Δίκη. καὶ τὸ [6] μαρτύριον,
Πανταλκῆς [7] ὁ Λακεδαιμόνιος ἀναστείλας διὰ τῆς
Σπάρτης ἐλθεῖν τοὺς ἐς Κύθηρα ἀπιόντας τῶν
περὶ τὸν Διόνυσον τεχνιτῶν, εἶτα καθήμενος ἐν
τῷ ἐφορείῳ ὑπὸ κυνῶν διεσπάσθη.

---

[1] ⟨αἱ⟩ add. H.
[2] Wesseling : Κορίαν.
[3] ταῦτα.
[4] ὑφορμοῦσαι τῇ πόλει.
[5] προειρημένῃ θαλάσσης ἐπικλύσει πολλῇ.
[6] καὶ τοῦδε.
[7] Παντεδίδας, Πανίηκλας etc.

emigrate. This, you know, is what they say happened at Helice,[a] for when the people of Helice
treated so impiously the Ionians who had come to
them, and murdered them at their altar, then it was
(in the words of Homer [*Od.* 12. 394]) that ' the gods
showed forth wonders among them.' For five days
before Helice disappeared all the mice and martens
and snakes and centipedes and beetles and every
other creature of that kind in the town left in a body
by the road that leads to Cerynea.[b] And the people
of Helice seeing this happening were filled with
amazement, but were unable to guess the reason.
But after the aforesaid creatures had departed, an
earthquake occurred in the night; the town col- Earthquake
lapsed; an immense wave poured over it, and Helice at Helice
disappeared, while ten Lacedaemonian vessels
which happened to be at anchor close by were
destroyed together with the city I speak of.

Justice at the same time uses animals as her
ministers to punish impious men. Witness the case
of Pantacles the Lacedaemonian [c] who, after pre-
venting some of the artists of Dionysus [d] who were
on their way to Cythera from passing through
Sparta, later, when seated upon the Ephor's throne,
was torn to pieces by dogs.

---

[a] In Achaia, about 1½ mi. from the Gulf of Corinth. In
373 B.C. delegates from Ionia came to beg for the statue of
Poseidon in Helice or at least for a plan of his temple and
altar, and at the very altar they were murdered by the people
of Helice. In the same year the town was destroyed by an
earthquake. See Frazer on Paus. 8. 24. 6.

[b] Hill-town, a short distance S of Helice.

[c] Pantacles is named as Ephor for the year 407 B.C. in two
interpolated passages of Xenophon, *Hell.* 1. 3. 1 and 2. 3. 10.

[d] Actors and musicians.

387

20. Ἐν Σικελίᾳ Ἀδρανός ἐστι πόλις, ὡς λέγει
Νυμφόδωρος, καὶ ἐν τῇ πόλει ταύτῃ Ἀδρανοῦ
νεώς, ἐπιχωρίου δαίμονος· πάνυ δὲ ἐναργῆ φησιν
εἶναι τοῦτον. καὶ τἆλλα μὲν ὅσα ὑπὲρ αὐτοῦ
λέγει, καὶ ὅπως ἐμφανής ἐστι καὶ ἐς τοὺς δεομένους [1] εὐμενής τε ἅμα καὶ ἵλεως, ⟨ἄλλοτε⟩ [2]
εἰσόμεθα· νῦν δὲ ἐκεῖνα εἰρήσεται. κύνες εἰσὶν
ἱεροί, καὶ οἵδε θεραπευτῆρες αὐτοῦ καὶ λατρεύοντές
οἱ, ὑπεραίροντες τὸ κάλλος τοὺς Μολοττοὺς κύνας
καὶ σὺν τούτῳ καὶ τὸ μέγεθος, χιλίων οὐ μείους
τὸν ἀριθμόν. οὐκοῦν οὗτοι μεθ' ἡμέραν μὲν
αἰκάλλουσί τε καὶ σαίνουσι τοὺς ἐς τὸν νεὼν καὶ τὸ
ἄλσος παριόντας, εἴτε εἶεν ξένοι εἴτε ἐπιχώριοι·
νύκτωρ δὲ τοὺς μεθύοντας ἤδη καὶ σφαλλομένους
κατὰ τὴν ὁδὸν οἶδε πομπῶν δίκην καὶ ἡγεμόνων
μάλα εὐμενῶς [3] ἄγουσι, προηγούμενοι ἐς τὰ
οἰκεῖα ἑκάστῳ,[4] καὶ τῶν μὲν παροινούντων τιμωρίαν ἀρκοῦσαν ἐσπράττονται· ἐμπηδῶσι γὰρ καὶ
τὴν ἐσθῆτα αὐτοῖς καταρρηγνύουσι, καὶ σωφρονί-
ζουσιν [5] ἐς τοσοῦτον αὐτούς· τούς γε μὴν πει-
ρωμένους λωποδυτεῖν διασπῶσι πικρότατα.

21. Κοχλίας δὲ ἄρα θαλάττιος ὁ ἐν τῇ Ἐρυθρᾷ
θαλάττῃ γινόμενος ὡραιότατος ἰδεῖν ἦν καὶ
μέγιστος· ἔστι μὲν γὰρ φοῖνιξ τὸ ἔλυτρον, ἔχει
δὲ καὶ ἕλικα [6] διηνθισμένην καὶ πεποικιλμένην
ὑπὸ τῆς φύσεως.[7] στέφανον ἂν εἴποις ὁρᾶν ἔκ
τινος πολυχροίας ἀνθῶν διαπλακέντα [8] πρασίνων

---

[1] δεομένους πρόχειρος.　　　[2] ⟨ἄλλοτε⟩ add. H.
[3] Schn : εὐγενῶς.　　　　　[4] ἕκαστον.
[5] Ges : σωφρονοῦσιν.

20. Adranus is a town in Sicily,[a] according to Nymphodorus, and in this town there is a temple to Adranus, a local divinity. And they say that he is there in very presence. And all that Nymphodorus tells of him besides, and how he shows himself and how kindly and favourable he is to his suppliants, we shall learn some other time. But now I shall give the following facts. There are sacred Hounds and they are his servants and ministers; they surpass Molossians in beauty and in size as well, and there are not less than a thousand of them. Now in the daytime they welcome and fawn upon visitors to the shrine and the grove, whether they be strangers or natives. But at night they act as escorts and leaders, and with great kindness conduct those who are already drunk and staggering along the road, guiding each one to his own house, while those who indulge in tipsy frolics they punish as they deserve, for they leap upon them and rip their clothes to pieces and chasten them to that extent. But those who are bent on highway robbery they tear most savagely.

Sacred Hounds in the temple of Adranus

21. There is, it seems, a marine snail which is born in the Red Sea and of great beauty and very large. Its shell is purple and its spiral has been decorated and made gay by Nature.[b] You would say you were looking at a garland subtly woven of

A Red Sea Snail

---

[a] On the SW slopes of mt Etna.
[b] This is the *Mitra papalis*, Gossen § 20.

---

[6] ἕλικα μεστήν MSS, καλλίστην *Jac.*
[7] φύσεως κόσμῳ δὲπεριττῷ.
[8] ποικίλως διαπλακέντα.

τε καὶ χρυσοειδῶν καὶ κινναβαρίνων, ἐναλλὰξ τῶν
χρωμάτων κατεσπαρμένων διαστήμασιν ¹ ἴσοις.

22. Τὸν δελφῖνα ἡ φύσις ἀεικίνητον εἰργάσατο,
ὥς φασι, καὶ πέρας τούτῳ τῆς κινήσεως τὸ καὶ
τοῦ βίου.² ὕπνου γοῦν δεόμενος μετεωρίσας τὸ
σῶμα καὶ ἀναπλεύσας ἐπ᾽ ἄκρον τὸ ὕδωρ, ὡς
ὁρᾶσθαι πᾶς, καταδαρθάνει τηνικάδε· ἄυπνος δὲ
καὶ ἄμοιρος τοῦδε τοῦ θεοῦ οὐδὲ οὗτός ἐστιν. ὅτε
γοῦν καθεύδει, ὠθεῖται ἐς βυθόν, ἕως ἂν ψαύσῃ ³
τῆς κάτω γῆς. ὅταν δὲ προσπελασθῇ αὐτῇ,
διυπνίζεται κρουσθεὶς πρὸς τὸ δάπεδον, εἶτα
ἀναδύνει. καὶ πάλιν ἐς ὕπνον ὑπαχθεὶς καὶ
νικώμενος τοῦ θεοῦ κατολισθάνει, καὶ αὖθις
ἀφυπνισθεὶς τῇ αὐτῇ κρούσει ἀναπλεῖ πάλιν. καὶ
πολλάκις δρᾷ τοῦτο, μεταξὺ ἡσυχίας καὶ ἐνεργείας
ὤν, οὐ μὴν ἐς ἀκινησίαν ἐκπίπτων παντελῆ ποτε.

23. Ἐν τῇ Ἐρυθρᾷ θαλάττῃ γίνεται ἰχθὺς
πλατὺς τὸ σχῆμα κατὰ τὴν βούγλωττον, ὥς φασι.
καὶ φολίδας μὲν οὐ σφόδρα τραχύς ἐστι προσαψα-
μένῳ, τὴν χρόαν δὲ ὑπόχρυσός ἐστι, μελαίναις τε
γραμμαῖς ἐς τὸ οὐραῖον ἀπὸ τῆς κεφαλῆς ἄκρας
καταγέγραπται. εἴποι τις ἂν αὐτὰς εἶναι χορδὰς
ἐντεταμένας· ἔνθεν τοι καὶ ⟨ὁ⟩ ⁴ ἰχθὺς αὐτὸς
κιθαρῳδὸς κέκληται. τὸ στόμα δὲ αὐτῷ συνίζει
καὶ ἔστι μέλαν ἰσχυρῶς, ζώνῃ γε μὴν κροκοειδεῖ
κατείληπται· πεποίκιλται δέ οἱ ἡ κορυφὴ διαφόρως
τῇ τε χρυσοειδεῖ αὐγῇ καὶ μέντοι καὶ μελαίναις
τισὶ περιγραφαῖς. καὶ πτερύγια χρυσοειδῆ ἔχει,

¹ τοῖς δ.   ² βίου τέλος.

flowers of varied hue, green and golden and ver-
milion, the colours alternating at equal intervals.

22. Nature, they say, has caused the Dolphin to The Dolphin
be in perpetual motion, and for the Dolphin motion motion
ends with the end of life. At any rate when in need
of sleep it rises and floats up to the surface so that
its whole body is visible, and then goes to sleep.
Even the Dolphin is not unsleeping or devoid of a
share of the god of sleep. At all events when it does
sleep it sinks into the depths until it touches the
bottom, and when it reaches it, it wakes on the
impact with the floor of the sea and rises again.
And again when overcome by sleep and subdued by
the god, down it sinks, and again when roused by the
impact as before, up it floats; and it does this time
after time, being half-way between repose and
activity, and yet never once does it lapse into
complete immobility.

23. In the Red Sea there occurs a flat-fish shaped The
like the sole, so they say. Its scales are not very fish
'Harper'
rough to the touch; its colour is golden, and from
head-tip to tail it is marked with black lines. One
might describe them as tense strings, which is the
reason why the fish itself is called the 'Harper.'[a]
Its mouth is compressed and is a deep black and is
enclosed in a saffron-coloured ring; its head is
variegated, gleaming like gold and with black lines.
It has fins like gold, but its tail is black except at the

[a] A species of *Chaetodont*, a brightly-coloured fish inhabiting
coral-reefs.

---

[3] ἕως ψαύσει.   [4] ⟨ὁ⟩ add. H.

μέλαινα δὲ αὐτῷ ἡ οὐρὰ πλὴν τῶν ἄκρων· ταῦτα
δὲ λευκὰ ἰσχυρῶς. καὶ ἄλλοι δὲ ᾄδονται κιθαρῳδοὶ
τίκτεσθαι.[1] καί εἰσι πορφυροῖ μὲν τὸ πᾶν σῶμα,
γραμμὰς ἐκ [2] διαστημάτων ἔχοντες χρυσᾶς· ζώνας
δ' ἔχουσιν ἐπὶ τῇ κεφαλῇ ἴοις τοῖς ἄνθεσι [3] παρα-
πλησίας, τὴν μὲν πρὸ τῶν ὀφθαλμῶν μέχρι τῶν
βραγχίων καθέρπουσαν, τὴν δὲ μετὰ τοὺς ὀφθαλμοὺς ἐς τὸ ἥμισυ τῆς κεφαλῆς προχωροῦσαν, τὴν
δὲ περιθέουσαν κατὰ τῆς δέρης ὡς ὅρμον.

24. Πάρδαλις δὲ ἰχθὺς ἐν τῇ Ἐρυθρᾷ φύεται
θαλάττῃ, ὡς οἱ θεασάμενοι λέγουσι, καὶ ἔοικε τὴν
χρόαν καὶ τὰ στίγματα τὰ περιφερῆ τῇ ὀρείῳ
παρδάλει. ὁ δὲ ὀξύρυγχος ὁ ἐνταῦθα γινόμενος
ἔχει μὲν πρόμηκες [4] τὸ στόμα, τοὺς δὲ ὀφθαλμοὺς
χρυσοειδεῖς, τὰ δὲ βλέφαρα αὐτῷ λευκά· τῷ δὲ
νώτῳ οἱ σημεῖά τε ἐπέστικται ὠχρά, καὶ πτέρυγες
αὐτῷ αἱ μὲν παρ' ἑκάτερα [5] μέλαιναι, αἱ δὲ
νωτιαῖαι λευκαί· καὶ ἡ οὐρὰ προμήκης τὸ σχῆμα,
τὴν δὲ χρόαν πράσινός ἐστι, μέσην δὲ αὐτὴν
διείληφε χρυσοειδὴς γραμμή.

25. Τῷ Πτολεμαίῳ τῷ δευτέρῳ, ὃν καὶ Φιλά-
δελφον καλοῦσι, βρέφος ἐλέφαντος [6] δῶρον ἐδόθη,
καὶ τῇ φωνῇ ἐνετράφη [7] τῇ Ἑλλάδι, καὶ λαλούντων συνίει. ἐπεπίστευτο δὲ πρὸ τοῦδε τοῦ ζῴου
τῆς Ἰνδῶν μόνης φωνῆς ἐπαΐειν τοὺς ἐλέφαντας.

26. Ἔοικε δὲ ἄρα καὶ ἐν τοῖς ἀλόγοις ὑπὸ τῆς
φύσεως προτιμᾶσθαι τὸ ἄρρεν. ἔχει γοῦν ὁ μὲν

---

[1] στικτοὶ τίκτεσθαι.   [2] δὲ ἐκ.

tip, and that is the purest white. And other kinds of Harper are said to occur: some are purple all over, with golden lines at intervals. They have rings the colour of gilliflowers on their head: one descends from below the eyes down to the gills, another extends from behind the eyes half-way down the head, and another encircles the neck like a necklace.

24. The Leopard-fish is native to the Red Sea, according to those who have seen it, and in its colour and circular markings resembles the leopard of the mountains.

The Oxyrhynchus, which occurs there, has an elongated mouth, eyes like gold, and white eyelids. There are pale markings on its back, but the fins on either side are black, while the dorsal fins are white. Its tail is oblong in shape and its colour is green, and a streak of gold bisects it.

25. Ptolemy the Second, also called Philadelphus, was presented with a young Elephant, and it was brought up where the Greek language was used, and understood those who spoke it. Up to the time of this particular animal it was believed that Elephants only understood the language spoken by the Indians.

26. It seems that among brute beasts also Nature has put the male above the female. At any rate

The Leopard-fish

The 'Oxyrhynchus' fish

Ptolemy II and his Elephant

The Male superior to the Female

---

³ τοῖς ἄ. del. Cobet.   ⁴ Ges: προμήκης, -κη.
⁵ παρ' ἑ.] πρῶται.
⁶ Gron: ἐλάφου, and below, ἐλάφους.
⁷ Schn: ἀνετράφη.

δράκων ὁ ἄρρην τὸν λόφον καὶ τὴν ὑπήνην,[1]
ὁ δὲ ἀλεκτρυὼν καὶ οὗτος ⟨τὸν⟩[2] λόφον καὶ τὰ
κάλλαια,[3] ὁ δὲ ἔλαφος[4] τὰ κέρατα, ⟨τὴν⟩[5]
χαίτην ὁ λέων, ὁ τέττιξ τὴν φωνήν.

27. Ὑπόθεσις μὲν τοῖς Ἀχαιοῖς καὶ τοῖς Τρωσὶ
τοῦ πολέμου ἡ Διὸς Ἑλένη φασί, καὶ Πέρσαις
πρὸς τοὺς Ἕλληνας Ἄτοσσα ἡ Δαρείου γυνὴ
ποθήσασα θεραπαίνας κτήσασθαι Ἀττικάς,[6] καὶ
τοῦ μακροῦ πολέμου τοῖς Ἕλλησι τὸ πινάκιον τὸ
κατὰ τῶν Μεγαρέων. Μάγνητας δὲ καὶ Ἐφεσίους
ἐς πόλεμον ἀκρὶς ἐξῆψε, περιστερὰ δὲ Χάονας καὶ
Ἰλλυριούς, Θηβαῖοι δ' ⟨οἱ⟩[7] ἐν Αἰγύπτῳ πρὸς
Ῥωμαίους ὑπὲρ κυνὸς πολεμῆσαι λέγονται.

28. Λέγει τις λόγος Πυθοχάρην τὸν αὐλητὴν
ἀναστεῖλαι λύκων ὁρμὴν αὐλήσαντα σύντονον καὶ
γενναῖον αὔλημα. μυιῶν δὲ πλῆθος ἀνέστησε
Μεγαρέας, Φασηλίτας δὲ σφῆκες, σκολόπενδραι δὲ
Ῥοιτιεῖς.

---

[1] ὑπήνην δασεῖαν.   [2] ⟨τόν⟩ add. H.
[3] Reiske: κάλλεα.   [4] Ges: ἐλέφας.
[5] ⟨τὴν⟩ add. Schn.   [6] Ἀττικὰς καὶ Ἰάδας.
[7] ⟨οἱ⟩ add. H.

---

[a] See Hdt. 3. 134.
[b] Pericles in 432 B.C. attempted to stop Megara from trading
in the Aegean, and so starve it into surrender. This was a
contributory cause of the Peloponnesian war.

the male Dragon has the crest and the beard; and the Cock too has the comb and the wattles; and the Stag has the horns, the Lion the mane, the male Cicada the voice.

27. The war between the Achaeans and Trojans was caused, they say, by Helen the daughter of Zeus; the war of the Persians against the Greeks was caused by Atossa the wife of Darius who had conceived a desire to obtain Athenian women for her service; [a] and the long war in Greece [b] was due to the proclamation directed against the people of Megara. The people of Magnesia [c] and of Ephesus were roused to war by a locust; the people of Chaonia [d] and of Moesia by a dove; and the people of Thebes in Egypt are said to have made war against the Romans because of a dog.[e]

Small causes of great wars

28. There is a story that Pythochares the piper repelled an attack of wolves by playing a loud and noble strain on his pipe. And a swarm of flies drove out the people of Megara, wasps the people of Phaselis,[f] and centipedes the people of Rhoeteum.[g]

Victor and vanquished

[c] Magnesia on the river Maeander rivalled Ephesus in importance, but was destroyed by the Ephesians in the middle of the 7th cent. B.C. The reference to a locust has not been explained.

[d] The Chaones were a powerful tribe in Epirus. The 'dove' may conceal a reference to the oracle at Dodona, whose priestesses were called 'doves'; cp. Hdt. 2. 57. But of a war between the Chaones and their northern neighbours the Illyrians nothing is known. Moesia lay some hundreds of miles N of Epirus beyond mt Haemus.

[e] Nothing is known of this.

[f] Town on the E coast of Lycia.

[g] Town NE of Troy on the Hellespont.

AELIAN

29. Πρόβατα ἄχολα ἐν τῷ Πόντῳ φασίν, ἐν δὲ τῇ Νάξῳ τῇ νήσῳ δίχολα.

30. Ὁ μέροψ ⟨ὁ⟩ [1] ὄρνις ταύτῃ τοι δοκεῖ δικαιότερος εἶναι τῶν πελαργῶν· οὐ γὰρ ἀναμένει γηράσαντας τρέφειν [2] τοὺς πατέρας, ἀλλ᾽ ἅμα τῷ φῦσαι τὰ ὠκύπτερα τοῦτο ἐργάζεται.[3]

31. Ἴδιον δὲ ἄρα τῶν ζῴων καὶ ἐκεῖνο ἀγαθόν. πρόνοιαν αὐτῶν οἱ [4] θεοὶ ποιοῦνται, καὶ οὔτε αὐτῶν καταφρονοῦσιν οὔτε [5] μὴν ὀλιγώρως ἔχουσιν. εἰ γὰρ καὶ ἀμοιρεῖ λόγου, ἀλλὰ γοῦν συνέσεως καὶ τῆς καθ᾽ ἑαυτὰ σοφίας οὐκ ἀτυχεῖ.[6] ὅπως οὖν [7] καὶ αὐτὰ φιλεῖται θεοῖς ἐρῶ, καὶ εἰ μὴ πολλὰ ἐκ πολλῶν, ὅσα δ᾽ οὖν ἀποχρήσει.[8] ἀνὴρ ἱππεύς,[9] Ληναῖος τὸ ὄνομα, ἵππον εἶχεν ἰδεῖν μὲν ὡραῖον, δραμεῖν δὲ ὤκιστον, τὸν δὲ θυμὸν ἀνδρειότατον· καὶ ἀγαθὸν μὲν ἐν ταῖς ἐπιδείξεσι τὴν ἱππείαν περιδραμεῖν τὴν δεδιδαγμένην, καρτερικὸν δὲ ἐν αὐτῷ τῷ πολέμῳ, καὶ διῶξαι ἔνθα ἦν καιρὸς καὶ ἀναχωρῆσαι ὅπου αὐτὸν ⟨ἡ⟩ [10] χρεία παρεκάλει πάνυ γεννικόν. οὐκοῦν ἐκ τούτων ἁπάντων ὁ μὲν κτῆμα ἦν ἀγαθόν, ὁ δὲ εὐκλεέστατος ἐν τοῖς ὁμοτέχνοις ἱππεῦσιν ἐδόκει. ὁ τοίνυν ἵππος ὁ τοιοῦτος τὴν ἱππικὴν ἀρετὴν θατέρῳ τοῖν ὀφθαλμοῖν τῷ δεξιῷ ὑπό τινος πληγῆς προσπεσούσης ὁρᾶν ἀδύνατος ἦν. οὐκοῦν ὁ Ληναῖος ὁρῶν ἑαυτοῦ σαλεύ-

---

[1] ⟨ὁ⟩ add. H.   [2] ἐκτρέφειν.
[3] ἐργάζεται καὶ ἔστι δικαιότερος καὶ εὐσεβέστερος ὀρνίθων ἁπάντων.
[4] καὶ οἱ.   [5] οὐδέ.
[6] ἔστιν ἀτυχῇ.   [7] δ᾽ οὖν.

29. They say that the Sheep of Pontus have no gall-bladder, whereas those on the isle of Naxos have two. _{The Sheep of Pontus and of Naxos}

30. The Bee-eater appears to be more dutiful than the stork, for this reason: it does not wait for its parents to grow old before it starts to feed them, but does so directly it grows its quill-feathers. _{The Bee-eater}

31. Here is another characteristic of animals and a good one. The gods take thought for them, neither looking down upon them nor reckoning them of small account. For although destitute of reasoning power, at any rate they possess understanding and knowledge proportionate to their needs. And I will explain how they are beloved of the gods, not by many examples taken from a multitude but by a sufficient number. _{Serapis restores a Horse's eye}

A cavalry officer of the name of Lenaeus owned a horse of fine appearance, very fleet of foot and of dauntless spirit; in displays it was good at running the course it had been taught; in war itself it was capable of endurance; and was quite excellent both in pursuit, when occasion arose, and in retreat, where necessity called for it. And in consequence of all this the horse was a valued possession, and the owner was accounted most fortunate by his fellow cavalrymen. Now the horse, with the excellent qualities I have described, in consequence of a blow which it received in its right eye was incapacitated for seeing. Accordingly Lenaeus seeing all his

---

[8] ἀποχρήσει τοσαῦτα. [9] *Jac*: ἱππεὺς τὴν στρατιάν.
[10] ⟨ἡ⟩ add. H.

AELIAN

ουσαν τὴν πᾶσαν ἐλπίδα ἐν τῷ τοῦ ἵππου τοῦ
εὐγενοῦς ἐκείνου πάθει, ἐπεὶ [1] καὶ ἡ ἀσπὶς ἡ
ἱππικὴ τὸν λαιὸν ὀφθαλμὸν οἱ ἔσκεπε τὸν μόνον
ὁρῶντα, ἐς τοῦ Σαράπιδος ἔρχεται, θεράπευμα [2]
ἀνάγων καὶ μάλα ἄηθες [3] τὸν ἵππον, καὶ δεῖται
τοῦ θεοῦ ὡς ὑπὲρ ἀδελφοῦ τινος ἢ υἱοῦ ὁ Ληναῖος
τοῦ ἵππου οἰκτεῖραι τὸν ἱκέτην, καὶ ταῦτα ἀδική-
σαντα οὐδὲ ἕν. εἶναι γάρ τοι [4] ἀνθρώπους σφίσι
κακῶν αἰτίους ἢ δράσαντάς τι ἀσεβὲς ἢ εἰπόντας
τι ἀπόφημον· ‘ἵππου δὲ’ ἔλεγε ‘ποία μὲν θεοσυλία,
φόνος δὲ τίς, βλασφημία δὲ πῶς ἢ πόθεν;’ ἐμαρ-
τύρατο δὲ τὸν θεὸν καὶ αὐτὸς ὡς οὐδεπώποτε
οὐδένα οὐδὲν ἀδικήσας, καὶ διὰ ταῦτα τὸν συστρατιώτην οἱ καὶ φίλον ἐδεῖτο τῆς ὀφθαλμίας ἀπαλλάξαι. [5]
ὁ δὲ οὐχ ὑπερορᾷ οὐδὲ ἐξεφαύλισε τὸν ἄλογόν τε
καὶ ἄφωνον ἰάσασθαι, ὢν τοσοῦτος θεός, καὶ διὰ
ταῦτα οἰκτείρει καὶ τὸν νοσοῦντα καὶ τὸν δεόμενον
ὑπὲρ αὐτοῦ, καὶ δίδωσιν ἴασιν μὴ καταιονεῖν μὲν
τὸν ὀφθαλμόν, πυριάσεσι δὲ αὐτὸν ἀλεαίνειν μεσού-
σης ἡμέρας ἐν τῷ τοῦ νεὼ περιβόλῳ. καὶ ταῦτα
μὲν ἐπράττετο, ἐρρώσθη δὲ τῷ ἵππῳ τὸ ὄμμα.
καὶ ὁ μὲν Ληναῖος χαριστήριά τε καὶ ζωάγρια
ἀπέθυεν, ὁ δὲ ἵππος ἐσκίρτα τε καὶ [6] ἐφριμάττετο
καὶ ἐδόκει μείζων τε καὶ ὡραιότερος, καὶ ἦν
φαιδρὸς καὶ τῷ βωμῷ προσθέων ἐκυδροῦτο, καὶ
μέντοι καὶ πρὸς τοῖς ἀναβαθμοῖς καλινδούμενος
ἑωρᾶτο τῷ θεῷ τῷ σωτῆρι χαριστήρια ἐκτίνων,
ᾗπερ οὖν ἔσθενεν.

[1] ἐπεὶ τὰ ἄλλα.   [2] θρέμμα.
[3] Reiske : ἀληθές MSS, adding ὥσπερ οὖν ἱερεῖον, del. H.
[4] τινων.   [5] ἀπαλλάξαι τὸν θεόν.
[6] τε καὶ περὶ τὸν νεών.

hopes anchored upon the condition of his noble horse
(the cavalry shield covered the left eye which alone
could see), went to the temple of Serapis bringing a
patient of a most unusual kind,—his horse, and, as
though he were pleading for a brother or a son,
implored the god for the horse's sake to have com-
passion on his suppliant, especially as it had done
no wrong. For men, he said, may bring misfortune
upon themselves either by some impious act or
some blasphemous speech. 'But what sacrilege,' he
exclaimed, ' or what murder has a horse committed,
and how and by what means has it blasphemed?'
And he called the god to witness that he himself had
never wronged any man, and for this reason he
implored the god to relieve his comrade-in-arms and
friend of its blindness. And the god, although so
mighty, did not neglect or scorn to heal the dumb
beast, and therefore took pity both on the sick
animal and on the man who besought him on its
behalf, and prescribed a cure, not by fomenting the
eye but by warming it with vapour baths at midday
in the temple precinct. So this was done and the
eye of the horse was restored. And Lenaeus
sacrificed thank-offerings and donations for its re-
covery, while the horse pranced and snorted and
seemed larger and more beautiful and was full of
joy, and speeding to the altar moved so proudly, and
as it rolled in front of the steps was seen to be giving
thanks with all its might to the god who had healed
it.

32. Ἐν ἀμπελῶνι [1] δὲ γεωργὸς εἰργάζετο τάφρον, ἵνα ἐμφυτεύσῃ [2] κλῆμα καλόν [3] τε καὶ εὐγενές· εἶτα τὴν σμινύην καταφέρων ὑποικουροῦσαν ἀσπίδα ἱερὰν καὶ ἀνθρώπων ἥκιστα ἐχθρὰν λαθὼν διέκοψε μέσην. καὶ τὴν γῆν διαξαίνων τὸ μὲν οὐραῖον βλέπων τῇ ψάμμῳ κατειλημένον,[4] τὸ δὲ ἡμίτομον τὸ ἐκ τῆς γαστρὸς ἐς τὴν δέρην ἀνιὸν ἔτι ἕρπον καὶ τοῦ λύθρου τοῦ διὰ τὴν τομὴν πεπληρωμένον, ἐκπλήττεται, καὶ ἔκφρων γενόμενος ἔς τε ὀρθὴν μανίαν καὶ ὡς τὰ μάλιστα ἰσχυρὰν ἐκφοιτᾷ. καὶ μεθ' ἡμέραν ἑαυτοῦ τε καὶ τοῦ λογισμοῦ ἦν ἀκράτωρ καὶ μέντοι καὶ νύκτωρ [5] ἦν παράφορος, καὶ ἐκ τοῦ λέχους ἀνεθόρνυτο καὶ ἔλεγε τὴν ἀσπίδα διώκειν, καὶ ὥσπερ οὖν ὁμοῦ τι τῷ δήγματι ὢν ἐκπληκτικώτατα ἐβόα καὶ ἐκάλει συμμάχους, καὶ μέντοι καὶ τῆς ἀνῃρημένης ὑπ' αὐτοῦ τὸ εἴδωλον ἔλεγεν ὁρᾶν βριμούμενόν τε καὶ ἀπειλοῦν, καὶ ὡμολόγει ποτὲ καὶ δεδῆχθαι, καὶ ὡς ὠδυνᾶτο ἐξ ὧν ᾤμωζεν [6] ἦν δῆλος. ἐπεὶ μέντοι ἡ νόσος πόρρω τοῦ χρόνου ἦν, οἱ προσήκοντές οἱ τὸν ἄνδρα ἐς τοῦ Σαράπιδος ἄγουσιν ἱκέτην, καὶ ἐδέοντο ἀναστεῖλαι καὶ ἀφανίσαι τῆς προειρημένης τὸ φάσμα. οἰκτείρει μὲν οὖν τὸν ἄνδρα ὁ θεὸς καὶ ἰᾶται· ὡς δὲ ἀτιμώρητος οὐκ ἔμεινεν ἡ ἀσπὶς εἴρηται καὶ πάνυ γε ἀποχρώντως.

33. Ταῶν δὲ Ἰνδικὸν δῶρον λαβὼν ὁ τῶν Αἰγυπτίων βασιλεύς, ταῶνων ἰδεῖν μέγιστόν τε καὶ ὡραιότατον, οὐκ ἀξιοῖ σὺν τοῖς ἀγελαίοις τρέφειν, ὡς οἰκίας ἄθυρμα αὐτὸν εἶναι ἢ γαστρὸς

---

[1] ἀμπέλῳ. [2] ἐμφυτευθῇ.
[3] καλὸν κλῆμα. [4] Cobet : κατειλημμένον MSS, H.

32. A husbandman was digging a trench in a vine- A sacred
yard in order to plant some fine, choice cutting, when $\frac{\text{Asp and its}}{\text{slayer}}$
he brought down his mattock upon a sacred Asp that
had its lair below the soil and was far from hostile to
man, and without knowing it cut the snake in half.
And as he was breaking up the soil he caught sight
of the tail involved in the sand, while the severed
portion from the belly upwards to the neck was still
crawling and covered with gore from the cut. He
was horror-struck, went out of his mind, and passed
into a state of real madness of the most acute
description. By day he lost control of himself and
of his reason; moreover at night he was in a state of
frenzy, and would leap out of bed saying that the
Asp was pursuing him, and as though he was on the
point of being bitten would utter the most horrifying
cries and shout for help. He would even say that he
saw the form of the snake which he had slain, angrily
threatening him; at times he avowed that he had
been bitten, and it was evident from his groans that
he was in pain. So when his affliction had lasted for
some time, his relations took him as a suppliant to the
temple of Serapisand implored the god to remove
and abolish the phantom of the aforesaid Asp. Well,
the god took pity on the man and cured him. But I
have described how the Asp had not to wait for its
revenge, and a very sufficient revenge too.

33. The King of Egypt was presented with a A sacred
Peacock from India, the largest and most magnificent Peacock
of its kind. He was unwilling to keep it along with
the common flock as a household pet or for eating,

---

⁵ νυκτός.　　　⁶ ᾤμωξεν.

χάριν, ἀλλὰ ἀνάπτει τῷ Πολιεῖ Διί, κρίνας ἀνάθημα
ἐπάξιον τῷ θεῷ τὸν ὄρνιν τὸν προειρημένον. ἐρᾷ
τοῦτον συλλαβεῖν ἄσωτος νεανίας καὶ πάνυ γε
πλούσιος καὶ ποιήσασθαι δεῖπνον· ἀεὶ γὰρ τῇ
γαστρὶ ἐχαρίζετο καὶ ἐξ ἁπάσης αἰτίας ὁ ἄσωτος
οὗτος, τὸ ποικίλον τῆς τροφῆς καὶ τὸ σὺν κινδύνῳ
πορισθὲν καὶ τὸ ἐωνημένον πολλῶν πόνων [1]
λαιμαργίας καὶ βδελυρίας ὑπερβολῇ κέρδος ἡγούμε-
νος ἐς ἡδονήν. μισθὸν οὖν τῆς θεοσυλίας ἁδρὸν
προτείνει τινὶ τῶν τοῦ θεοῦ θεραπευτήρων, καὶ
ὑπισχνεῖται καὶ ἄλλον. ὁ δὲ ἐλπίδι κουφισθεὶς
ματαίᾳ, ἔνθα ᾔδει τὸν ὄρνιν αὐλιζόμενον ἐλθὼν
ἐπεχείρει συλλαβεῖν καὶ τῷ πλουσίῳ κομίσαι.
καὶ ἐκεῖνον μὲν οὐχ ὁρᾷ, μεγίστην δὲ ἀσπίδα
ὀρθὴν εἶδε καὶ ἐπ' αὐτὸν τεθυμωμένην. καὶ τὰ
μὲν πρῶτα ἔδεισε καὶ ὑπαπῆλθεν,[2] ἐγκειμένου δὲ
τοῦ ἀσώτου καὶ παρορμῶντος ὁ ⟨μὲν⟩ [3] ὑπηρέτης
ἐπὶ τὸν ταῶν ἦλθεν, ὁ δὲ ἀνωτέρω [4] ᾄξας καὶ
ἑαυτὸν τοῖς πτεροῖς μετεωρίσας καὶ ἀρθεὶς κοῦφος
οὔτε ἐπὶ τι τῶν ἱερῶν δένδρων ἐκάθισεν οὔτε ἐπ'
ἄλλον μετέωρόν τε καὶ ὑψηλὸν χῶρον, ἀλλὰ ἐπὶ τὸ
μέσον τοῦ νεώ,[5] καὶ ἐς αὐτοὺς ἀτρέπτως ἑώρα,
οἷον ὑποφαίνων ὅτι ἄρα τῆς ἐπιβουλῆς τῆς ⟨ἐξ⟩ [6]
ἐκείνων κρείττων πέφυκε, καὶ οὐκ ἔστιν αὐτοῖς
ἑλεῖν αὐτόν. οὐκοῦν ἐπεὶ μηδὲν ἤνυστο, ὅπερ ἦν
προδοὺς ἀργύριον ὁ ἄσωτος ἀπῄτει λαβεῖν, ὁ δὲ
οὐκ ἀπεδίδου λέγων ποιῆσαι μὲν τὰ ἑαυτοῦ πάντα,
ἀδύνατος δὲ εἶναι θεῶν κτῆμα ὑφελέσθαι. οἷα

---

[1] πολλῶν ὤνων V, πολλοῦ τρόπῳ most MSS.
[2] ἀπῆλθεν H.   [3] ⟨μέν⟩ add. H.
[4] ἀμφοτέρων.
[5] ἐπὶ τι μέσον τοῦ θεοῦ τοῦ νεώ.

but attached it to the temple of Zeus Protector of the City, judging the aforesaid bird to be an offering worthy of the god. This bird a dissolute youth of considerable wealth longed to capture and to make a meal of, for he habitually indulged his appetite on any and every pretext, and in his extravagant gluttony and depravity he regarded variety of food and what had been acquired by dangerous means and what had been purchased at the cost of immense trouble as an accession to his pleasure. Accordingly he offered one of the attendants on the god a substantial bribe to commit sacrilege, and promised him a further sum besides. And the man elated by a vain hope went to the spot where he knew the bird lodged and tried to lay hands on it and bring it to his rich patron. But the bird he did not see: what he did see was a huge asp reared up in anger against him. At first he was afraid and made off, but when the dissolute man insisted and urged him on, the attendant went to get the Peacock. But the bird sprang up out of reach and raising itself lightly through the air on its wings, settled not upon one of the sacred trees nor upon any other lofty and high spot but upon the centre of the temple, and surveyed them with an unflinching eye as though to show that it was too clever for their designs and that it was not to be caught. Accordingly since the attendant had accomplished nothing, the dissolute man demanded the money, which he had already given, back again; but the other refused, saying that he had carried out his orders but was unable to steal what belonged to the gods. As was natural, a quarrel arose over the

---

[6] ⟨ἐξ⟩ add. H.

οὖν εἰκὸς ἦν ἔρις ὑπὲρ τούτων καὶ βοὴ ἤδη, καὶ
ἤκουον πολλοί· εἶτα ἄνεισιν ὁ ἐπὶ πᾶσιν ἱερεύς,
καὶ ἐρωτᾷ τῆς ἐν τῷ νεῷ φιλονικίας τὴν ὑπόθεσιν,
καὶ ἐλέγχουσιν ἀλλήλους. καὶ ὁ μὲν πλούσιος
ἀπειλαῖς καὶ βλασφημίαις καὶ λοιδορίαις αἰκισθεὶς
ἀπαλλάττεται. καὶ ὄρνιθος ἄλλου καταπιὼν ὀστοῦν
καὶ ὀδυνώμενος τὸν βίον κατέστρεψεν ἀλγεινότατα,
τὸν δὲ ὑπηρέτην τὸν κακὸν οἷα δήπου ἱερόσυλον ὁ
τῆς πόλεως ἁρμοστὴς ἐκόλασε, τὸν δὲ ὄρνιν οὔτε
ζῶντα οὔτε νεκρὸν ἐθεάσαντο, ἀλλὰ ἑκατὸν ὡς
λόγος ἔτη διαβιώσας εἶτα ἠφανίσθη.

34. Καὶ ἐκεῖνο δὲ ἔοικε τούτῳ καὶ ὁμολογεῖ.
Κίσσος ὄνομα θεραπεύων τὸν Σάραπιν ἰσχυρῶς,
ἐπιβουλευθεὶς ὑπὸ τῆς πρότερον μὲν ἐρωμένης
ὕστερον δὲ γαμετῆς, καὶ ᾠὰ ὄφεως φαγών,
ὠδυνᾶτο καὶ ἑαυτοῦ κακῶς εἶχε, καὶ ἐπίδοξος
τεθνήξεσθαι ἦν. δεῖται δὲ τοῦ θεοῦ, ὁ δὲ προσέταξε
πρίασθαι μύραιναν ζῶσαν, καθεῖναι δὲ τὴν χεῖρα
ἐς τὸ ζωγρεῖον.[1] καὶ ὁ Κίσσος πείθεται καὶ
καθῆσιν, ἡ δὲ ἐμφῦσα εἴχετο, ἀποσπωμένη δὲ καὶ
τὴν νόσον τὴν ἐν τῷ νεανίᾳ συναπέσπασεν.[2]
ὑπηρέτις μὲν δὴ θεοῦ θεραπείας ἡ μύραινα αὕτη
γενομένη ἐς[3] ἀκοὴν τὴν ἡμετέραν ἀφίκετο.

35. Χρύσερμον δὲ[4] ἐπὶ Νέρωνος αἷμα ἀνεμοῦντα
καὶ τηκόμενον ἤδη, αἷμα ταύρου πιόντα ἰάσατο ὁ
αὐτὸς οὗτος θεός. ἐγὼ δὲ λέγω ταῦτα, ὅτι ἐς
τοσοῦτον ἄρα τὰ ζῷα θεοφιλῆ ἐστιν, ὡς καὶ ὑπὸ
τῶν θεῶν σώζεσθαι, καὶ σώζειν ἐκείνοις βουλομέ-

---

[1] ζωγριον
[2] συνέσπασεν.
[3] καὶ εἰς.
[4] τε.

affair and presently there was shouting, and many people heard the noise. Next, the chief priest arrived and enquired what was the reason of this wrangling in the temple, and the men began to accuse one another. And the rich man, outraged by threats, blasphemy, and abuse, took his departure, and after swallowing the bone of another bird was in pain and died in agonies, while the wicked attendant was punished by the governor of the city for sacrilege. As for the bird, it was not seen either alive or dead, but the story goes that after living for a hundred years it disappeared.

34. The following story too is like the above and concurs with it. One Cissus by name, a devoted servant of Serapis, was the victim of a plot on the part of a woman whom he had once loved and later married: he ate some eggs of a snake, which caused him pain; he was in a grievous state and in danger of death. But he prayed to the god, who bade him buy a live Moray and thrust his hand into the creature's tank. Cissus obeyed and thrust in his hand. And the Moray fastened on and clung to him, but when it was pulled off it pulled away the sickness from the young man at the same time. It was because this Moray was a minister of the god's healing power that the tale reached my hearing. *A victim of poisoning saved by Serapis*

35. And this same god in the days of Nero cured Chrysermus who was vomiting blood and already beginning to waste away, by means of a draught of bull's blood. And I mention these facts because animals are so dearly beloved by the gods that their lives are saved by them, and when the gods desire, *Cures wrought by Serapis*

AELIAN

νων ἑτέρους. ἀτὰρ οὖν καὶ Βάσιλιν [1] τὸν Κρῆτα ἐς
νόσον φθίσεως ἐμπεσόντα ἐξάντη τοῦ τοσούτου
κακοῦ ὅδε ὁ θεὸς εἰργάσατο ὀνείων κρεῶν γευσάμε-
νον. καὶ προσέπεσε γενέσθαι αὐτῷ κατὰ [2] τὸ
ὄνομα τοῦ ζῴου· ἔφατο γὰρ ὀνησιφόρον οἱ
ταύτην ἔσεσθαι τὴν θεραπείαν καὶ ἴασιν. καὶ
ὑπὲρ μὲν τούτων ἀπόχρη καὶ ταῦτα.

36. Ἴδια δὲ ἄρα τῶν ζῴων καὶ ἐκεῖνα.[3] αἱ
ἵπποι ἐς ἁρματηλασίαν ἐπιτηδειόταται [4] εἶναι
πιστεύονται. πυνθάνομαι δὲ τοὺς ἄνδρας τοὺς
πωλευτικοὺς λέγειν ὅτι ἄρα χαίρουσιν ἵπποι
λουτρῷ τε καὶ ἀλοιφῇ. ὅτι δὲ καὶ μύρῳ ἐχρίοντο
ἵπποι, Σημωνίδης [5] ἐν τοῖς ἰάμβοις λέγει. Πέρσαι
δὲ μετὰ τὴν Κύρου μάχην τὴν ἐν Λυδίᾳ καμήλους
τοῖς ἵπποις συντρέφουσι, τὸ δέος τῶν ἵππων τὸ ἐκ
τῶν καμήλων ἐς αὐτοὺς ἐξαπτόμενον ἐκβάλλειν
πειρώμενοι τῇ συντροφίᾳ.

37. Καλεῖται δὲ σελάχια ὅσα οὐκ ἔχει λεπίδας·
εἴη [6] δ' ἂν μύραινα γόγγρος νάρκη τρυγὼν βοῦς
γαλεός· . . . [7] δελφὶς φάλλαινα φώκη. ταῦτα δὲ
ἄρα μόνα τῶν ἐνύδρων ζῳοτοκεῖ. μαλάκια δὲ
καλεῖται ὅσα [8] ἀνόστεά ἐστι [9]· καὶ εἴη ἂν πολύπους
σηπία [10] τευθὶς ἀκαλήφη. ταῦτά τοι καὶ αἵματος
ἄμοιρα καὶ σπλάγχνων ἐστί. μαλακόστρακα δὲ
ἀστακοὶ καρίδες καρκίνοι πάγουροι· ἀποδύεται δὲ

---

[1] Βάθυλιν, -ηλυν, -ελιν.
[2] παρά.
[3] ἐκεῖνα λέγεται.
[4] ἐπιτηδειότεραι.
[5] Σιμωνίδης MSS, edd.
[6] εἶεν MSS.
[7] Lacuna : ⟨κήτη δέ⟩ add. Jac.
[8] ὅσα τῶν ἐνύδρων.
[9] Jac : εἰσι.

406

they save others. It was this god (Serapis) who
when Basilis the Cretan fell into a wasting disease,
rid him of this terrible complaint by causing him to
eat the flesh of an ass. And the result was in
accordance with the name of the beast, for the god
said that this treatment and remedy would be of
ass-istance to him.

On these topics enough has been said.

36. Here are further peculiarities of animals. *The Horse*
Mares are believed to be most suitable for drawing
chariots. And I learn that trainers assert that
horses delight in being washed and anointed. And
Semonides in his iambics [*fr.* 7. 57 D] says that
horses were even rubbed with perfume. And the
Persians, since the battle which Cyrus fought in
Lydia,[a] keep camels together with their horses, and
attempt by so doing to rid horses of the fear which
camels inspire in them.

37. Fishes that have no scales are called ' carti- *Various*
laginous ': for example, the moray, the conger-eel, *genera of*
the torpedo, the sting-ray, the horned-ray, the dog- *world*
fish; ⟨' cetaceans '⟩, the dolphin, the whale, the
seal; these are the only aquatic creatures that are
viviparous. ' Cephalopod mollusca ' is the name
given to those that have no bones: for example, the
octopus, the cuttlefish, the squid, the sea-anemone;
these have no blood and no intestines. ' Crustacea,'
lobsters, prawns, crabs of all kinds;[b] these slough

[a] He defeated Croesus, King of Lydia, 546 B.C.
[b] See 9.6 note c.

10 σηπία ⟨τεῦθος⟩ τευθίς add. *Wellmann.*

AELIAN

καὶ τὸ γῆρας ταῦτα. ὀστρακόδερμα δὲ ὄστρεα
πορφύραι κήρυκες στρόμβοι ἐχῖνοι κάραβοι. καρ-
χαρόδοντα δὲ [1] λύκος κύων λέων πάρδαλις. ταῦτά
τοι [2] καὶ σαρκῶν ἐσθίει. ἀμφώδοντα [3] δὲ ἄνθρωπος ἵππος ὄνος, ἅπερ οὖν ⟨καὶ⟩ [4] πιμελὴν ἔχει.
συνόδοντα δὲ βοῦς πρόβατον αἴξ. χαυλιόδοντα δὲ
ὗς ὁ ἄγριος σπάλαξ· τὸν γὰρ ἐλέφαντα οὔ φημι
ὀδόντας ἔχειν ἀλλὰ κέρατα. ἔντομα δὲ σφὴξ
μέλιττα· λέγουσι δὲ μηδὲ πνεύμονας ἔχειν ταῦτα.
ἀμφίβια δὲ ἵππος ποτάμιος ἐνυδρὶς κάστωρ
κροκόδιλος. φολιδωτὰ δὲ σαῦρος σαλαμάνδρα
χελώνη κροκόδιλος ὄφις· ταῦτα δὲ καὶ τὸ γῆρας
ἀποδύεται πλὴν κροκοδίλου καὶ χελώνης. μώνυχα
δὲ ἵππος ὄνος. δίχηλα δὲ βοῦς ἔλαφος [5] αἴξ οἶς
χοῖρος. πολυσχιδῆ δὲ ἄνθρωπος κύων. στεγανό-
ποδα ⟨δὲ⟩ [6] καὶ πλατυώνυχα κύκνος χήν.[7] γαμψώ-
νυχα δὲ [8] ἱέρακες ἀετοί. τὴν δὲ [9] τῶν ἄλλων
ζῴων ἰδιότητα ἀλλαχοῦ εἶπον.

38. Φιλότεκνον δὲ ἄρα ζῷον ἦν καὶ ὁ χηναλώπηξ,
καὶ ταὐτὰ τοῖς πέρδιξι δρᾷ. καὶ γὰρ οὗτος πρὸ
τῶν νεοττῶν ἑαυτὸν κυλίει, καὶ ἐνδίδωσιν ἐλπίδα
ὡς θηράσοντι αὐτὸν τῷ ἐπιόντι· οἱ δὲ ἀποδιδράσκουσιν ἐν [10] τῷ τέως. ὅταν δὲ πρὸ ὁδοῦ γένωνται,
καὶ ἐκεῖνος ἑαυτὸν τοῖς πτεροῖς ἐλαφρίσας ἀπαλλάττεται.

[1] δὲ στρογγύλους ἔχοντα τοὺς ὀδόντας.
[2] μέντοι.
[3] ἀμφώδοντα H.
[4] ⟨καὶ⟩ add. H.
[5] Ges: ἐλέφας.
[6] ⟨δέ⟩ add. H.
[7] χὴν, δερμόπτερος δὲ νυκτερίς.
[8] δὲ οἷς καὶ τὸ ῥάμφος ἐπικαμπές.
[9] Reiske: δὲ ἄλλην.
[10] οἱ νεοττοὶ ἐν.

408

their 'old age.' 'Testacea,' oysters, purple shell-fish, whelks, trumpet-shells,[a] sea-urchins, crayfish. 'Saw-toothed' animals are the wolf, the dog, the lion, the leopard; these, you know, are carnivorous. Incisor-teeth in both jaws are found in man, horses, and asses, and these creatures have fat. Animals whose upper and lower teeth meet evenly are the ox, the sheep, the goat. Animals with projecting teeth, the wild boar, the blind-rat; the elephant, I maintain, has horns, not teeth. Insects, the wasp, the bee; these are even said to have no lungs. 'Amphibians,' the hippopotamus, the otter, the beaver, the crocodile. Scaley creatures, the lizard, the salamander, the tortoise, the crocodile, the snake; and these also, with the exception of the tortoise and the crocodile, slough their 'old age.' Animals with uncloven hoofs, the horse, the ass; cloven-hoofed animals, the ox, the stag, the goat, the sheep, the pig. Creatures with toes, men and dogs. Web-footed and flat-nailed creatures, the swan, the goose. Creatures with crooked talons, hawks and eagles. I have mentioned elsewhere the distinguishing marks of other animals.

38. It seems that the Egyptian Goose also is devoted to its offspring and behaves as partridges do. For it also rolls on the ground in front of its young and affords its pursuer the hope of catching it; meantime the chicks make their escape. And when they are some distance away, the parent also takes wing and is off.

The Egyptian Goose

[a] Κήρυξ and στρόμβος appear to be synonyms for 'whelk,' and both were used as conchs or trumpets.

39. Λέγουσι δὲ Αἰγύπτιοι τὸν ἱέρακα ζῶντα μὲν καὶ ἔτι περιόντα θεοφιλῆ ὄρνιν εἶναι, τοῦ βίου δὲ ἀπελθόντα καὶ μαντεύεσθαι καὶ ὀνείρατα ἐπιπέμπειν, ἀποδυσάμενον τὸ σῶμα καὶ ψυχὴν γεγενημένον γυμνήν. λέγουσι δὲ Αἰγύπτιοι [1] καὶ τρίποδα ἱέρακα παρ' αὐτοῖς φανῆναί ποτε, καὶ ὑγιᾶ [2] δοκοῦσι [3] λέγειν τοῖς πεπιστευκόσιν.

40. Πέρδικες οἱ Παφλαγόνες δικάρδιοί εἰσιν, ὥσπερ οὖν Θεόφραστος λέγει. καὶ Θεόπομπος λέγει τοὺς ἐν Βισαλτίᾳ λαγὼς διπλᾶ ἥπατα ἔχειν ἕκαστον. λέγει δὲ Ἀπίων, εἰ μὴ τερατεύεται, καὶ ἐλάφους νεφροὺς τέτταρας ἔχειν κατά τινας τόπους. λέγει δὲ ὁ αὐτὸς καὶ κατ' Ἀτώθιδα [4] τὸν Μήνιδος [5] δικέφαλον γέρανον φανῆναι, καὶ εὐθηνῆσαι τὴν Αἴγυπτον· καὶ ἐπ' ἄλλου βασιλέως τετρακέφαλον ὄρνιν, καὶ πλημμυρῆσαι τὸν Νεῖλον ὡς οὔποτε, καὶ καρπῶν ἀφθονίαν γενέσθαι καὶ εὐποτμίαν ληίων θαυμαστήν. τετράκερων δὲ ἔλαφον Νικοκρέων ὁ Κύπριος ἔσχε, καὶ ἀνέθηκε Πυθοῖ καὶ ὑπέγραψε

σῆς ἕνεκεν, Λητοῦς τοξαλκέτα κοῦρ', ἐπινοίας
τήνδ' ἕλε Νικοκρέων τετράκερων ἔλαφον.

καὶ μέντοι καὶ τετράκερω πρόβατα ἐν τῷ τοῦ Διὸς τοῦ Πολιέως ἦν καὶ τρίκερω. ἐγὼ δὲ καὶ πεντάποδα βοῦν ἱερὸν ἐθεασάμην, ἀνάθημα τῷ θεῷ τῷδε ἐν τῇ πόλει τῇ Ἀλεξανδρέων τῇ μεγάλῃ, ἐν

---

[1] οἱ Αἰγύπτιοι. [2] ὑγιῆ.
[3] Ges : δοκοῦσιν εἶναι.
[4] Bunsen : κατὰ τὸν Οἴνιδα MSS, H.
[5] M. βασιλεύοντα.

39. The Egyptians say that the Hawk while alive The Hawk and active is beloved of the gods, and when it has departed this life and shed its body and become a disembodied spirit, it prophesies and sends dreams. And the Egyptians say that a Hawk with three legs once appeared among them, and believers accept the statement as sound.

40. The Partridges of Paphlagonia have two hearts, Freaks of according to Theophrastus [*fr.* 182]. And Theo- Nature pompus says that Hares in Bisaltia have each of them a double liver. And Apion says—unless he is romancing—that the Stags in certain districts have four kidneys. And the same writer states that in the time of Atothis *a* son of Menis there appeared a Crane with two heads, and that there was prosperity in Egypt; and in the reign of another King there appeared a bird with four heads, and the Nile over-flowed as never before and the fruits were abundant and the crops flourished marvellously. Nicocreon of Cyprus possessed a Deer with four horns; this he gave as an offering at Delphi and wrote beneath it:

' It was thy doing, O son of Leto, mighty archer, that Nicocreon captured this four-horned deer.'

Moreover there were even Sheep with four horns and with three horns in the temple of Zeus, the Guardian of the City. And I myself have seen *b* a sacred Ox with five feet which was an offering to this god in the great city of Alexandria, in the far-famed

a Atothis (or Ath-) was the second king of the First Dynasty, fl. *c.* 3140 B.C.; he built the palace at Memphis.
b See vol. I, p. xij, note 2.

AELIAN

τῷ ἀδομένῳ [1] τοῦ θεοῦ ἄλσει, ἔνθα περσέαι
σύμφυτοι σκιὰν περικαλλῆ καὶ ψῦξιν [2] ἀπεδείκνυντο.
καὶ ἦν μόσχος ἐνταῦθα τὴν χρόαν κηρῷ προσει-
κασμένος, καὶ ἐπὶ τοῦ ὤμου πόδα ἀπηρτημένον
εἶχε περίεργον [3] μὲν ὅσα ἐπιβῆναι, τέλειον δὲ
ὅσα ἐς πλάσιν. καὶ ταῦτα μὲν δοκεῖ τῇ φύσει
ὁμολογεῖν οὐ πάνυ τι, [4] ἐγὼ δὲ ὅσα ἐς ἐμὴν ὄψιν
τε καὶ ἀκοὴν ἀφίκετο εἶπον.

[1] καλουμένῳ.
[2] ὄψιν.
[3] συνεργόν MSS, ἀσύνεργον Ges.
[4] πάντῃ.

412

grove of the god, where the persea-trees close-planted afforded the loveliest shade and coolness. And there was a Calf with the colour of wax, and it had a foot attached to its shoulder which was superfluous for walking although it was perfectly formed. True, these phenomena appear far from conformity to nature, but I have reported what I myself have seen and heard.

PRINTED IN GREAT BRITAIN BY
RICHARD CLAY AND COMPANY, LTD.,
BUNGAY, SUFFOLK.

THE LOEB CLASSICAL LIBRARY

VOLUMES ALREADY PUBLISHED

Latin Authors

AMMIANUS MARCELLINUS. Translated by J. C. Rolfe. 3 Vols. (*3rd Imp., revised.*)

APULEIUS: THE GOLDEN ASS (METAMORPHOSES). W. Adlington (1566). Revised by S. Gaselee. (*8th Imp.*)

S. AUGUSTINE: CITY OF GOD. 7 Vols. Vol. I. G. E. McCracken.

ST. AUGUSTINE, CONFESSIONS OF. W. Watts (1631). 2 Vols. (Vol. I. 7th Imp., Vol. II. 6th Imp.)

ST. AUGUSTINE, SELECT LETTERS. J. H. Baxter. (*2nd Imp.*)

AUSONIUS. H. G. Evelyn White. 2 Vols. (*2nd Imp.*)

BEDE. J. E. King. 2 Vols. (*2nd Imp.*)

BOETHIUS: TRACTS and DE CONSOLATIONE PHILOSOPHIAE. Rev. H. F. Stewart and E. K. Rand. (*6th Imp.*)

CAESAR: ALEXANDRIAN, AFRICAN and SPANISH WARS. A. G. Way.

CAESAR: CIVIL WARS. A. G. Peskett. (*6th Imp*)

CAESAR: GALLIC WAR. H. J. Edwards. (*11th Imp.*)

CATO: DE RE RUSTICA; VARRO: DE RE RUSTICA. H. B. Ash and W. D. Hooper. (*3rd Imp.*)

CATULLUS. F. W. Cornish; TIBULLUS. J. B. Postgate; PERVIGILIUM VENERIS. J. W. Mackail. (*13th Imp.*)

CELSUS: DE MEDICINA. W. G. Spencer. 3 Vols. (Vol. I. 3rd Imp. revised, Vols. II. and III. 2nd Imp.)

CICERO: BRUTUS, and ORATOR. G. L. Hendrickson and H. M. Hubbell. (*3rd Imp.*)

[CICERO]: AD HERENNIUM. H. Caplan.

CICERO: DE FATO; PARADOXA STOICORUM; DE PARTITIONE ORATORIA. H. Rackham (With De Oratore. Vol. II.) (*2nd Imp.*)

CICERO: DE FINIBUS. H. Rackham. (*4th Imp. revised.*)

CICERO: DE INVENTIONE, etc. H. M. Hubbell.

CICERO: DE NATURA DEORUM and ACADEMICA. H. Rackham. (*3rd Imp.*)

CICERO: DE OFFICIIS. Walter Miller. (*7th Imp.*)

CICERO: DE ORATORE. 2 Vols. E. W. Sutton and H. Rackham. (*2nd Imp.*)

CICERO: DE REPUBLICA and DE LEGIBUS; SOMNIUM SCIPIONIS. Clinton W. Keyes. (*4th Imp.*)

CICERO: DE SENECTUTE, DE AMICITIA, DE DIVINATIONE. W. A. Falconer. (*6th Imp.*)

CICERO: IN CATILINAM, PRO FLACCO, PRO MURENA, PRO SULLA. Louis E. Lord. (*3rd Imp. revised.*)

CICERO: LETTERS TO ATTICUS. E. O. Winstedt. 3 Vols.
(Vol. I. 7*th Imp.*, Vols. II. and III. 4*th Imp.*)
CICERO: LETTERS TO HIS FRIENDS. W. Glynn Williams. 3
Vols. (Vols. I. and II. 4*th Imp.*, Vol. III. 2*nd Imp. revised.*)
CICERO: PHILIPPICS. W. C. A. Ker. (4*th Imp. revised.*)
CICERO: PRO ARCHIA, POST REDITUM, DE DOMO, DE HARUS-
PICUM RESPONSIS, PRO PLANCIO. N. H. Watts. (3*rd Imp.*)
CICERO: PRO CAECINA, PRO LEGE MANILIA, PRO CLUENTIO,
PRO RABIRIO. H. Grose Hodge. (3*rd Imp.*)
CICERO: PRO CAELIO, DE PROVINCIIS CONSULARIBUS. PRO
BALBO. R. Gardner.
CICERO: PRO MILONE, IN PISONEM, PRO SCAURO, PRO FONTEIO.
PRO RABIRIO POSTUMO, PRO MARCELLO, PRO LIGARIO, PRO
REGE DEIOTARO. N. H. Watts. (3*rd Imp.*)
CICERO: PRO QUINCTIO, PRO ROSCIO AMERINO, PRO ROSCIO
COMOEDO, CONTRA RULLUM. J. H. Freese. (3*rd Imp.*)
CICERO: PRO SESTIO, IN VATINIUM. R. Gardner.
CICERO: TUSCULAN DISPUTATIONS. J. E. King. (4*th Imp.*)
CICERO: VERRINE ORATIONS. L. H. G. Greenwood. 2 Vols.
(Vol. I. 3*rd Imp.*, Vol. II. 2*nd Imp.*)
CLAUDIAN. M. Platnauer. 2 Vols. (2*nd Imp.*)
COLUMELLA: DE RE RUSTICA. DE ARBORIBUS. H. B. Ash,
E. S. Forster and E. Heffner. 3 Vols. (Vol. I. 2*nd Imp.*)
CURTIUS, Q.: HISTORY OF ALEXANDER. J. C. Rolfe. 2 Vols.
(2*nd Imp.*)
FLORUS. E. S. Forster and CORNELIUS NEPOS. J. C. Rolfe.
(2*nd Imp.*)
FRONTINUS: STRATAGEMS and AQUEDUCTS. C. E. Bennett and
M. B. McElwain. (2*nd Imp.*)
FRONTO: CORRESPONDENCE. C. R. Haines. 2 Vols. (3*rd
Imp.*)
GELLIUS, J. C. Rolfe. 3 Vols. (Vol. I. 3*rd Imp.*, Vols. II. and
III. 2*nd Imp.*)
HORACE: ODES and EPODES. C. E. Bennett. (14*th Imp.
revised.*)
HORACE: SATIRES, EPISTLES, ARS POETICA. H. R. Fairclough.
(9*th Imp. revised.*)
JEROME: SELECTED LETTERS. F. A. Wright. (2*nd Imp.*)
JUVENAL and PERSIUS. G. G. Ramsay. (8*th Imp.*)
LIVY. B. O. Foster, F. G. Moore, Evan T. Sage, and A. C.
Schlesinger and R. M. Geer (General Index). 14 Vols. (Vol.
I. 5*th Imp.*, Vol. V. 4*th Imp.*, Vols. II.–IV., VI. and VII.,
IX.–XII. 3*rd Imp.*, Vol. VIII., 2*nd Imp. revised.*)
LUCAN. J. D. Duff. (4*th Imp.*)
LUCRETIUS. W. H. D. Rouse. (7*th Imp. revised.*)
MARTIAL. W. C. A. Ker. 2 Vols. (Vol. I. 5*th Imp.*, Vol. II.
4*th Imp. revised.*)
MINOR LATIN POETS: from PUBLILIUS SYRUS TO RUTILIUS
NAMATIANUS, including GRATTIUS, CALPURNIUS SICULUS,
NEMESIANUS, AVIANUS, and others with " Aetna " and the
" Phoenix." J. Wight Duff and Arnold M. Duff. (3*rd
Imp.*)

2

OVID: THE ART OF LOVE and OTHER POEMS. J. H. Mozley. (*4th Imp.*)
OVID: FASTI. Sir James G. Frazer. (*2nd Imp.*)
OVID: HEROIDES and AMORES. Grant Showerman. (*7th Imp.*)
OVID: METAMORPHOSES. F. J. Miller. 2 Vols. (Vol. I. 11*th Imp.*, Vol. II. 10*th Imp.*)
OVID: TRISTIA and EX PONTO. A. L. Wheeler. (*4th Imp.*)
PERSIUS. Cf. JUVENAL.
PETRONIUS. M. Heseltine, SENECA APOCOLOCYNTOSIS. W. H. D. Rouse. (*9th Imp. revised.*)
PLAUTUS. Paul Nixon. 5 Vols. (Vol. I. 6*th Imp.*, II. 5*th Imp.*, III. 4*th Imp.*, IV. and V. 2*nd Imp.*)
PLINY: LETTERS. Melmoth's Translation revised by W. M. L. Hutchinson. 2 Vols. (*7th Imp.*)
PLINY: NATURAL HISTORY. H. Rackham and W. H. S. Jones. 10 Vols. Vols. I.–V. and IX. H. Rackham. Vols. VI. and VII. W. H. S. Jones. (Vol. I. 4*th Imp.*, Vols. II. and III. 3*rd Imp.*, Vol. IV. 2*nd Imp.*)
PROPERTIUS. H. E. Butler. (*7th Imp.*)
PRUDENTIUS. H. J. Thomson. 2 Vols.
QUINTILIAN. H. E. Butler. 4 Vols. (Vols. I. and IV. 4*th Imp.*, Vols. II. and III. 3*rd Imp.*)
REMAINS OF OLD LATIN. E. H. Warmington. 4 vols. Vol. I. (ENNIUS AND CAECILIUS.) Vol. II. (LIVIUS, NAEVIUS, PACUVIUS, ACCIUS.) Vol. III. (LUCILIUS and LAWS OF XII TABLES.) (*2nd Imp.*) (ARCHAIC INSCRIPTIONS.)
SALLUST. J. C. Rolfe. (*4th Imp. revised.*)
SCRIPTORES HISTORIAE AUGUSTAE. D. Magie. 3 Vols. (Vol. I. 3*rd Imp. revised*, Vols. II. and III. 2*nd Imp.*)
SENECA: APOCOLOCYNTOSIS. Cf. PETRONIUS.
SENECA: EPISTULAE MORALES. R. M. Gummere. 3 Vols. (Vol. I. 4*th Imp.*, Vols. II. and III. 3*rd Imp.*)
SENECA: MORAL ESSAYS. J. W. Basore. 3 Vols. (Vol. II. 4*th Imp.*, Vols. I. and III. 2*nd Imp. revised.*)
SENECA: TRAGEDIES. F. J. Miller. 2 Vols. (Vol. I. 4*th Imp.* Vol. II. 3*rd Imp. revised.*)
SIDONIUS: POEMS AND LETTERS. W. B. Anderson. 2 Vols. (Vol. I. 2*nd Imp.*)
SILIUS ITALICUS. J. D. Duff. 2 Vols. (Vol. I. 2*nd Imp.* Vol. II. 3*rd Imp.*)
STATIUS. J. H. Mozley. 2 Vols. (*2nd Imp.*)
SUETONIUS. J. C. Rolfe. 2 Vols. (Vol. I. 7*th Imp.*, Vol. II. 6*th Imp. revised.*)
TACITUS: DIALOGUES. Sir Wm. Peterson. AGRICOLA and GERMANIA. Maurice Hutton. (*7th Imp.*)
TACITUS: HISTORIES AND ANNALS. C. H. Moore and J. Jackson. 4 Vols.. (Vols. I. and II. 4*th Imp.* Vols. III. and IV. 3*rd Imp.*)
TERENCE. John Sargeaunt. 2 Vols. (Vol. I. 8*th Imp.*, Vol. II. 7*th Imp.*)
TERTULLIAN: APOLOGIA and DE SPECTACULIS. T. R. Glover. MINUCIUS FELIX. G. H. Rendall. (*2nd Imp.*)
VALERIUS FLACCUS. J. H. Mozley. (*3rd Imp. revised.*)

VARRO: DE LINGUA LATINA. R. G. Kent. 2 Vols. (*3rd Imp. revised.*)
VELLEIUS PATERCULUS and RES GESTAE DIVI AUGUSTI. F. W. Shipley. (*2nd Imp.*)
VIRGIL. H. R. Fairclough. 2 Vols. (Vol. I. 19*th Imp.*, Vol. II. 14*th Imp. revised.*)
VITRUVIUS: DE ARCHITECTURA. F. Granger. 2 Vols. (Vol. I. 3*rd Imp.*, Vol. II. 2*nd Imp.*)

Greek Authors

ACHILLES TATIUS. S. Gaselee. (*2nd Imp.*)
AELIAN: ON THE NATURE OF ANIMALS. 3 Vols. Vols. I. and II. A. F. Scholfield.
AENEAS TACTICUS, ASCLEPIODOTUS and ONASANDER. The Illinois Greek Club. (*2nd Imp.*)
AESCHINES. C. D. Adams. (*3rd Imp.*)
AESCHYLUS. H. Weir Smyth. 2 Vols. (Vol. I. 7*th Imp.*, Vol. II. 6*th Imp. revised.*)
ALCIPHRON, AELIAN, PHILOSTRATUS LETTERS. A. R. Benner and F. H. Fobes.
ANDOCIDES, ANTIPHON, Cf. MINOR ATTIC ORATORS.
APOLLODORUS. Sir James G. Frazer. 2 Vols. (*3rd Imp.*)
APOLLONIUS RHODIUS. R. C. Seaton. (*5th Imp.*)
THE APOSTOLIC FATHERS. Kirsopp Lake. 2 Vols. (Vol. I. 8*th Imp.*, Vol. II. 6*th Imp.*)
APPIAN: ROMAN HISTORY. Horace White. 4 Vols. (Vol. I. 4*th Imp.*, Vols. II.–IV. 3*rd Imp.*)
ARATUS. Cf. CALLIMACHUS.
ARISTOPHANES. Benjamin Bickley Rogers. 3 Vols. Verse trans. (*5th Imp.*)
ARISTOTLE: ART OF RHETORIC. J. H. Freese. (*3rd Imp.*)
ARISTOTLE: ATHENIAN CONSTITUTION, EUDEMIAN ETHICS, VICES AND VIRTUES. H. Rackham. (*3rd Imp.*)
ARISTOTLE: GENERATION OF ANIMALS. A. L. Peck. (*2nd Imp.*)
ARISTOTLE: METAPHYSICS. H. Tredennick. 2 Vols. (*4th Imp.*)
ARISTOTLE: METEOROLOGICA. H. D. P. Lee.
ARISTOTLE: MINOR WORKS. W. S. Hett. On Colours, On Things Heard, On Physiognomies, On Plants, On Marvellous Things Heard, Mechanical Problems, On Indivisible Lines, On Situations and Names of Winds, On Mellissus, Xenophanes, and Gorgias. (*2nd Imp.*)
ARISTOTLE: NICOMACHEAN ETHICS. H. Rackham. (6*th Imp. revised.*)
ARISTOTLE: OECONOMICA and MAGNA MORALIA. G. C. Armstrong; (with Metaphysics, Vol. II.). (*4th Imp.*)
ARISTOTLE: ON THE HEAVENS. W. K. C. Guthrie. (3*rd Imp. revised.*)
ARISTOTLE: ON THE SOUL, PARVA NATURALIA, ON BREATH. W. S. Hett. (*2nd Imp. revised.*)

4

ARISTOTLE: ORGANON—Categories, On Interpretation, Prior Analytics. H. P. Cooke and H. Tredennick. (3rd Imp.)
ARISTOTLE: ORGANON—Posterior Analytics, Topics. H. Tredennick and E. S. Forster.
ARISTOTLE: ORGANON—On Sophistical Refutations.
On Coming to be and Passing Away, On the Cosmos. E. S. Forster and D. J. Furley.
ARISTOTLE: PARTS OF ANIMALS. A. L. Peck; MOTION AND PROGRESSION OF ANIMALS. E. S. Forster. (4th Imp. revised.)
ARISTOTLE: PHYSICS. Rev. P. Wicksteed and F. M. Cornford. 2 Vols. (Vol. I. 2nd Imp., Vol. II. 3rd Imp.)
ARISTOTLE: POETICS and LONGINUS. W. Hamilton Fyfe; DEMETRIUS ON STYLE. W. Rhys Roberts. (5th Imp. revised.)
ARISTOTLE: POLITICS. H. Rackham. (4th Imp. revised.)
ARISTOTLE: PROBLEMS. W. S. Hett. 2 Vols. (2nd Imp. revised.)
ARISTOTLE: RHETORICA AD ALEXANDRUM (with PROBLEMS. Vol. II.). H. Rackham.
ARRIAN: HISTORY OF ALEXANDER and INDICA. Rev. E. Iliffe Robson. 2 Vols. (3rd Imp.)
ATHENAEUS: DEIPNOSOPHISTAE. C. B. Gulick. 7 Vols. (Vols. I.–IV., VI. and VII. 2nd Imp., Vol. V. 3rd Imp.)
ST. BASIL: LETTERS. R. J. Deferrari. 4 Vols. (2nd Imp.)
CALLIMACHUS: FRAGMENTS. C. A. Trypanis.
CALLIMACHUS, Hymns and Epigrams, and LYCOPHRON. A. W. Mair; ARATUS. G. R. Mair. (2nd. Imp.)
CLEMENT of ALEXANDRIA. Rev. G. W. Butterworth. (3rd Imp.)
COLLUTHUS. Cf. OPPIAN.
DAPHNIS AND CHLOE. Thornley's Translation revised by J. M. Edmonds; and PARTHENIUS. S. Gaselee. (4th Imp.)
DEMOSTHENES I.: OLYNTHIACS, PHILIPPICS and MINOR ORATIONS. I.–XVII. AND XX. J. H. Vince. (2nd Imp.)
DEMOSTHENES II.: DE CORONA and DE FALSA LEGATIONE. C. A. Vince and J. H. Vince. (3rd Imp. revised.)
DEMOSTHENES III.: MEIDIAS, ANDROTION, ARISTOCRATES, TIMOCRATES and ARISTOGEITON, I. AND II. J. H. Vince (2nd Imp.)
DEMOSTHENES IV.–VI.: PRIVATE ORATIONS and IN NEAERAM. A. T. Murray. (Vol. IV. 3rd Imp., Vols. V. and VI. 2nd Imp.)
DEMOSTHENES VII.: FUNERAL SPEECH, EROTIC ESSAY, EXORDIA and LETTERS. N. W. and N. J. DeWitt.
DIO CASSIUS: ROMAN HISTORY. E. Cary. 9 Vols. (Vols. I. and II. 3rd Imp., Vols. III.–IX. 2nd Imp.)
DIO CHRYSOSTOM. J. W. Cohoon and H. Lamar Crosby. 5 Vols. (Vols. I.–IV. 2nd Imp.)
DIODORUS SICULUS. 12 Vols. Vols. I.–VI. C. H. Oldfather. Vol. VII. C. L. Sherman. Vols. IX. and X. R. M. Geer. Vol. XI. F. Walton. (Vol. I. 3rd Imp., Vols. II.–IV. 2nd Imp.)
DIOGENES LAERTIUS. R. D. Hicks. 2 Vols. (5th Imp.).
DIONYSIUS OF HALICARNASSUS: ROMAN ANTIQUITIES. Spelman's translation revised by E. Cary. 7 Vols. (Vols. I.–V. 2nd Imp.)

EPICTETUS. W. A. Oldfather. 2 Vols. (*3rd Imp.*)
EURIPIDES. A. S. Way. 4 Vols. (Vols. I. and IV. *7th Imp.*, Vol. II. *8th Imp.*, Vol. III. *6th Imp.*) Verse trans.
EUSEBIUS: ECCLESIASTICAL HISTORY. Kirsopp Lake and J. E. L. Oulton. 2 Vols. (Vol. I. *3rd Imp.*, Vol. II. *5th Imp.*)
GALEN: ON THE NATURAL FACULTIES. A. J. Brock. (*4th Imp.*)
THE GREEK ANTHOLOGY. W. R. Paton. 5 Vols. (Vols. I.–IV. *5th Imp.*, Vol. V. *3rd Imp.*)
GREEK ELEGY AND IAMBUS with the ANACREONTEA. J. M. Edmonds. 2 Vols. (Vol. I. *3rd Imp.*, Vol. II. *2nd Imp.*)
THE GREEK BUCOLIC POETS (THEOCRITUS, BION, MOSCHUS). J. M. Edmonds. (*7th Imp. revised.*)
GREEK MATHEMATICAL WORKS. Ivor Thomas. 2 Vols. (*3rd Imp.*)
HERODES. Cf. THEOPHRASTUS: CHARACTERS.
HERODOTUS. A. D. Godley. 4 Vols. (Vol. I. *4th Imp.*, Vols. II. and III. *5th Imp.*, Vol. IV. *3rd Imp.*)
HESIOD AND THE HOMERIC HYMNS. H. G. Evelyn White. (*7th Imp. revised and enlarged.*)
HIPPOCRATES and the FRAGMENTS OF HERACLEITUS. W. H. S. Jones and E. T. Withington. 4 Vols. (Vol. 1. *4th Imp.*, Vols. II.–IV. *3rd Imp.*)
HOMER: ILIAD. A. T. Murray. 2 Vols. (*7th Imp.*)
HOMER: ODYSSEY. A. T. Murray. 2 Vols. (*8th Imp.*)
ISAEUS. E. W. Forster. (*3rd Imp.*)
ISOCRATES. George Norlin and LaRue Van Hook. 3 Vols. (*2nd Imp.*)
ST. JOHN DAMASCENE: BARLAAM AND IOASAPH. Rev. G. R. Woodward and Harold Mattingly. (*3rd Imp. revised.*)
JOSEPHUS. H. St. J. Thackeray and Ralph Marcus. 9 Vols. Vols. I.–VII. (Vol. V. *4th Imp.*, Vol. VI. *3rd Imp.*, Vols .I.–IV. and VII. *2nd Imp.*)
JULIAN. Wilmer Cave Wright. 3 Vols. (Vols. I. and II. *3rd Imp.*, Vol. III. *2nd Imp.*)
LUCIAN. A. M. Harmon. 8 Vols. Vols. I.–V. (Vols. I. and II. *4th Imp.*, Vol. III. *3rd Imp.*, Vols. IV. and V. *2nd Imp.*)
LYCOPHRON. Cf. CALLIMACHUS.
LYRA GRAECA. J. M. Edmonds. 3 Vols. (Vol. I. *5th Imp.* Vol. II *revised and enlarged*, and III. *4th Imp.*)
LYSIAS. W. R. M. Lamb. (*3rd Imp.*)
MANETHO. W. G. Waddell: PTOLEMY: TETRABIBLOS. F. E. Robbins. (*3rd Imp.*)
MARCUS AURELIUS. C. R. Haines. (*4th Imp. revised.*)
MENANDER. F. G. Allinson. (*3rd Imp. revised.*)
MINOR ATTIC ORATORS (ANTIPHON, ANDOCIDES, LYCURGUS, DEMADES, DINARCHUS, HYPEREIDES). K. J. Maidment and J. O. Burtt. 2 Vols. (Vol. I. *2nd Imp.*)
NONNOS: DIONYSIACA. W. H. D. Rouse. 3 Vols. (*2nd Imp.*)
OPPIAN, COLLUTHUS, TRYPHIODORUS. A. W. Mair. (*2nd Imp.*)
PAPYRI. NON-LITERARY SELECTIONS. A. S. Hunt and C. C. Edgar. 2 Vols. (*2nd Imp.*) LITERARY SELECTIONS. (Poetry). D. L. Page. (*3rd Imp.*)

6

PARTHENIUS. Cf. DAPHNIS AND CHLOE.
PAUSANIAS: DESCRIPTION OF GREECE. W. H. S. Jones. 5
Vols. and Companion Vol. arranged by R. E. Wycherley.
(Vols. I. and III. *3rd Imp.*, Vols. II., IV. and V. *2nd Imp.*)
PHILO. 10 Vols. Vols. I.–V.; F. H. Colson and Rev. G. H.
Whitaker Vols. VI.–IX.; F. H. Colson. (Vols. I.–II., V.–
VII., *3rd Imp.*, Vol. IV. *4th Imp.*, Vols. III., VIII., and IX.
2nd Imp.)
PHILO: two supplementary Vols. (*Translation only.*) Ralph
Marcus.
PHILOSTRATUS: THE LIFE OF APPOLLONIUS OF TYANA. F. C.
Conybeare. 2 Vols. (Vol. I. *4th Imp.*, Vol. II. *3rd Imp.*)
PHILOSTRATUS: IMAGINES; CALLISTRATUS: DESCRIPTIONS.
A. Fairbanks. (*2nd Imp.*)
PHILOSTRATUS and EUNAPIUS: LIVES OF THE SOPHISTS.
Wilmer Cave Wright. (*2nd Imp.*)
PINDAR. Sir J. E. Sandys. (*8th Imp. revised.*)
PLATO: CHARMIDES, ALCIBIADES, HIPPARCHUS, THE LOVERS,
THEAGES, MINOS and EPINOMIS. W. R. M. Lamb. (*2nd
Imp.*)
PLATO: CRATYLUS, PARMENIDES, GREATER HIPPIAS, LESSER
HIPPIAS. H. N. Fowler. (*4th Imp.*)
PLATO: EUTHYPHRO, APOLOGY, CRITO, PHAEDO, PHAEDRUS.
H. N. Fowler. (11*th Imp.*)
PLATO: LACHES, PROTAGORAS, MENO, EUTHYDEMUS. W. R. M.
Lamb. (*3rd Imp. revised.*)
PLATO: LAWS. Rev. R. G. Bury. 2 Vols. (*3rd Imp.*)
PLATO: LYSIS, SYMPOSIUM GORGIAS. W. R. M. Lamb. (*5th
Imp. revised.*)
PLATO: REPUBLIC. Paul Shorey. 2 Vols. (Vol. I. *5th Imp.*,
Vol. II. *4th Imp.*)
PLATO: STATESMAN, PHILEBUS. H. N. Fowler; ION. W. R. M.
Lamb. (*4th Imp.*)
PLATO: THEAETETUS and SOPHIST. H. N. Fowler. (*4th Imp.*)
PLATO: TIMAEUS, CRITIAS, CLITOPHO, MENEXENUS, EPISTULAE.
Rev. R. G. Bury. (*3rd Imp.*)
PLUTARCH: MORALIA. 14 Vols. Vols. I.–V. F. C. Babbitt.
Vol. VI. W. C. Helmbold. Vol. VII. P. H. De Lacey and
B. Einarson. Vol. X. H. N. Fowler. Vol. XII. H.
Cherniss and W. C. Helmbold. (Vols. I.–VI. and X. *2nd Imp.*)
PLUTARCH: THE PARALLEL LIVES. B. Perrin. 11 Vols.
(Vols. I., II., VI., VII., and XI. *3rd Imp.*, Vols. III.–V. and
VIII.–X. *2nd Imp.*)
POLYBIUS. W. R. Paton. 6 Vols. (*2nd Imp.*)
PROCOPIUS: HISTORY OF THE WARS. H. B. Dewing. 7 Vols.
(Vol. I. *3rd Imp.*, Vols. II.–VII. *2nd Imp.*)
PTOLEMY: TETRABIBLOS. Cf. MANETHO.
QUINTUS SMYRNAEUS. A. S. Way. Verse trans. (*3rd Imp.*)
SEXTUS EMPIRICUS. Rev. R. G. Bury. 4 Vols. (Vol. I. *4th
Imp.*, Vols. II. and III. *2nd Imp.*)
SOPHOCLES. F. Storr. 2 Vols. (Vol. I. 10*th Imp.* Vol. II. 6*th
Imp.*) Verse trans.

STRABO: GEOGRAPHY. Horace L. Jones. 8 Vols. (Vols. I., V., and VIII. 3rd Imp., Vols. II., III., IV., VI., and VII. 2nd Imp.)
THEOPHRASTUS: CHARACTERS. J. M. Edmonds. HERODES, etc. A. D. Knox. (3rd Imp.)
THEOPHRASTUS: ENQUIRY INTO PLANTS. Sir Arthur Hort, Bart. 2 Vols. (2nd Imp.)
THUCYDIDES. C. F. Smith. 4 Vols. (Vol. I. 5th Imp., Vols. II. and IV. 4th Imp., Vol. III., 3rd Imp. revised.)
TRYPHIODOBUS. Cf. OPPIAN.
XENOPHON: CYROPAEDIA. Walter Miller. 2 Vols. (Vol. I. 4th Imp., Vol. II. 3rd Imp.)
XENOPHON: HELLENICA, ANABASIS, APOLOGY, and SYMPOSIUM. C. L. Brownson and O. J. Todd. 3 Vols. (Vols. I. and III 3rd Imp., Vol. II. 4th Imp.)
XENOPHON: MEMORABILIA and OECONOMICUS. E. C. Marchant (3rd Imp.)
XENOPHON: SCRIPTA MINORA. E. C. Marchant. (3rd Imp.)

IN PREPARATION

Greek Authors

ARISTOTLE: HISTORY OF ANIMALS. A. L. Peck.
PLOTINUS: A. H. Armstrong.

Latin Authors

BABRIUS AND PHAEDRUS. Ben E. Perry.

DESCRIPTIVE PROSPECTUS ON APPLICATION

London
Cambridge, Mass.

WILLIAM HEINEMANN LTD
HARVARD UNIVERSITY PRESS